T0133342

DECEMBER 11, 2016

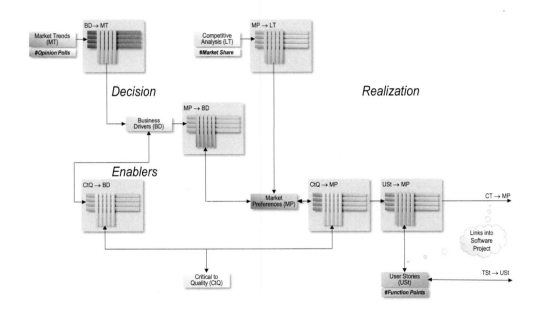

MANAGING COMPLEXITY
UNCOVER THE MYSTERIES WITH SIX SIGMA TRANSFER FUNCTIONS

THOMAS MICHAEL FEHLMANN
EURO PROJECT OFFICE AG
Zurich, Europe

WHY THIS BOOK?

Today's rising complexity in society and technology can only be managed by appropriate means. One tool that has become highly important are the transfer functions used in Six Sigma, but also in search algorithms lie Google's PageRank, or Big Data analytics.

This book is organized into four parts

Part 1)   Transfer Functions and what they are good for

Part 2)   Applications to Lean Six Sigma for Software

Part 3)   Applications to product and project management

Part 4)   Appendices – a short introduction into mathematical topics

All the examples shown are implemented in Excel in the Microsoft Office 2016 version and freely available to readers of this book. The examples in this book, including technical information needed to run the examples, are available to readers of this book. Interested readers must send an e-Mail to info@e-p-o.com with a proof of purchase and a valid e-Mail address. Access is personal, encrypted and protected. This is necessary since the examples contain open VBA code that otherwise can be compromised.

ACKNOWLEDGEMENT

Important contributions originate from discussions and workshops with colleagues from the Software Metrics and the Quality Function Deployment communities; especially, but not limited to, Luigi Buglione, Eberhard Kranich, Sylvie Trudel and the AHP working group of the German QFD Institute.

TABLE OF CONTENTS

PART ONE:

TRANSFER FUNCTIONS

AND

QUALITY FUNCTION DEPLOYMENT

CHAPTER 1: LEAN SIX SIGMA

Short introduction into Six Sigma: DMAIC, Process Control, Defect Elimination, Design for Six Sigma. Quality Function Deployment is one of the key methods for predicting controls in Six Sigma. Design for Six Sigma and Quality Function Deployment predict process responses. How does this work?

Why do controls influence the response and not vice-verso? Six Sigma analysis links controls to responses in a cause-effect relationship. Six Sigma is about reducing variations in processes – be it production or development by strictly focusing on controls that affect the response. However, that raises a bunch of questions: what means process? What means tolerance for variations? In addition, how to design a process, such that it is possible to predict how many times exceptions do occur?

All that is supposed to be "lean" – avoiding 無駄 *muda, waste. However, what means "lean"? Does it have a mathematical foundation, just as Six Sigma with mathematical statistics has?*

1-1 SIX SIGMA – REDUCING VARIATION

Six Sigma orientates products and services towards measurable customer values. It provides several tools and methods to eliminate and prevent defects whereby anything, which diminishes customer values, is termed a defect.

1-1.1 WHAT MAKES A SIX SIGMA APPROACH SO SPECIAL?

A Six Sigma approach has not much to do with bugs found by engineers: non-adherence to specification for instance does not automatically imply that customers see this as a defect. Specifications may be wrong, or not provide any value to a customer, and then non-adherence to specification is not a defect. This makes defect counting difficult, especially in software, as only the customer value counts. Mistakes that engineers classify as minor may end up as major defects in customer's perception, and vice-versa. Customers might ignore even major failures of software if this does not hurt their business values.

Six Sigma for Software liaises with Requirements Elicitation and Engineering practices (the process areas RD and REQM in CMMI (Siviy, et al., 2008) rather than with Reviews and Testing (VER and VAL). Defect containment focuses on finding missed requirements early enough. Six Sigma metrics are goal-oriented whereas traditional software metrics such as bug counting are solution-oriented. Six Sigma thinking is much nearer to agile approaches than to anything else. To make processes measurable, Six Sigma requires translation between the different views of customers and engineers, of users and administrators, or developers and testers. Transfer functions provide such translations.

1-1.2 A SIMPLE SIX SIGMA CASE

Statistics use multidimensional vector spaces for modeling statistical behavior. The following short excursion may help readers to understand statistical thinking.

Consider the problem of throwing a ball into some target. For instance, the target may be a basket (Figure 1-1). Metrics for reaching the goal are simple: the ball must fly into the target basket, which may have any position in the three-dimensional space. Three degrees of freedom exist for positioning the basket - distance, direction and height. The tolerance range is the diameter of the basket's opening. Touching the basket's ring is acceptable if the ball still falls into the basket. This is the goal side of the problem. The solution side is intrinsic.

Figure 1-1: A Spring Canon

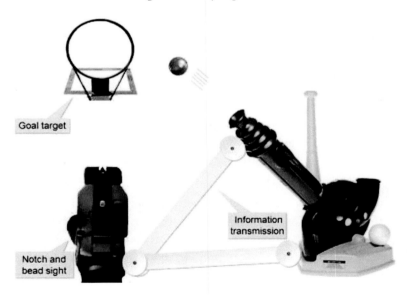

Goal target

Information transmission

Notch and bead sight

The investigated solution is a cannon that transports the ball with a spring into a well-definable direction. Greek researcher of the 3rd century BC already developed a ballistic theory to understand how the spring cannon puts the ball into target. Although mostly forgotten, this work was mostly done at the University of Alexandria (Russo, 2004).

There are two degrees of freedom for the spring cannon: One is the angle of the eject device, the other is the force impact on the ball. The direction of the spring cannon is no longer free; it depends from the already fixed position of the target basket.

There are two additional degrees of freedom needed to build the ball-throwing system. One is the accuracy of the optical system for measuring distance and direction of the target basket. Imagine a system with notch and bead sights. The other degree of freedom is the accuracy of the information transmission mechanism between the optical system and the throwing device; for instance, a control shaft transmitting information from the optical system to the spring cannon. Together, these four dimensions control the spring canon.

The function that maps these four solution space dimensions into the three space dimensions defining the target's position transfers a solution with four degrees of freedom into a target result with three degrees of freedom. The solution space must always have more, or at least equally many, dimensions than the goal space. Only then the system is flexible enough to meet goal space requirements. Other solutions might exist with more or even with less degree of freedom. The one just described with four dimensions might just be the most effective conceivable system.

Unlike the engineer who tries to find a mechanism transforming optical information into a spring cannon angle and a spring force impact according his ballistic theory, the Six Sigma transfer function looks at all four degrees of freedom for the controls that influence the outcome. The Six Sigma view is for system capability. Six Sigma is concerned with statistical variations of the controls. *Design for Six Sigma* (DfSS) targets at solving at design stage or at redesigning an existing process or product; DfSS uses *Quality Function Deployment* (QFD) for finding the proper controls. With an engineered device at hand, it is possible to measure the variations of the four control dimensions, controlling all four degrees of freedom *Chapter 4: Quality Function Deployment* will explain how to calculate optimum controls.

In either case, the result is a matrix with four solution control columns and three goal rows; each matrix cell records the impact that this solution dimension has on the respective goal dimension. For instance, a variation in the optical system may have substantial impact on the flight of the projected ball because adjusting the vertical orbit is more critical than the target direction. If in contrary the information coupling between the optical system and the spring cannon can use sufficient numerical accuracy, then no impact on targeting might be observed.

1-2 THE SECRETS OF SIX SIGMA

The basic idea behind Six Sigma is *Do It Right the First Time*.

This sounds easy and self-explanatory. However, it is not. Neither it is clear what means "right" nor what is "the first time". Right is something if results are within meets tolerance limits. How to find out what the right tolerance is? What should the right tolerance be when doing something the first time? Moreover, what means "first time"? Is Six Sigma applicable for once-upon-a-time events only, or does it apply for repeatable processes?

Six Sigma originates from making the outcome of repeatable processes predictable by systematically reducing variation. It has a statistical context: if you run a process and its output (the response) varies, you get a potentially infinite sequence of data points in the response space that statistical techniques can analyze. If the process is mechanical, as most manufacturing processes, then response distribution often is normal; i.e., it varies around a mean value μ that marks the ideal response for the process.

The *Standard Deviation* σ for a data set $x_1, x_2, \ldots, x_n$ of n normally distributed random values is the square root of the total *Variance* between all observed process responses, see equation (1-1):

$$\sigma = \sqrt{\frac{\sum_{i=1}^n (x_i - \mu)^2}{n}}, \text{where } \mu = \frac{\sum_{i=1}^n x_i}{n} \tag{1-1}$$

Given a random normal distribution, the probabilities for the response samples constitute the *Gaussian Normal Distribution*. The standard deviation marks the distance on the horizontal axis between the mean, μ, and the curve's inflection point, as represented in Figure 1-2.

A normal distribution is indicative for the presence of random influence factors, including measurement errors, which affect the responses. Variations go equally in all directions. The probability that an event occurs within the normal distribution area is one. No response will ever arrive outside the normal distribution. That makes a normal distribution an attractive model for Six Sigma. If the normal distribution area fits within lower and upper specification limit, we did it right the first time.

Next to the standard deviation σ, there is a popular process performance index C_{pk} indicating whether process responses fall in a range specified by lower and upper specification limit LSL and USL.

$$C_{pk} = min \left[\frac{USL - \mu}{3\sigma}, \frac{\mu - LSL}{3\sigma} \right], \sigma, \mu \text{ defined as in (1-1)} \tag{1-2}$$

A Six Sigma quality process has $C_{pk} = 2$, because if USL and LSL both are equal 6σ and the process center around mean value $\mu = 0$, the minimum of the pair is $6\sigma/3\sigma = 2$, while ordinary process capabilities are in between 1 and 2. However, experience has shown that processes usually do not perform as well in the long term as they do in the short term. Thus, the number of sigma intervals fitting between mean response and specification limit may well drop over time, compared to an initial short-term study. An empirically based 1.5 sigma shift accounts for real-life increase in process variation over time (Creveling, et al., 2003). Experience has shown that processes usually do not perform as well in the long term as they do in the short term. Consequently, a process fitting 6σ between the process mean and the nearest specification limit in a short-term study will in the long term fit $4.5 * \sigma$ instead of $6.0 * \sigma$. This happens either because the process mean will shift over time, or because the long-term standard deviation of the process will exceed the observed short-term standard deviation, or for both reasons, compare Figure 1-2.

Figure 1-2: Gaussian Normal Distribution

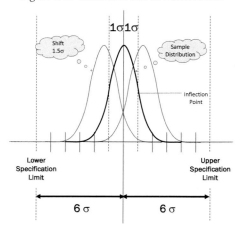

Hence, the widely-accepted definition of a Six Sigma process is a process that produces 3.4 defective parts per million opportunities (DPMO) (Creveling, et al., 2003). This is because a process that is normally distributed will have 3.4 parts per million beyond a point that is 4.5 standard deviations above or below the mean (one-sided capability study).

The 3.4 DPMO of a Six Sigma process corresponds to 4.5 sigma; namely six minus the 1.5 sigma shift that was introduced for dealing with long-term variation. This prevents underestimation of the defect levels that likely occur in real-life operation. The Six Sigma level of a process corresponds to its yield.

Table 1-3: Percentage Yield of Processes (long term) according El-Haik (El-Haik, 2005)

Sigma Level	Sigma (1.5σ shift)	DPMO	Percent Defective	Percentage Yield
1	-0.5	691'462	69%	31%
2	0.5	308'538	31%	69%
3	1.5	66'807	6.7%	93.3%
4	2.5	6'210	0.62%	99.38%
5	3.5	233	0.023%	99.977%
6	4.5	3.4	0.00034%	99.99966%

Table 1-3 gives long-term DPMO values and their correspondence to short-term sigma levels and yields. These figures assume that the process mean will shift by 1.5 sigma toward the side with the critical specification limit. In other words, they assume that after the initial study determining the short-term sigma level, process capability will turn out to be less than the short-term capability value. Defect percentages indicate only defects exceeding the specification limit to which the process mean is nearest. Defects beyond the far specification limit are not included in the percentages.

However, today's world is full of production series whose size is just one. The most stringent example is writing software. Reproducing software by legal copying is never the problem and is defect-free. However, not only software, any problem solving and decision-making activity is a one-time undertaking. How should Six Sigma help to reduce variations, if no variation exists? How should the process become predictable without reducing variation? Many attempts were made in the last decennia to make use of Six Sigma techniques, for instance to deal with software defects. Among the best approaches is that of Fenton, Neil, and Marquez, which applies Bayesian networks for prediction (Fenton, et al., 2008).

This book presents a new approach developed during more than twenty years of applying Six Sigma to software development. It uses Six Sigma statistical techniques, adding *Transfer Functions* from signal theory and *Eigenvector* solutions to the Six Sigma toolbox. This chapter introduces the notion of transfer functions conceptually; for the mathematical foundations, see *Chapter 2: Transfer Functions*. Although most of the examples originate from software development, the approach is valid for all kind of problem solving, including product development and placement, decision metrics, decision methods in politics and economics, and might even help financial industry doing better investments and adapt strategy.

The approach is straightforward: since we do not have data point series, we need goals for a process and a means to predict what the process will deliver. Software development is a knowledge acquisition journey and not civil engineering. Thus, software engineers need goals not plans. Moreover, they need a compass telling them whether they are underway towards these goals or whether they are going to miss them.

Six Sigma transfer functions do this. A transfer function models a process, and the process needs not producing series. In this respect, transfer functions extend the traditional Six Sigma toolbox considerably, and make it applicable to once-upon a time undertakings such as developing a new software product. Even with repeatable methodical components, it is not easily possible to apply statistical process control to software development. Decision processes, as needed in management, or in any other complex undertaking, are also not following statistical behavior. Nevertheless, you can do decisions, and develop software, right the first time. The Six Sigma *Define – Measure – Analyze – Implement – Control* cycle, the famous DMAIC, if read with care, simply explains what a transfer function is.

1-2.1 DEFINE THE GOALS OF THE PROCESS

Defining goals for process is not as straightforward as it might seem. Most organizations find it quite difficult to agree on goals; most often, it requires a difficult decision process by itself. The Six Sigma toolbox offers a wide selection of methods how to define goals; this book lists and explains the most important ones; see *Chapter 4: Quality Function Deployment* and other chapters referenced therein. As other process improvement initiatives, Six Sigma largely focuses on customer's needs as the goal, at least for commercial organizations. Nevertheless, methods such as Net Promoter® Score (*Chapter 5: Voice of the Customer by Net Promoter®*) and New Lanchester Strategy (*Chapter 11: Application to Product Management*) allow handling situations where the notion of customers is not so obvious, where the market's voices are rather obscure, potential customers difficult to guess and their opinion impossible to measure directly.

Figure 1-4: Define Level of Process Response

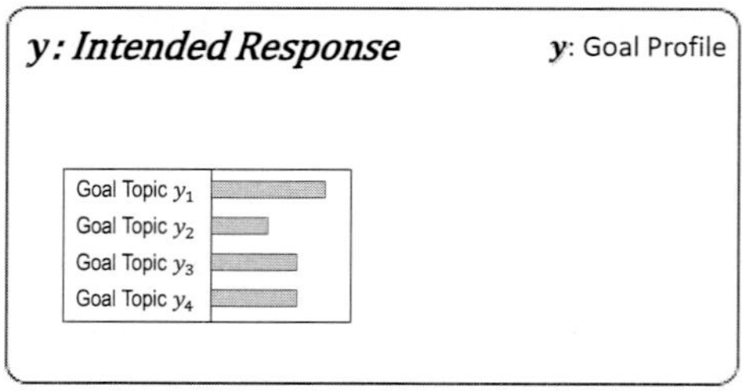

In either case, goals come with *Weights* reflecting their relative importance; see Figure 1-4. A *Goal Profile* displays a relative set of weights, assigning each goal topic a weight in comparison to other goal topics. Describing a profile is possible by reference to the percentage of the total for each component; then the components are dimensionless numbers between 0 and 1, or 0% and 100%. The sum of weights totals to 1, or 100%. In *Chapter 2: Transfer Functions*, vectors in a linear vector space will represent such goal profiles rather than weights in percent; this has a few distinct advantages, and many consequences, as we will see later.

1-2.2 MEASURE THE PROCESS

Before and while executing a process, measurement of data pertaining to the process allows making goals measurable. Process measurements must include parameters that allow measuring goal topics, and thus determine their respective weights. Otherwise, it is impossible to measure whether the process ever meets its goals.

With user-defined upper and lower specification limits, it is necessary to measure the response to know whether it is within specification limit or not, as symbolized in Figure 1-5. Measuring is straightforward once physical entities are involved, but become increasingly difficult if service or software process responses are measured.

For instance, the response of a development process for a software product should be sales revenue, which is easily measurable but difficult to attribute to some specific goal topic. It could relate to software functional size, to its quality, its low defect density or else; thus, it is necessary to measure the software process and separate it from the sales process.

Figure 1-5: Measure Level of Process Response

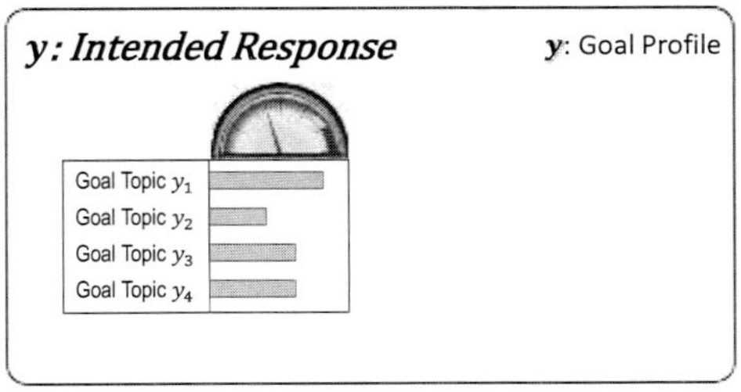

The amount of measurements needed interacts with the Analysis step. To a certain extent, the Measure and the Analysis phase in DMAIC work in parallel. Preliminary analysis induces new measurements, and any such measure improves analysis capacity. This book discusses measurement aspects in *Chapter 3: What is AHP?*, *Chapter 5: Voice of the Customer by Net Promoter®*, *Chapter 6: Functional Sizing* and *Chapter 10: Software Testing and Defect Density Prediction*.

1-2.3 ANALYZE THE PROCESS

The outcome of the process A matches the expected response, called goal profile in Figure 1-5. How to assess whether the process A is a Six Sigma quality process, with known variation and responses within upper and lower limits? The controls are unknown. A function A^{T} predicting the controls is needed for investigation and analysis. The Analysis phase of the DMAIC methods aims at finding the prediction function A^{T}.

Most measurements in Six Sigma follow the *Design of Experiments* (DoE) paradigm (Creveling, et al., 2003). It means that a few candidate controls are selected that are tested by experiment whether they have impact on the process response. A DoE measurement might happen once per each pair of response component and candidate control, resulting in a matrix of measurements.

The measurements reference in Six Sigma are the controls; see Figure 1-6. DoE always refers to some specific controls in effect. Often, in practice, this is quite clear; however, sometimes it is not. If it is not, the measurement phase and the subsequent analyze phase interact and go in parallel. Analyzing the measurements should allow understanding how the process works. What are the reasons for variations observed in the process? What is needed to make it compliant to specification limits?

The main problem with measurements at this stage is measuring against what. Controls affect the process such that responses keep within the Six Sigma range. Finding these controls is the most critical step in Six Sigma.

Figure 1-6: Analyze Level of Process Response

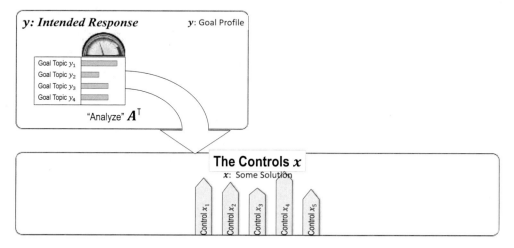

Many controls exist that have little or no impact on response, but some are critical. Usually, when analyzing, there are many candidate controls to consider.

1-2.4 Implement the Process

With m different response aspects (a "response profile") and n controls, one has $n *$ m measurement points. This creates a matrix, the *generic Six Sigma matrix*, linking controls to responses. The reverse of the generic Six Sigma measurements matrix defines a *Transfer Function* that explains the observed signals (responses) by suitable causes (controls), shown as A in Figure 1-7. Thus, the transfer function explains the causes for the effects observed.

For instance, for analyzing a software process, the generic Six Sigma matrix is very simple. It assesses which software requirement affects which business driver, or what functionality. The ultimate responses from the software process are something making the software product attractive to buyers, or users. Features that create value drive software business. Software components may be features or services that influence business drivers; identifying these is simple and mostly straightforward. Measuring the software process is simpler than many other processes outside the manufacturing floor, especially when adopting an agile development approach as shown in *Chapter 8: Lean & Agile Software Development*.

Figure 1-7: Implement Level of Process Response

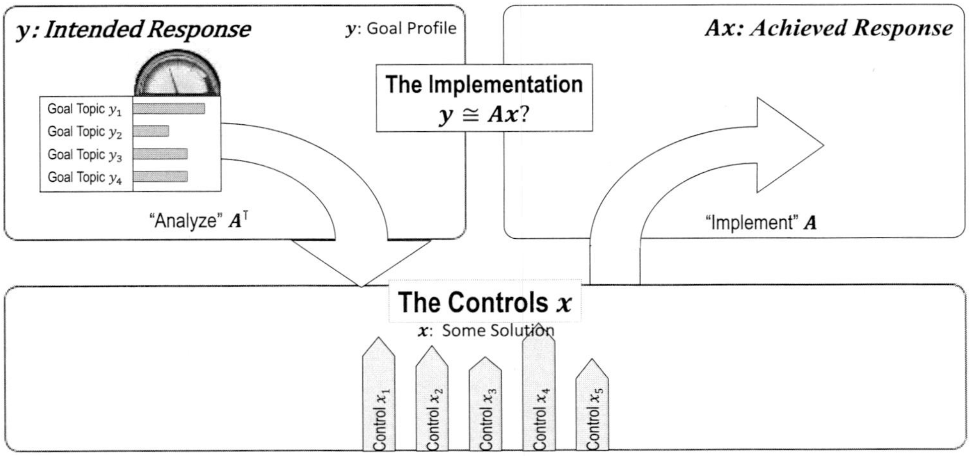

Agile software teams use measurements to identify which functional and non-functional tasks meet the expectations of the customer and its strategic business goals that drive that software development project (Buglione, et al., 2011). This can be done interactively, visualized on pin walls, involving the team. It causes almost no additional cost because it is part of the agile software development process.

More complex but still feasible is measuring and analyzing cost drivers to understand total cost, for instance for estimating projects. Examples include complex software projects, but also undertakings such as building a new airport from scratch – where each change decision affects almost all cost drivers such as environmental protection, safety and security. The key to successful cost prediction is to understand, how changes in one cost driver affects all the other costing factors. The Six Sigma transfer functions introduced in *Chapter 2: Transfer Functions*, and *Chapter 12: Effort Estimation for ICT Projects* do cost estimations. In Figure 1-7, Ax represents the response of the process.

1-2.5 CONTROL

The measurements done per each pair of goal topic and control represents the current process design, based on the controls provided.

Figure 1-8: Control Level of Process Response

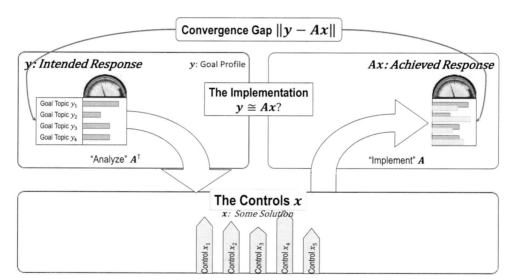

Controlling the process might involve changing the controls, add others, and replace some, influencing the coupling between controls and responses. Weak couplings between team size and, say, work output – since increasing the team size does not give you more value – can be changed by adding structure to the team; e.g., by creating competitive small sub-teams with clearly separated tasks, and the coupling could increase close to normal. By adding quality enhancements, it might be possible to close the gap towards customer's expectations, potentially at higher cost.

There is not much difference between implementing a new process, and improving an existing one. Sometimes, some of the controls turn out to be superfluous. Removing controls reduces cost. However, removing controls also decreases the degree of free-dom needed to make your process capable to deliver right the first time. With less controls, it might become difficult to predict the process response reliably. Less controls means less possibility to influence the software development process.

There is an optimum selection of controls. An optimum has balanced controls with equal impact, however on different response topics. Ideally, there is one control per response topic, but such an ideal case is seldom, and it is not what we consider high complexity. Complexity originates from unwanted and uncontrolled side effects.

Because of measurements and analysis, controls have a profile, like the response pro-file. Let x be the control profile and Ax the process response. While y denotes the profile of the expected response, the optimum selection of controls aims at $Ax = y$. The difference between Ax and y is the *Convergence Gap*, visualized in Figure 1-8 and mathematically introduced in *Chapter 2: Transfer Functions*.

The convergence gap is the single most important quality criterion for a Six Sigma process. It tells whether the controls found are good enough to explain the observed, or expected, behavior of the response. Moreover, measurements errors or wrong con-trol selection might affect the analysis. However, from a large convergence gap it is not easily derivable what causes the gap. Mathematically speaking, the information content of the convergence gap is not specific enough to advise a fix for a bad process.

More information is available when looking at the difference between Ax and y per goal topics. Each goal topic, if negative, indicates a lack of support by the process' implementation; positive goal topics indicate over-fulfillment. Even so, it is not gen-erally possible to derive whether measurement errors or a bad selection of controls causes the gap. Only a domain expertise does help. Consequently, transfer functions are a constructive approach to problem solving, not a general problem solver. General problem solvers do not exist; see (Engeler, 1995); however, constructive approaches approximating the solution are often available.

Thus, reducing the convergence gap is a task left to domain experts. Only a small con-vergence gap indicates that the process is under control. For instance, if cost drivers control the capital spending process, the control profile defines the allocated budget to each organization units that represent the controls. If in contrary, the controls rep-resent user stories in a software development process, the control profile defines their priorities. In any case, the control profile is how you numerically control the process such that its response falls within the expected response range. When executing the process while keeping controls within the range that keeps the response within the goal profile, the process response "is under control".

1-3 THE THREE FUNDAMENTAL SIX SIGMA METRICS

To reduce variations, Six Sigma takes the following three fundamental kinds of measurements into consideration:

1. Response profile measurement; e.g., profiles of customer's needs or strategic goals. Such measurements are often based on measuring people's opinion, be it executives' or consumers' opinion. The typical approach to such measurements is the *Analytic Hierarchy Process* (AHP);

2. Transfer function measurements, measuring the coupling between controls and responses, typically one measurement per cell of the controls/responses matrix. Here the widest range applies for the kind of measurement applied; from consensus-based workshops to physical measurements and regression analysis based on a series of measurements for response and control profiles;

3. Sometimes, it is possible to compare control profile with measurements of solution profiles. Then the process controls compare with the solution profile by direct measurements. This is the optimal case because the difference between the goal profile y and the effective response profile $y' = Ax'$ for a measurement x' of solution profile x seconds the convergence gap. Moreover, the difference between the achieved solution profile $y_A = Ax$ and the effective response profile y' indicates process capability. These three gaps are called the *Three Six Sigma Metrics:*

$$\text{Convergence Gap:} \qquad \|y - y_A\| \qquad \text{where } y_A = Ax \qquad (1\text{-}3)$$

$$\text{Effective Gap:} \qquad \|y - y'\| \qquad \text{where } y' = Ax' \qquad (1\text{-}4)$$

$$\text{Process Capability:} \qquad \|y_A - y'\| \qquad \text{where } y' = Ax' \text{ and } y_A = Ax \qquad (1\text{-}5)$$

Process capability refers to the implementation of the process, targeted at solution profile x; however, implemented was x' not x. The three Six Sigma metrics refer to responses not controls. Transfer functions provide a measurement theory based on direct and indirect measurements of controls. Detecting exoplanets and other physical measurement methods use exact that approach – under condition that calibration exists with measurement units.

In practice, controls are not always easily measurable; in this case, only the convergence gap is effectively measurable. If there is no direct measurement method available for the controls, the solution profile x' remains unknown, $y' = Ax'$ cannot be calculated, and consequently two of the three Six Sigma metrics remain theoretical.

1-4 LEAN SIX SIGMA

Lean has become a buzzword for manufacturing and now as well for various other industries including software development. Agile Software Development is no longer good enough; neither is Six Sigma. Nowadays, it is *Lean & Agile Software Development* and *Lean Six Sigma*, even if the tools and methods did not change much. The concept of Lean is appealing and attractive, and easy to communicate and generate consensus. Thus, the lean mindset is an asset when managing complexity. Because of this, *Lean* is integral to this book's topics.

1-4.1 LEAN IN MANUFACTURING

Lean in manufacturing is associated with no 無駄 (Muda), "no waste", or in positive terms, optimum use of material resources, effort and energy. George in (George, 2002) and (George, 2010), has defined Muda as *Transportation, Inventory, Motion, Waiting, Overproduction, Over-Processing*, and *Defects* – abbreviated by TIMWOOD. Manufacturing processes generate value, at least for the immediate "next customer" in the value chain. This sounds simpler as it is. Many manufacturing processes suffer not only from needless process steps that only add cost or are done for traditional reasons only. Some of these "waste" steps are for eliminating risks or increase process capability, eliminating variations from processes. Does risk-avoidance introduce over-processing or not? Lean and Six Sigma are at first glance contradictory. So, how can Lean and Six Sigma brought together? How much risk and variations can a manufacturing process tolerate?

Essential for Lean Six Sigma is the customer viewpoint. Eliminating variation from a process makes sense only if somebody, the customer, is ready to pay for it. Otherwise, if a process is widely within tolerance limits but costs more than another process that is lax but still capable, why should the customer pay more?

If Six Sigma process engineering adds stuff into a process that needlessly decreases variation, this is waste produced, Muda, and therefore not lean. Thus, criteria are needed that can determine whether a process is lean or not.

In short, Lean Six Sigma is Six Sigma statistical process control with a clear goal profile setting. Thus, adding "Lean" to "Six Sigma" means requiring that the transfer function mapping the Six Sigma controls – controlling process variation – to the customer's needs – defining the goal profile of the process response – becomes minimal with respect to the number and complexity of controls needed.

1-4.2 LEAN SIX SIGMA IN SOFTWARE

Producing software is not a mass production process, not even an industrial engineering discipline. Software development is knowledge acquisition and information knowledge deployment into mechanical and commercial systems. Six Sigma for Software is not only defect density reduction in the implementation of requirements to software; it must take defective requirements and incomplete knowledge deployment into consideration as well. Muda in software is different from in manufacturing: transportation, inventory and motion are irrelevant, while waiting, overproduction and defects become even more critical. Over-processing is a rather rare phenomenon in software development.

It is virtually impossible to do Six Sigma for Software without some Lean component. Defects are a concern for all software artifacts ever written or run, and waiting for software being ready for operation is a major concern in almost all software development projects. The only sources for requirements are users and customers of the software, or of the software-driven device. Even in special domains as highly technical and high-reliability software; e.g., in the medical or military domain, Lean is part of every Six Sigma approach even if risk avoidance dominates all other Muda concerns.

On the other hand, the software development process might also suffer from not being lean. One major issue in many projects is that features are implemented that are not necessary for the customer. The reason might be convenience, avoiding technical debt, or simply not understanding customer's needs. However, if a customer does not request a feature, it does not mean that this feature is wasted. Customer requirements are usually all but complete; features might be needed because they make it easier to implement required features.

1-4.3 LEAN CONTROLS IN SOFTWARE

Mapping controls that limit variations of the process is the aim of the method known as *Quality Function Deployment* (QFD). The most renowned QFD transfer function is the *House of Qualities* (HoQ). The HoQ maps process characteristics, or qualities, to the process response, well-known as *Customer's Needs* (Herzwurm & Schockert, 2006); demanded, and expected, qualities by customers. If the customer is not present in person; e.g., for software product development, often the term *Business Drivers* (Denney, 2005) is used instead of customer's needs.

A *Lean Design for Six Sigma* means mapping design characteristics to software product features. It allows choosing design characteristics that optimize qualities and product features. This is a QFD transfer function; design characteristics are the controls.

While the criteria for Lean Six Sigma in manufacturing seem not difficult to identify, given the TIMWOOD criteria and the ability to measure key performance indicators in manufacturing easily, for software the only relevant reference measurement are customer's needs. Customer's needs represent business value for the user or the customer of software. Value decreases with functional defects delivered, or missing functionality, usability or attractiveness. Sometimes, excess functionality also can decrease value of software. Lean Six Sigma for Software includes requirements elicitation, defect density prediction, usability metrics and functional effectiveness.

1-4.4 MAKING TRANSFER FUNCTIONS LEAN

There is a simple approach to lean with transfer functions. Google made it popular with its famous search engine based on the PageRank methods (Gallardo, 2007). As with any measurement expected to establish the choice of controls, the transfer function A must explain the observed response y by the profile found, denoted in Figure 1-9 by x_E. The convergence gap must be small; i.e., the achieved response $y_E = Ax_E$ must be close enough to y.

Figure 1-9: DMAIC with Eigenvector Solution for Process Response

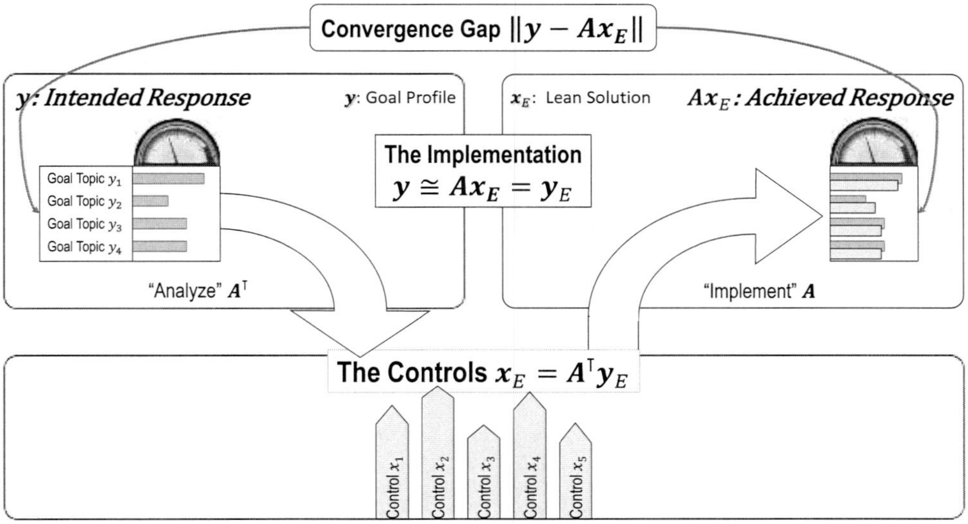

Assume there is some "oracle" denoted as $A^\mathsf{T}y$ guessing a profile $x_E = A^\mathsf{T}y$ for any chosen set of controls. *Chapter 2: Transfer Functions* will explain how to construct such oracles. This oracle $x_E = A^\mathsf{T}y$ predicts the control profile $x_E = \langle x_1, x_2, \dots, x_n \rangle$, enabling users to select and prioritize controls, such that they can start with the most important first and fine-tune with the least expensive ones. Traditionally, QFD aims for this; however, the methods used for finding such a profile for x_E often rely more on experienced moderators than on mathematical reasoning and evidence.

If the oracle $A^\mathsf{T} y$ is not perfect, as happens for oracles from time to time, the convergence gap tells the difference. If the oracle is perfect, then the process becomes lean.

Thus, if such an oracle exists for the transfer function A, the selected controls can deliver a process response $y_E = Ax_E$ close enough to the desired response y. Then, the process repeats without variation. The convergence gap is an indicator for Lean, like the standard deviation sigma for the degree of variation in a process.

1-5 LINKING RESPONSES TO CONTROLS IN QFD

In QFD, we need goals that are measurable by *A Ratio Scale* (Michell, 1986). Traditionally, the scale ran continuously over numbers between zero and five. Throughout this book, we prefer using a ratio scale 0 to 1, allowing profiles represented by unit length vectors. This has advantages for calculating convergence gap and comparing profiles, and it supports the use of vector profiles that define the vector's direction.

1-5.1 RESPONSES

Responses constitute vectors in the *Response Space*, a linear vector space whose dimension is the number of criteria compared within the response. Sample Six Sigma process responses are customer's needs, the dimension of the linear vector space are the various topics mentioned in the list of customer's needs. Value criteria for the customer do not consist of a ranking only but have assigned a ratio scale value on the 0 to 1 scale.

Comparing one profile with another works well with normalized profiles. Older attempts to QFD have used maximum metrics instead of normalization to vector length. Strictly speaking, it compared ranking only. When using ratio scales, comparison is based on recognizing their nature as (statistical) event vectors. To compare direction not length, the vectors must have equal length. For instance, the vectors $\langle 1,1,1,3,1 \rangle$ and $\langle 0.3,0.3,0.3,0.9,0.3 \rangle$ point to the same direction but have different lengths by a factor of $3/0.9 = 3.33$. See equation (2-23) how to compute the length of a vector. Thus, if you use transfer functions, you must do it right.

Traditional QFD has a long history of practitioners who applied bad mathematics to QFD matrices because they did not understand that they were doing statistical investigations in a multi-dimensional vector space. The direction of the process response modeled in QFD matters, not its length. The profiles always have relative values, one criterion is three times as important as the second criteria, and the third and fourth are somewhere in between. Teams doing QFD optimize the direction of the response vector, by adding or removing weights to the controls, until the system delivers the required response. Translating relative ratio scales into absolute values; e.g., for ranking, for cost priority or other constraints, is simple.

1-5.2 CONNECTING CAUSE TO EFFECT

It is the role of the transfer function A to model that behavior as a mapping between the solution controls and the response. This mapping is linear in most of the time, and always for QFD. The reason why, is because quality is linear – there are no jump in quality; however, there might be jumps in perception. Quality increases together with value; but customers may perceive high value as a delighter and react differently. They become loyal customers and do not compare any longer with competition because they feel comfortable and secure with a product or service. This is a result of passing a threshold, not because quality is overwhelming.

QFD often goes together with the New Lanchester strategy (see *section 11-3.1* in *Chapter 11: Application to Product Management*) that explains such paradigm shifts, but also can predict similar shifts for the future. Adding controls to a QFD is costly but can be very rewarding, especially if others did not think of.

Choice of topics for the goal response is critical, but selection of control topics is even more rewarding. Goal topics are often not freely selectable, because they must meet something that customers recognize. Moreover, make customers recognize something they do not know is difficult. On the other hand, solution controls are more freely selectable, and are the main source of product innovation. Controls influence the response and not vice-versa, but people think this unbelievably difficult to understand. A mental barrier in humans fixes them thinking towards solutions that once worked, rather than to the goals that are worth aiming at and blocks trying other controls.

1-5.3 LEAN LINKS BETWEEN CONTROLS AND RESPONSES

As Saaty says, humans are not logical creatures: "Most of the time we base our judgments on hazy impressions of reality, and then use logic to defend our conclusions" (Saaty, 1990). Therefore, many couplings between controls and responses come with little objective evidence and clumsy explanations. It is helpful to consider each cell in a generic Six Sigma matrix means work effort for somebody or something. In case of hardware, a cell represents the physical cause-effect coupling between two topics. The work needed for the coupling may consist of physical force between hardware components. In case of service or software development, the work effort needed for the coupling typically consist of payable human work. Thus, it may be waste, or it might create value. Traditional QFD looked at linking controls to responses by means of correlation. The cells represented physical constraints linking responses to controls. In today's Six Sigma, correlation constitutes from the work effort needed to make the system do what the customer expects.

Consequently, QFD matrices with lots of controls and fully completed cells are costly and suspect to waste. First, with less controls your process becomes less complex. Second, the more cells you should keep under control, the more difficult it is to keep cross-effects under control. Ideally, controls are independent from each other. Even if they are not, some degree of independence is always preferable. Independence is easily recognizable from the matrix because it becomes sparse.

Controls that need the least links to responses are preferable. The emptier cells in the matrix, the less costly links exist between controls and responses. Moreover, the impact on responses is less suspect to unstable behavior. Unstable means if controls slightly change, the impact should be predictable and small as well. *Chapter 2: Transfer Functions* will explain the mathematics behind this. The reasoning behind minimizing the number of links between controls and responses are the costs.

Contrary to older approaches, design in *Design for Six Sigma* (DfSS) means not only the selection of controls but also the choice of links. A matrix is like a working plan, a design, for a service or a software. In hardware, the matrix is still representing physical dependencies and constraints, unless the hardware has become intelligent by the Internet of Things. Petersen (Petersen & Wohlin, 2010) shows how measurements make matrix cell links lean.

1-6 CONCLUSIONS

Lean is a very strong concept, although not yet fully arrived in Six Sigma and process improvement. For solution profiles in transfer functions, we have now a concept that explains in simple terms what lean means. You even can measure how lean you are, by measuring coupling totals in its matrix representation. Comparing total cost with priorities creates insight into the benefits for Lean. *Chapter 12: Effort Estimation for ICT Projects* elaborates this approach.

CHAPTER 2: TRANSFER FUNCTIONS

Transfer functions are the most innovative mathematical invention of the 20th century. Although transfer function ever existed, until this time by lack of a theory people applied trial and error methods, sometimes also by regression analysis, to solve the problems.

This chapter presents the theory needed to understand transfer functions and what they contribute to solving complex problems. It contains some references to basic mathematics. Samples explain how it works.

2-1 INTRODUCTION

What was the most impressing invention in the 20th century? During the 20th century, many inventions from the 19th century, such as trains, cars, phones, electricity, matured. Planes were invented in the 19th century as well but only took off in the early years of the 20th century; some 20th century achievements like nuclear bombs and power plants probably are no longer that impressive, as they once seemed to be.

Figure 2-1: Transfer Functions in Various Disciplines

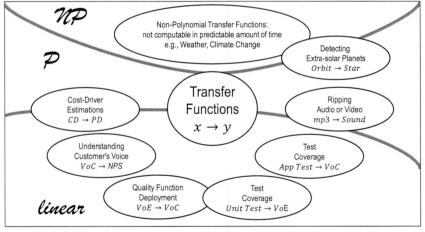

However, a few inventions changed the world, much more than cars and railways ever could do (Figure 2-1). Sure, Ada Lovelace, the first programmer, and Charles Babbage conceived digital data encoding and computing in the 19th century as well; however, the proof that the Fast Fourier Transfer (FFT) is not NP-complete made the deepest impact on humankind probably since invention of cooking by a fire (Cooley & Tukey, 1964). It led to the possibility of transferring audio signals in predictable time into digital code. Previously, sounds were analogous oscillations of the atmosphere, or of electro-magnetic potentials, caused by beings or loudspeakers. The FFT did allow describing with digital numbers what the original oscillations where. Thus, it became possible to build chips that transformed audio signals into storable digital code and later video signals as well. All the music and entertainment stuff that today makes up for a trillion-euro industry started with that invention. Today it populates computers, laptops, phones and television sets connecting by a global Internet.

The development of the FFT procedure is the result of long years of solid research in pure mathematics, especially in linear algebra. FFT transforms audio signals captured by a microphone as electrical pulses into digital numbers. To be more precise, the FFT algorithm selects the Fourier base functions in the functional vector space that model the analogous signal and represents the signal by the coordinates of the unit vector in this space. This works like a linear vector space. However, the base units are functions, not points in space. Finding controls for a transfer function to explain the observed response resembles the FFT procedure.

Eigenvector theory is another achievement of linear algebra that changed the world; however, not before the start of the 21st century. For instance, eigenvector theory is employed in the *Analytic Hierarchy Process* (AHP) developed by Saaty (Saaty, 1990) [175] to solve multi-criteria decision problems, in the football teams ranking method proposed by (Keener, 1993) or, more recently, for powering Google's PageRank algorithm. The name 'PageRank' refers to Larry Page, one of the founders of Google. The Eigenvector theory is now widely in use and paved the way to the Internet age.

2-2 TRANSFER FUNCTIONS – MATHEMATICAL FOUNDATIONS

As explained in *Chapter 1: Lean Six Sigma*, the goal of the Six Sigma methodology is to improve an existing process, product or service by eliminating or considerably decreasing the number of defects relating to the process, product or service output, which does not fulfill predefined customer specifications. Fundamental assumptions of the Six Sigma methodology are that the process, product or service is measurable or become measurable by suitable measurement methods, and observations include an enough representative defect data to initiate and to conduct the necessary improvement. Hence, Six Sigma is a retrospective or reactive methodology. In contrast, the *Design for Six Sigma* (DfSS) methodology is proactive, since it applies when a completely new measurable process, product or service is being designed or an existing measurable process, product or service is under redesign, with the aim that stated customer specifications are met within a predefined Six Sigma quality level.

DfSS is proactive by utilizing the transfer functions concept or transfer functions model that identifies and reveals the relationship between the input parameters or *controls* of the considered process, product or service, and the corresponding output termed the *response*. Evidently, the input control parameters are subject to variations, so that the response becomes predictable when altering selected input controls. Thus, the transfer functions concept is of interest for limiting variations against *Critical-To-Quality* (CTQ) or *Critical-To-Satisfaction* (CTS) requirements of the future process, product or service. A well-defined transfer function predicts the performance and quality of a product, process or service prior to a potentially cost intensive implementation of a prototype. Literature references include (Creveling, et al., 2003), El-Haik and Shaout (El-Haik & Shaout, 2010), Fehlmann and Kranich (Fehlmann & Kranich, 2011-2), Hu and Antony (Hu & Antony, 2007), Maass and McNair (Maass & McNair, 2010), and Yang and El-Haik (Yang & El-Haik, 2009).

2-2.1 WHAT IS A TRANSFER FUNCTION?

A *Transfer Function* f maps the input parameters $x_1, \ldots, x_n$ to the output y of a product, process or service, mathematically depicted as $y = f(x_1, \ldots, x_n)$. Following the usual Six Sigma terminology, the parameters $x_1, \ldots, x_n$ are *Process Controls*, denoted by the real-valued vector $\boldsymbol{x} = (x_1, \ldots, x_n)$. The output $y \in \mathbb{R}$ is the (single) *Process Response*. The transfer function f is a *Linear Single Response Transfer Function*. In case a transfer function f provides *Multiple Process Responses* denoted by the real-valued column vector $\boldsymbol{y} = \langle y_1, \ldots, y_m \rangle$, $\boldsymbol{x}$ is termed the *Solution Profile*, and $\boldsymbol{y}$ the *Response Profile* of the transfer function f. To avoid writing columns vectors vertically throughout this book, the writing convention $\boldsymbol{y} = \langle y_1, \ldots, y_m \rangle$ applies for representing column vectors.

$$\boldsymbol{y} = \begin{pmatrix} y_1 \\ y_2 \\ \ldots \\ y_m \end{pmatrix} = \langle y_1, \ldots, y_m \rangle \tag{2-1}$$

Of interest are multiple response transfer functions of the form $f : \mathbb{R}^n \to \mathbb{R}^m$. If f is a linear function; i.e., $f(\lambda \boldsymbol{x}) = \lambda f(\boldsymbol{x})$ for all $\lambda \in \mathbb{R}$, then the transfer functions f can be represented by a matrix $\boldsymbol{A} \in \mathbb{R}^{m \times n}$ such that

$$\boldsymbol{y} = \boldsymbol{A}\boldsymbol{x} \text{ with } \boldsymbol{A} \in \mathbb{R}^{m \times n}, \ \boldsymbol{x} \in \mathbb{R}^n, \ \boldsymbol{y} \in \mathbb{R}^m \tag{2-2}$$

Then f is termed a *Linear Multiple Response Transfer Function*. The term $\boldsymbol{A}\boldsymbol{x}$ represents a matrix vector multiplication, so that each response component y_j calculates as

$$y_j = \sum_{i=1}^{n} a_{ij} x_i, \, 1 \leq j \leq m \qquad (2\text{-}3)$$

To illustrate these notions considering the defect containment process as part of a software development process, see, for instance, Kan (Kan, 2004). Defect containment provides a retrospective insight into the defect distribution per stage of a project artefact, or other deliverable. The associated defect containment matrix reveals the relationship between the defect injection stages in the rows of the matrix and the defect detection and removal stages. The stages constitute the columns of the matrix; see Figure 2-2.

Since a defect does not exist prior to its inception, only the diagonal and the upper triangular part of the defect containment matrix are relevant. The diagonal matrix entries reflect the number of defects, which are detected in-stage; i.e., such defects are removed in the same stage in which they were detected. The strict upper triangular matrix entries represent those defects, which are detected out-of-stage, meaning that such defects escape or leak the stage they were injected in.

Figure 2-2: The Defect Containment Matrix

$$A = \begin{array}{c} 1 \\ 2 \\ 3 \\ 4 \\ 5 \end{array} \begin{pmatrix} a_{11} & a_{12} & a_{13} & a_{14} & a_{15} \\ & a_{22} & a_{23} & a_{24} & a_{25} \\ & & a_{33} & a_{34} & a_{35} \\ & & & a_{44} & a_{45} \\ & & & & a_{55} \end{pmatrix}$$

If the defect detection and removal cost per stage i are of interest, then Figure 2-2 reflects a single response transfer function. Each a_{ij} denotes the number of defects injected in stage i, and detected and removed in stage j; each x_j denotes the (a-priori known average) cost of defect detection and removal in stage j, and y_i the aggregated cost of each stage i where defects were injected in.

2-2.2 DERIVING SINGLE RESPONSE TRANSFER FUNCTIONS

A transfer function f may be simple like that in equation (2-3), polynomial or a composition of other (transfer) functions. If experimental or historical data are available and a transfer function is not obvious or pre-defined, the function can be derived by means of Design for Six Sigma (DfSS) tools. For instance, *Design of Experiments*, see e.g. Montgomery (Montgomery, 2009), or *Response Surface Methodology*; see e.g., Myers,

Montgomery and Anderson-Cook (Myers, et al., 2009) yield theoretical details and practical results.

The goal of *Multiple Linear Regression Analysis* – see Yan and Su (Yan & Su, 2009) – is to model the relationship between the independent process controls $x_1, \ldots, x_n$ and the single response y.

Multiple regression procedures estimate this relationship by utilizing a linear function of the form

$$y = \beta_0 + \sum_{i=1}^{n} \beta_i x_i + \varepsilon_i \tag{2-4}$$

where the factors β_i are termed *Regression* or *Profile Coefficients*. These coefficients reflect the (independent) linear contribution of each independent process control x_i to the response y. The constant β_0 is called the intercept of the hyperplane and ε denotes an error component which is approximately normally distributed with zero expected value and variance σ^2; i.e., $\varepsilon \sim N(0, \sigma^2)$.

Note that the regression line in equation (2-4) depicts the best possible prediction of the response y for the process controls $x_1, \ldots, x_n$.

2-3 SINGLE RESPONSE TRANSFER FUNCTIONS

In general, a sensitivity analysis explores the relative impact or the significant effect a change of a single process control x_k has on the single response y.

2-3.1 SENSITIVITY ANALYSIS

A widely-applied tool to conduct a sensitivity analysis is the *Local Method* which is based on taking partial derivatives of a given transfer function with respect to the process controls, typically one control at a time, and on scrutinizing the values of these partial derivatives.

2-3.1.1 MATHEMATICAL BACKGROUND

Consider an arbitrary transfer function $f: \mathbb{R}^n \to \mathbb{R}$ with $y = f(x)$, $x \in \mathbb{R}^n$. Assume that f and its derivatives are continuous, whereby the second and higher derivatives are excluded from further calculations, since these derivatives are assumed to be sufficiently small. Furthermore, the target value of the response y is denoted by

$$\tau_y = f(\tau) = f\left(\tau_{x_1}, \ldots, \tau_{x_n}\right) \tag{2-5}$$

where $\tau = \langle \tau_{x_1}, \dots, \tau_{x_n} \rangle$ is termed the *Target* or *Goal Profile* which, from a customer point of view, represents the optimal solution profile.

By the assumptions given in the previous paragraph, a first order Taylor series expansion around the target τ_y approximates the transfer function by means of

$$y = f(\boldsymbol{x}) = \tau_y + \sum_{i=1}^{n} \frac{\partial f}{\partial x_i}\left(x_i - \tau_{x_i}\right) \tag{2-6}$$

See, for instance, Denniston (Denniston, 2006) and Phadke (Phadke, 1989).

Setting

$$\Delta y = y - \tau_y \quad \text{and} \quad \Delta x_i = x_i - \tau_{x_i} \tag{2-7}$$

results in a more manageable form of equation (2-6):

$$\Delta y = \sum_{i=1}^{n} \frac{\partial f}{\partial x_i} \Delta x_i \tag{2-8}$$

To calculate the expected value and variance of y, apply the *Delta Method* – e.g., Montgomery (Montgomery, 2009) – to equation (2-6), respectively:

$$\mu_y = \sum_{i=1}^{n} \frac{\partial f}{\partial x_i} \mu_{\Delta x_i} \qquad \text{and} \qquad \sigma_y^2 = \sum_{i=1}^{n} \frac{\partial f}{\partial x_i} \sigma_{\Delta x_i}^2 \tag{2-9}$$

2-3.1.2 LOCAL SENSITIVITY ANALYSIS

The terms $\partial f / \partial x_i$ in equations (2-6), (2-8), and (2-9) are termed *Sensitivity Coefficients* and are themselves functions of the process controls values. The *Local Sensitivity* of an arbitrary process control x_i and its associated target value τ_{x_i} is equal to

$$S_{\Delta x_i} = \left(\left(\frac{\partial f}{\partial x_i} \right)^2 \sigma_{\Delta x_i}^2 \right) \sigma_y^{-2} \times 100\% \tag{2-10}$$

Local sensitivity analysis provides insight into the effects changes of the process controls have on the response y. Thus, such an analysis helps to decide which process controls are critical. To conduct such a local sensitivity analysis, the *one-at-a-time measurement* procedure can be performed. During this procedure, each process control is modified by a certain percent amount, for instance $\pm 20\%$, or by its positive or negative standard deviation; i.e., $\pm SD$ and the corresponding change in the response y is measured.

Assume that x_k is the actual modified process control. Hence, the remaining process controls x_i, $(1 \leq i \leq n, \ i \neq k)$ are treated as noises. In view of equation (2-8):

$$\Delta y = \frac{\partial f}{\partial x_k} \Delta x_k + \sum_{i \neq k} \frac{\partial f}{\partial x_i} \Delta x_i = \frac{\partial f}{\partial x_k} \Delta x_k + \varepsilon_k \tag{2-11}$$

It is assumed that ε_k is a random variable with expected value zero – compare Denniston (Denniston, 2006) – meaning that the sum of the (sensitivity coefficients weighted) expected values of the process controls x_i $(1 \leq i \leq n, \ i \neq k)$ is on target:

$$\mu_{\varepsilon_k} = \sum_{i \neq k} \frac{\partial f}{\partial x_i} \mu_{\Delta x_i} = 0 \tag{2-12}$$

In view of equation (2-9), this implies that

$$\mu_y = \tau_y + \frac{\partial f}{\partial x_k} \mu_{\Delta x_k} \tag{2-13}$$

The variance of ε_k, defined in equation (2-11), is

$$\sigma_{\varepsilon_k}^2 = \sum_{i \neq k} \left(\frac{\partial f}{\partial x_i} \right)^2 \sigma_{\Delta x_i}^2 \tag{2-14}$$

so, that the variance σ_y^2 of the response y with respect to the considered process control x_k is equal to

$$\sigma_y^2 = \sigma_{\Delta x_k}^2 + \sigma_{\varepsilon_k}^2 = \sigma_{\varepsilon_k}^2 \tag{2-15}$$

since Δx_k is a constant with $\sigma_{\Delta x_k}^2 = 0$.

2-4 MULTIPLE RESPONSE TRANSFER FUNCTIONS

Single response transfer functions $f: \mathbb{R}^n \to \mathbb{R}$ with $y = f(x)$, $x \in \mathbb{R}^n$ have been investigated. However, how to treat multiple response transfer function of the form $f: \mathbb{R}^n \to \mathbb{R}^m$ with, $x \in \mathbb{R}^n$, $y \in \mathbb{R}^m$? Especially in the case when the transfer function f is a linear mapping? As already defined, x is termed the solution profile and y the response profile. Sometimes, x is called *Cause*, and y the *Effect* of the transfer function. The corresponding *Target* or *Goal Profile* is the real-valued column vector $\tau_y = \langle \tau_{y_1}, \tau_{y_2} \dots, \tau_{y_m} \rangle$.

Of special interest, are the questions: (a) In which way can a response profile y be determined such that it approximates the user-defined target or goal profile $\tau_y \in \mathbb{R}^m$?

and (b) How can the quality of this approximation be measured? Answers to these questions will be given in the following subsections.

2-4.1 LINEAR ALGEBRA PRELIMINARIES

As will be seen later, linear mappings in form of multiple responses, transfer functions play an outstanding role in the QFD context. To help readers who want to understand these topics more in detail, this book includes a brief linear algebra background on linear mappings and their relationships to matrices in the appendix *(Appendix A: Linear Algebra in a Nutshell)*. For further details, consult the literature e.g., (Lang, 1973), (Roman, 2007), or (Kressner, 2005).

2-4.1.1 LINEAR MAPS

Let V and W be two vector spaces over the field of real numbers $\mathbb{R}$. The mapping $f: V \to W$ is a *linear map* or a *homomorphism* if $f(\alpha x + \beta y) = \alpha f(x) + \beta f(y)$ for all $\alpha, \beta \in \mathbb{R}$. One can easily show that if $f: V \to W$ and $g: W \to Z$ are two linear forms then the composition $g \circ f: V \to Z$ is a linear map. The *endomorphism* $id_V: V \to V$ is always linear. When defined, the inverse f^{-1} of a linear map $f: V \to V$ is linear.

Let V and W be two vector spaces over field of real numbers $\mathbb{R}$. The set of all linear mappings $f: V \to W$ is the *dual space*, denoted by

$$V^* = \{f: V \to W | f \ \ is \ \ linear\} \tag{2-16}$$

The elements of V^* are termed *linear forms*. The *dual linear form* $f^*: W^* \to V^*$ is defined by $f^*(\varphi) = \varphi \circ f$ for all $\varphi \in W^*$. Obviously, $\varphi \circ f \in V^*$.

Since the definition of a dual space is not self-evident, here is a practical explanation. Let $V = \mathbb{R}^n$ and $W = \mathbb{R}$. Then, for instance, an element of $(V)^* = (\mathbb{R}^n)^*$ can be defined by an arbitrary linear mapping $f: \mathbb{R}^n \to \mathbb{R}$ with

$$(x_1, \dots, x_n) \to \sum_{i=1}^{n} a_i x_i, \quad a_i \in \mathbb{R} \tag{2-17}$$

The summation term on the right side of equation (2-17) reflects the left-hand side of one row of a system of linear equations. Hence, the dual space $(\mathbb{R}^n)^*$ can be identified with the space of all row vectors of length n.

2-4.1.2 LINEAR MAPS AND MATRICES

To generalize equation (2-17), let $f_A: \mathbb{R}^n \to \mathbb{R}^m$ be a linear form. There exists a unique $m \times n$ matrix $A \in \mathbb{R}^{m \times n}$ with $f_A(x) = Ax$ and $x \to Ax$ for all $x \in \mathbb{R}^n$.

Hence,

$$f_A(x) = Ax = \begin{pmatrix} a_{11} & a_{12} & \cdots & a_{1n} \\ a_{21} & a_{22} & \cdots & a_{2n} \\ \vdots & \vdots & \ddots & \vdots \\ a_{m1} & a_{m2} & \cdots & a_{mn} \end{pmatrix} (x_1, x_2, ..., x_n) = \begin{pmatrix} \sum_{i=1}^{n} a_{1i} x_i \\ \vdots \\ \sum_{i=1}^{n} a_{mi} x_i \end{pmatrix} \quad (2\text{-}18)$$

reveals how the jth component of Ax is calculated.

In view of the definition of a dual space, let $f: V \rightarrow W$ be a linear mapping and A be the unique matrix representation of f. The *transpose matrix* A^T is the matrix representation of the dual linear functional $f^*: V^* \rightarrow W^*$. So, if $f_A: \mathbb{R}^n \rightarrow \mathbb{R}^m$ is represented by a unique $m{\times}n$ matrix $A \in \mathbb{R}^{m \times n}$, then $f_{A^\mathsf{T}}: \mathbb{R}^m \rightarrow \mathbb{R}^n$ is uniquely characterized[1] by the matrix $A^\mathsf{T} \in \mathbb{R}^{n \times m}$.

Let $f_A: \mathbb{R}^n \rightarrow \mathbb{R}^m$ and $f_B: \mathbb{R}^k \rightarrow \mathbb{R}^n$ be two linear forms described by a unique $m{\times}n$ matrix $A \in \mathbb{R}^{m \times n}$ and by a unique $k{\times}n$ matrix $B \in \mathbb{R}^{n \times k}$, respectively. The composition $f_A \circ f_B := B(A(x)) = f_{A \cdot B}$ uniquely defines matrix multiplication $A \cdot B \in \mathbb{R}^{m \times k}$. In other words, the composition of two linear forms results in a matrix multiplication of the matrices, yielding another linear form.

Using the notations of the previous paragraphs, let $f_A: \mathbb{R}^n \rightarrow \mathbb{R}^m$ and $f_{A^\mathsf{T}}: \mathbb{R}^m \rightarrow \mathbb{R}^n$ be two linear forms with associated matrices $A \in \mathbb{R}^{m \times n}$ and $A^\mathsf{T} \in \mathbb{R}^{n \times m}$, respectively. The matrix $AA^\mathsf{T} \in \mathbb{R}^{m \times m}$ describes the composition $f_A \circ f_{A^\mathsf{T}}$. Since $AA^\mathsf{T} = (AA^\mathsf{T})^\mathsf{T}$, this matrix is symmetric and – in most cases – positive definite; i.e., $x^\mathsf{T} AA^\mathsf{T} x > 0$, for all $x \in \mathbb{R}^n \setminus \{0\}$.

2-4.1.3 EIGENVALUES AND EIGENVECTORS

Let $\varphi \in W^*$ denote an endomorphism over the field of real numbers $\mathbb{R}$. $\lambda \in \mathbb{R}$ is called an *eigenvalue* of φ if a $v \in V$ exists with $\varphi(v) = \lambda v$. The non-zero vector $v \in V \setminus \{0\}$ is termed a *principal eigenvector* of φ if λ is the maximum eigenvalue among all eigenvectors. The vector space

$$Eig(A, \lambda) := \{x \in \mathbb{R}^n | Ax = \lambda x\}, \ A \in \mathbb{R}^{n \times n} \quad (2\text{-}19)$$

is called the *eigenspace* of A with respect to the eigenvalue λ. Consequently, the eigenspace $Eig(A, \lambda)$ is a vector subspace spanned by all associated eigenvectors x.

---

[1] This is a simplified representation of the correct form of the dual linear form $f_{A^\mathsf{T}}: (\mathbb{R}^m)^* \rightarrow (\mathbb{R}^n)^*$ since $(\mathbb{R}^m)^* \simeq \mathbb{R}^m$ and $(\mathbb{R}^n)^* \simeq \mathbb{R}^n$.

If $A \in \mathbb{R}^{n \times n}$ is a diagonal or an upper/lower triangular matrix, then the main diagonal entries a_{ii}, $1 \le i \le n$, are the eigenvalues of the matrix A. The eigenvectors of a diagonal matrix A are the n-dimensional standard basis vectors.

If the matrix $A \in \mathbb{R}^{n \times n}$ is symmetric, and then A has exactly n (not necessarily distinct) real eigenvalues, and there exists a set of n real eigenvectors, one for each eigenvalue, which are mutually orthogonal; i.e., the n real eigenvectors are linearly independent even if the eigenvalues are not distinct. Furthermore, for each symmetric matrix A, there exists an orthogonal matrix P with $P^{\mathsf{T}} P = P P^{\mathsf{T}} = I$ such that the matrix $D := P^{\mathsf{T}} A P$ is diagonal, and I is the unity matrix, consisting with 1's on the diagonal only. The eigenvalues of A are equal to the entries $d_{ii}, 1 \le i \le n$ of the diagonal matrix D, and the corresponding eigenvectors of A are the columns of P. If the symmetric matrix A is positive definite, then all eigenvalues of A are real and positive. In this case, the entries of the diagonal matrix D are also real and positive.

2-4.1.4 THEOREM OF PERRON

Let A be an $n \times n$ matrix with positive entries $a_{ij} > 0$, $1 \le i, j \le n$. Then,

a) there exists a (simple) eigenvalue $\lambda > 0$ such that $Av = \lambda v$, whereby $v > 0$ denotes the corresponding eigenvector;

b) λ is bigger (in modulus) than the other eigenvalues, and

c) any other positive eigenvector of A is a multiple of v.

Such an eigenvector is a *Principal Eigenvector*. A generalization of this theorem is the *Perron-Frobenius Theorem*. The details including the proof of this theorem are in *Appendix C-3*. The consequences are that matrices with positive entries always have a principal eigenvector and therefore matrices A with negative coefficients whose combination with the transform AA^{T} is no longer positive-definite have no real solution for $y = Ax$. See also Saaty and Özdemir (Saaty & Özdemir, 2003) and Saaty's explanation for the principal eigenvector (Saaty, 2003).

2-4.2 SOLVING MULTIPLE RESPONSE LINEAR TRANSFER FUNCTIONS

In the previous sections, it was pointed out that the linear form $f_A: \mathbb{R}^n \to \mathbb{R}^m$ can be uniquely represented by a $n \times m$ matrix $A \in \mathbb{R}^{m \times n}$ such that $x \to Ax$ with $y = Ax$, $x \in \mathbb{R}^n, y \in \mathbb{R}^m$ Hence, the linear functional f_A is a multiple response linear transfer function.

2-4.2.1 CALCULATING AN APPROXIMATION y TO THE TARGET PROFILE τ_y

From a mathematical point of view, the equation $y = Ax$ can be interpreted in several ways. At first, y is the result of the matrix-vector multiplication of the matrix A and the vector x. Secondly, the matrix A represents a transformation matrix, which transforms a vector x into the vector y. Thirdly, $y = Ax$ reflects a system of linear equations, a solution x is looked for.

In many application contexts, the most interesting case is to investigate various characteristics of the transformation A itself. At this point, the theory of eigenvectors and eigenvalues comes into play; see e.g., Kressner (Kressner, 2005). Outstanding application examples of this theory are the calculation of Google's PageRank, see Gallardo (Gallardo, 2007), and Langville and Meyer (Langville & Meyer., 2006), and Saaty's Analytic Hierarchy Process (AHP), see Saaty (Saaty, 1990), (Saaty & Peniwati, 2008). The following two chapters – *Chapter 3: What is AHP?* and *Chapter 4: Quality Function Deployment* – will elaborate further on this.

To explain eigenvectors and eigenvalues, let $S \in \mathbb{R}^{n \times n}$ be an arbitrary square matrix. From Linear Algebra (see *Appendix A: Linear Algebra in a Nutshell*), it is well known that almost all vectors $x \in \mathbb{R}^n$ reverse directions when the vectors are multiplied by the matrix S; see e.g., Lang (Lang, 1973), and Roman (Roman, 2007). A non-zero vector x is called an *Eigenvector* of the matrix S, if x and the vector Sx are in the same direction; i.e., eigenvectors are the directions which are invariant under the transformation S.

The *Eigenvalue* λ reveals whether the vector Sx remains unchanged ($\lambda = 1$), is reversed in direction ($\lambda < 0$), is shrunk ($0 < |\lambda| < 1$), or stretched ($|\lambda| > 1$). Thus, the fundamental equation to solve an eigenvector respectively an eigenvalue problem is

$$Sx = \lambda x \qquad\qquad (2\text{-}20)$$

A natural question arises: How can the equation above (2-20) help to solve $y = Ax$, in order to calculate a solution profile $x \in \mathbb{R}^n$ with respect to a linear multiple response transfer function $A \in \mathbb{R}^{m \times n}$ and a response profile $y \in \mathbb{R}^m$?

Following Fehlmann and Kranich (Fehlmann & Kranich, 2011-2), a response profile y can be determined by solving the following eigenvector respectively eigenvalue problem:

$$AA^\mathsf{T} y = \lambda y \qquad\qquad (2\text{-}21)$$

Obviously, the matrix $AA^\mathsf{T} \in \mathbb{R}^{m \times m}$ is symmetric; i.e. $AA^\mathsf{T} = (AA^\mathsf{T})^\mathsf{T}$. It is well-known from the theory of eigenvalues that this matrix has exactly m (not necessarily) distinct real eigenvalues. There exists a set of m real eigenvectors, one for each eigenvalue, which are mutually orthogonal and thus linear independent, even in the case when the eigenvalues are not distinct, see *Appendix C-3: The Perron-Frobenius Theorem*, and the literature. In most cases, the matrix AA^T in (2-21) is positive definite; i.e., $x^\mathsf{T} AA^\mathsf{T} x > 0$ for all $x \in \mathbb{R}^n \setminus \{0\}$, and therefore has a *Principal Eigenvector* y. This eigenvector plays an important role in the Analytic Hierarchy Process (AHP), which is a supporting tool of QFD (Saaty & Peniwati, 2008).

2-4.2.2 CALCULATING THE EIGENVECTOR

The idea is to revert *Cause* and *Effect*, shown in Figure 2-3. First, transpose the matrix and calculate the combined symmetrical square matrix AA^T. According the Theorem of Perron-Frobenius, this symmetrical square matrix has a principal eigenvector y_E with $y_E = AA^Ty_E$. Using that eigenvector, the solution for $Ax_E = y_E$ is $x_E = A^Ty_E$. $\|y - y_E\|$ is the convergence gap; $x_E = A^Ty_E$ are called the *Eigencontrols* of A. If y happens to be near an eigenvector of A, x is an approximate solution for $Ax = y$. However, solutions do not exist for all transfer functions A.

Figure 2-3: How the Eigenvector is calculated (Jacobi Iterative Method)

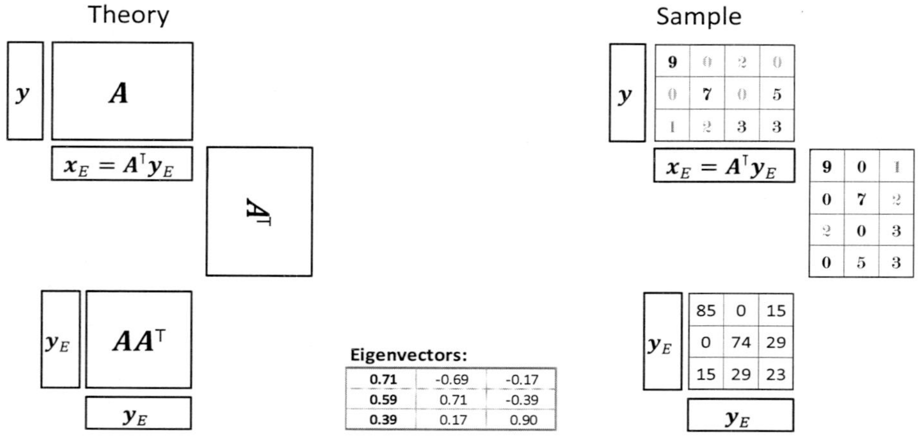

In Figure 2-3, the principal eigenvector is highlighted and the only one that is positive. There are many methods available for calculating eigenvectors; the most popular, and best suitable for our complex problem solving techniques, is the *Jacobi Iterative Method*, applicable for symmetric positive matrices. The method yields all three existing eigenvectors. Most mathematical packages contain eigenvector calculation methods; for Microsoft Excel, a free open source tool is available (Volpi & Team, 2007). For more information on this tool, consult the book and web site of Robert de Levie (Levie, 2012). Most statistical packages contain eigenvector calculation methods; e.g., the *R Project* (The R Foundation, 2015).

2-4.2.3 A Measure for the Closeness of y to τ_y

Let the current response profile y approximate the predefined target profile τ_y. Then the Euclidian norm

$$\|\Delta y\| = \sqrt{\sum_{i=1}^{m} (\Delta y_i)^2} \tag{2-22}$$

with $\Delta y_i = (y - \tau_y)_i$, $1 \leq i \leq m$, is termed the *Convergence Gap* and reveals the quality of the approximation of y to τ_y. If this gap fulfills a predefined convergence criterion, then y is (sufficiently) close to τ_y. In this case, $x = A^{\mathsf{T}} \tau_y$ calculates the approximate solution profile x, solving $y = Ax$.

2-4.2.4 Normalization of Vectors

When vectors differ not in length but in direction, the length of the vector difference still is significant. Adding or subtracting two vectors varies with its lengths; its direction depends on its length. However, when doing Six Sigma, we need to compare two profile vectors regarding direction.

Therefore, using the Euclidian norm for bringing vectors to equal lengths:

$$\|y\| = \sqrt{\sum_{i=1}^{m} y_i^2} \tag{2-23}$$

Thus, the Euclidian norm is appropriate for multiple response linear transfer functions while the traditional maximum norm used in QFD – taking the maximum difference between vector components – is not. Given a vector $y = \langle y_1, \ldots, y_m \rangle$ of dimension m, its normalized variant is

$$y' = \frac{y}{\|y\|} = \langle \frac{y_1}{\|y\|}, \frac{y_2}{\|y\|}, \ldots, \frac{y_m}{\|y\|} \rangle \tag{2-24}$$

This explains why certain practices, especially comparisons, made in traditional QFD approaches lead to *bad mathematics* (Mazur, 2014).

2-4.3 SOLVING $y = Ax$

If the convergence gap $\|\Delta y\|$ defined in (2-22) fulfills a predefined convergence criterion, the determination of the solution profile x with respect to the response profile y and the matrix A is self-evident in view of equation (2-21) with $\lambda = 1$ and $y = Ax$:

$$AA^\mathsf{T}y = y = Ax \implies x = A^\mathsf{T}y \tag{2-25}$$

Evidently, $A^\mathsf{T} \in \mathbb{R}^{n \times m}$ is a dual multiple response transfer function that predicts the solution profile x by $x = A^\mathsf{T}y$. Therefore, it is termed the *Prediction Function* for that solution profile. The solution is stable in the sense that when repeatedly applying the process represented by the transfer function A, the response y remains always the same.

$$y = AA^\mathsf{T}y = AA^\mathsf{T}AA^\mathsf{T}y = AA^\mathsf{T}AA^\mathsf{T}AA^\mathsf{T}y = \cdots \tag{2-26}$$

In *Chapter 3: What is AHP?* it is shown how to calculate eigenvectors for AHP.

2-4.4 PROFILES AND WEIGHTS

Statistical thinking means looking at events as vectors in a multidimensional vector space, instead of maximizing favored topic. The difference can become substantial. In modern QFD, *Weights* and *Profiles* must be clearly distinguished. A *Weight* is a percentage of its total importance; e.g., topics 1 to 5 show different importance in terms of relative weights. The total of all weights is 100%. A *Profile* in contrary is a vector of length 1. Their vector components are not relative weights.

Weights allow distributing a 100% budget to different solution elements. The sum of weights $w_1 + w_2 + \cdots + w_n$ is 100%. Weights convert into profiles by equation (2-27)

$$y_i = \frac{w_i}{\|w\|}, \text{ where } \|w\| = \sqrt{\sum w_i^2} \tag{2-27}$$

On the other hand, given some profile $y = \langle y_1, y_2, \ldots, y_n \rangle$, this easily converts into weights $\langle w_1, w_2, \ldots, w_n \rangle$ using equation (2-28)

$$w_i = \frac{y_i}{|y|} * 100\%, \text{ where } |y| = \sum y_i \tag{2-28}$$

A *Weight Vector* $w = \langle w_1, w_2, \ldots, w_n \rangle$ is the information sought from the technical solution x of a problem of the form $y = Ax$, because this is what the finance manager needs when allocating the budget for the technical solution. Consequently, while profiles are needed for the mathematics, results always are shown as weights as well.

Figure 2-4 demonstrates how two priority profiles (2-29) and (2-30) in five dimensions sum up into a combined profile vector (2-31). This happens in the three columns, labeled with *Weights*, respectively *Profiles*. In the rows, weight vectors transform into profile vectors and back again. The second and the fourth column below the arrows contain the intermediate calculation steps of the transformation *Weights* → *Profiles* → *Weights*. In the left column, the two weight vectors are summed up. This yields a significantly different result than summing up two profiles in the middle column and transform the sum back to a weight vector, as done in the third row. Summing up weight vectors is bad mathematics.

Figure 2-4: Adding Weights versus Adding Profiles

	Weights	→	Profiles	→	Weights	
Topic 1	5%	0.25%	0.06	0.06	5%	
Topic 2	90%	81.00%	1.00	1.00	90%	
Topic 3	2%	0.04%	0.02	0.02	2%	
Topic 4	1%	0.01%	0.01	0.01	1%	
Topic 5	2%	0.04%	0.02	0.02	2%	
	100%	90.19%	1.00	1.11	100%	Weight & Profile 1
	plus↓	→	plus↓	→	Weights	
Topic 1	35%	12.25%	0.68	0.68	35%	
Topic 2	5%	0.25%	0.10	0.10	5%	
Topic 3	10%	1.00%	0.19	0.19	10%	
Topic 4	30%	9.00%	0.58	0.58	30%	
Topic 5	20%	4.00%	0.39	0.39	20%	
	100%	51.48%	1.00	1.94	100%	Weight & Profile 2
	sum↓		sum↓	→	Weights	
Topic 1	40%	0.74	0.48	0.48	24%	
Topic 2	95%	1.10	0.72	0.72	36%	
Topic 3	12%	0.22	0.14	0.14	7%	
Topic 4	31%	0.59	0.39	0.39	19%	
Topic 5	22%	0.41	0.27	0.27	13%	
	200%	1.52	1.00	2.01	100%	Sum of Profiles 1+2
	norm↓	→	Profiles	→	Weights	≠
Topic 1	20%	4.00%	0.36	0.36	20%	
Topic 2	48%	22.56%	0.86	0.86	48%	
Topic 3	6%	0.36%	0.11	0.11	6%	
Topic 4	16%	2.40%	0.28	0.28	16%	
Topic 5	11%	1.21%	0.20	0.20	11%	
	100%	55.26%	1.00	1.81	100%	Sum of Weights 1+2
			0.23			Convergence Gap

In detail, when represented as a vector, the weights look as follows:

$$\langle 5\%, 90\%, 2\%, 1\%, 2\% \rangle \tag{2-29}$$

The Euclidian length of vector (2-29) is $\sqrt{5\%^2 + 90\%^2 + 2\%^2 + 1\%^2 + 2\%^2} = 90.19\%$, or after squaring, $\sqrt{0.25\% + 81.00\% + 0.04\% + 0.01\% + 0.04\%} = 90.19\%$; see the second column of Figure 2-4. Since importance might correlate to the budgeted amount that is being spend for these topics, weight percentage matters.

Take as an example another weight vector, say the second part in Figure 2-4:

$$\langle 35\%, 5\%, 10\%, 30\%, 20\% \rangle \qquad (2\text{-}30)$$

Clearly, its Euclidian length is 51.48%. Thus, the two vectors (2-29) and (2-30) have unequal length. Adding them by components – keeping its representation as percent – yields $\langle 40\%, 95\%, 12\%, 31\%, 22\% \rangle$. This corresponds to the weight vector

$$\langle 20\%, 48\%, 6\%, 16\%, 11\% \rangle \qquad (2\text{-}31)$$

seen in the fourth row of Figure 2-4. However, since the first weight vector was longer, it contributes more to the sum vector (2-31) than the second does. Unless there is some special business reason for it, such an unequal treatment can have disastrous results, for instance spending investment money in the wrong places.

Converting both vectors to profiles by dividing their components through its length, yields two vectors of equal length but pointing into two different directions in the vector space. Adding the components yields the profile $\langle 0.48, 0.72, 0.14, 0.39, 0.27 \rangle$ with unit length $\| \langle 0.48, 0.72, 0.14, 0.39, 0.27 \rangle \| = 1$. This profile again converts into weights by normalizing the sum of components to 100%.

The result is the weight vector

$$\langle 24\%, 36\%, 7\%, 19\%, 13\% \rangle \qquad (2\text{-}32)$$

seen in the third row of Figure 2-4. This obviously is not the same vector as (2-31). The convergence gap – or difference in the $\mathcal{L}_2$ norm, see below – is 0.23. For a vector of unit length, this corresponds to one quarter of its length, thus vector (2-32) points to another direction. The reason for the difference is the length inequality. In practice, this difference is difficult to detect. selecting extremely different weight vectors demonstrates the case.

As a conclusion, while it is safe comparing, adding, subtracting, or multiplying vector profiles as needed in a QFD, it is bad mathematics if you compare, add or subtract weight vectors.

2-4.5 REGULARIZATION

Regularization, in mathematics and statistics and particularly in the fields of machine learning and inverse problems, refers to a process of introducing additional information to solve an ill-posed problem or to prevent over-fitting. In statistics and machine learning, regularization methods are used for model selection, to prevent over-fitting by penalizing models with extreme parameter values. The most common variants in machine learning are $\mathcal{L}_1$ and $\mathcal{L}_2$ regularization, introducing a 'distance' between two events (in statistics) or two states (in machine learning). Regularization for transfer functions uses the statistical way with the $\mathcal{L}_2$ norm. In mathematics, the $\mathcal{L}p$ spaces are known as function spaces defined using a natural generalization of the distance norm for finite-dimensional vector spaces. Say n is the dimension of the vector space; then, the p-norm definition is

$$\mathcal{L}p(\langle x_1, x_1, \dots, x_n\rangle) = \sqrt[p]{x_1^p + x_2^p + \cdots + x_n^p} \tag{2-33}$$

for any $p = 1, 2, \dots$ In measurement practice, $p = 1$ and $p = 2$ play the most prominent roles, representing distance in the $\mathcal{L}_1$ and $\mathcal{L}_2$ spaces.

Statistics rely on the $\mathcal{L}_2$ norm, predictive analysis also. Economic models rely on the $\mathcal{L}_1$ norm.

$$\mathcal{L}_1(\langle x_1, x_2, \dots, x_n\rangle) = \sqrt[1]{x_1^1 + x_2^1 + \cdots + x_n^1} = |x_1| + |x_2| + \cdots + |x_n| \tag{2-34}$$

$$\mathcal{L}_2(\langle x_1, x_2, \dots, x_n\rangle) = \sqrt[2]{x_1^2 + x_2^2 + \cdots + x_n^2} = \|\langle x_1, x_2, \dots, x_n\rangle\| \tag{2-35}$$

The weights obviously follow the $\mathcal{L}_1$ norm while profiles observe the $\mathcal{L}_2$ norm. Statistics rely on the $\mathcal{L}_2$ norm, while economics prefer the $\mathcal{L}_1$ norm. In short, therefore statistics and economics do not always work well together, adding just another complexity level to manage. Throughout this book, we use the $\mathcal{L}_2$ norm. References include (Bektas & Sisman, 2010), (Wu, 2004) and (Fehlmann, 2003).

The $\mathcal{L}_1$ norm of a mathematical object x (2-34) usually has the denotation $|x|$ – representing the absolute value of a real number for instance – while the common convention for the $\mathcal{L}_2$ norm (2-35) usually is $\|x\|$. This book adheres to this convention.

2-5 CONCLUSIONS

The main result of this chapter reveals how to apply multiple response linear transfer functions, eigenvalue theory and dual space linear forms. Because matrices represent multiple response transfer functions, it is advantageous to introduce and utilize the (algebraic) dual space and corresponding linear forms to be able to determine a solution profile. Building the composition of a multiple response linear transfer function and its associated linear form results in a symmetric and positive definite matrix. Eigenvalues and eigenvectors are then computed and the convergence gap is adopted for measuring how close the solution of the eigenvector problem is to a given target profile.

CHAPTER 3: WHAT IS AHP?

Rating and Ranking always was an important task. Who wants to be picked last for teams in the gym class?

Likewise, manufacturers of goods like to know what makes its product attractive, why people buy it. Thus, they need ranking.

The Analytic Hierarchy Process (AHP) is the calculation of priority vectors based on geometric mean instead of arithmetic average. AHP uses ratio scales, thus the AHP allows measuring entities like stones, or effort without referring to reference objects, even if only fuzzy information is available.

A sample application shows how AHP supports project estimation.

3-1 INTRODUCTION

The power of *Linear Algebra* is not yet widely used for business decisions – despite the widespread use of *Eigenvector Theory* for instance in Google search (Langville & Meyer., 2006). One useful application of linear algebra to decision-making based on incomplete knowledge is the *Analytic Hierarchy Process* (AHP) that Thomas Saaty devised some twenty year ago, and that has been widely used in business decision-making and medical diagnostics since (Saaty, 2003).

One reason for the weak popularity of linear algebra might simply be ignorance. Vector spaces, although everybody experiences it in out three-dimensional instance, are less directly linked to business success than, for instance, proportional calculations or compound interest. As the traffic incidence frequency suggests, human seem less prone to correctly assess braking distance than grasp a cost advantage. Cost is a matter of ordered numbers, while ordering in a three-dimensional vector space is already something painful.

Nevertheless, the world is not one-dimensional. Most complexity arises from multidimensional structures that affect the status and pending decisions in an unprecedented way, and often surprising, to some extent. Thus, the question raised is how to determine a ranking for complex, interacting matters?

3-1.1 A POPULAR RATING METHOD USING BAD MATHEMATICS

Methods used for ranking are a quagmire of different approaches taken from various disciplines, some with scientific touch, and others simply because they seem practical.

One noteworthy misleading example is the popular paired comparison process. For instance, Cohn (Cohn, 2005) uses it to size software based on Story Points. Paired comparison is better than linear arrangements for prioritization because it considers cross-dependencies. However, paired comparison yields inconsistent results, as can be seen from Figure 3-1 when prioritizing seven alternatives.

Figure 3-1: Inconsistent Paired Comparison – Expert says: A7 = A5 – but overall A5 < A7!

Requirements	A1.	A2.	A3.	A4.	A5.	A6.	A7.	Importance of requirement (Points total)	Weighting of requirement (%)	Ranking of requirement
A1.		1/2	1/2	0	1/2	1	0	2.5	12%	5
A2.	1/2		1	1/2	0	1	0	3.0	14%	4
A3.	1/2	0		0	0	1	0	1.6	8%	6
A4.	1	1/2	1		1	1/4	0	3.8	18%	3
A5.	1/2	1	1	0		1	1/2	4.0	19%	2
A6.	0	0	0	3/4	0		0	0.8	4%	7
A7.	1	1	1	1	1/2	1		5.5	26%	1
Total								21.1	100%	

In this case, an expert has made a possibly wrong judgment about the requirement A7, which is overall most important. For instance, when comparing the row A1 with the column A2, the fractions mean

0 = A1 is less important than A2
½ = A1 is as important as A2
1 = A1 is more important than A2

Intermediary values such as ¼ or ¾ are also valid. In one case (circled in Figure 3-1) the expert considered A7 equally important as A5. The misjudgment leads to a change in ranking the requirements A4 and A5, a change that is hardly detectable without proper means and metrics. However, if such variation remains undetected, the impact and cost incurred for this methodical weakness can become substantial, later.

The reason why traditional paired comparison fails to deliver consistent results is the same as exemplified in section *2-4.4 Profiles and Weights*. Missing regularization leads to dependencies from the requirements importance total, as collected in the first column after the matrix in Figure 3-1. While comparing pairwise seems a good idea, the evaluation of the comparison makes the difference.

3-1.2 THE METHODOLOGY OF AHP

When utilizing the AHP procedure to solve complex decision problems, the decision makers pass through three different phases: The *Design Phase*, the *Evaluation Phase*, and the *Validation Phase*.

The first step in the design phase is to describe in detail the specific characteristics of the underlying decision problem and then to define the goal of that problem. In a second step, decision makers must structure its characteristics hierarchically into criteria, sub-criteria and alternatives.

In the subsequent evaluation phase, the criteria and alternatives of each hierarchy level compare pairwise with respect to their higher-level elements. When the evaluations of each level are completed, a consistency check should not reveal contradiction to the single evaluations when aggregated to an overall result. This process repeats for the whole hierarchy.

Finally, the validation phase consists of a sensitivity analysis to ensure that the final evaluation result is stable.

The basic steps in the solution of a multi-criteria decision problem using AHP are:

- Step 1: Define the goal of the decision by answering the questions
    1. " What do I want to decide?"
    2. " For what purpose?" and
    3. " What are the alternatives?"
- Step 2: Structure the decision problem into a hierarchy of categories or attributes and decision criteria.
- Step 3: Conduct a pairwise comparison of the decision criteria in each category.
- Step 4: Calculate the priorities and a consistency index.
- Step 5: Evaluate alternatives per priorities identified.

The advantage of dividing complex decision problems into smaller pieces is obvious. Naïve decision makers are tempted using the paired comparison process and nevertheless split the problem; however, combining two paired comparisons is dangerous except when using sound mathematical foundation. This is Saaty's contribution to decision making.

3-1.3 IMPROVING WITH AHP

Thomas Saaty proposed already in 1980 (Saaty & Alexander, 1989) the better way of evaluating paired comparison who termed his approach Analytic Hierarchy Process (AHP). He applies *Eigenvector Theory* to enforce decision consistency (Saaty, 2003).

AHP also uses the upper triangle part of a of a quadratic rectangular matrix, like paired comparison, but looks at the matrix as a representation for a linear mapping in a multidimensional vector space $\mathbb{R}^n$ over the real numbers (the ratios). AHP uses geometrical distances (square root of sum of coefficient's squares, annotated by $\mathcal{L}2$) instead of arithmetical distance (sum of absolute coefficients measure $\mathcal{L}1$) as with paired comparison. Thus, the diagonal elements have value 1, whereas paired comparison has 0 on the diagonal, see Figure 3-1, and the lower triangle part of the matrix contains the reciprocal values; i.e.

$$\alpha_{i,j} = \frac{1}{\alpha_{j,i}} \tag{3-1}$$

for all elements $\alpha_{i,j}$ in the matrix A, $1 \leq i,j \leq n$, as opposed to paired comparison that uses

$$\alpha_{i,j} = 1 - \alpha_{j,i} \tag{3-2}$$

in the lower triangle.

Saaty explains the AHP method in (Saaty & Peniwati, 2008) with a nice example involving comparing weights of stones. Focusing on one property alone (namely the weight of the stones) allows measuring the ratios between the stones without referring to a standard weight measurement unit such as kilogram. Let $x = \langle x_1, x_2, \ldots, x_n \rangle$ denote the unknown weights of the n stones. Now compare the stones pairwise and insert their ratio

$$a_{i,j} = {x_i}/{x_j}, 1 \leq i,j \leq n \tag{3-3}$$

in each matrix cell. It is obvious that the matrix becomes reciprocal.

$$A = \begin{pmatrix} x_1/x_1 & x_1/x_2 & \cdots & x_1/x_n \\ x_2/x_1 & x_2/x_2 & \cdots & x_2/x_n \\ \vdots & \vdots & \ddots & \vdots \\ x_n/x_1 & x_n/x_2 & \cdots & x_n/x_n \end{pmatrix} \tag{3-4}$$

The ratio matrix A has rank one, because every row is a multiple of the first row. This in turn has important theoretical consequences. All eigenvalues of the linear mapping A except one are zero. The sum of the eigenvalues of a matrix is equal to its trace; the sum of the diagonal elements, and in this case, the trace of A is equal to n. Therefore, n is the largest, or *principal*, eigenvalue of A. Thomas Saaty observed that A yields a measurement scale for the unknown weights.

By vector-multiplying each column of A with x, the solution of $Ax = \lambda x$ consists of positive entries and is unique up to a multiplicative constant λ. It is called the *Principal Right Eigenvector* of A; λ is called its *Eigenvalue*. After normalization of A, $Ax = x$ holds. The vector difference $\|Ax - x\|$ is zero; this is the ideal case because the weight ratio matrix is consistent.

Consistency means that for all elements $a_{i,j}$ of the square matrix A with n rows and columns holds

$$a_{i,j} = a_{i,k} * a_{k,j} \text{ for all } k \text{ in the interval } 1 \leq i < k < j \leq n \qquad (3\text{-}5)$$

The ratio $a_{i,j}$ between element i and element j of Saaty's stone comparison exercise reflects the decisions taken on the stone ratios in between. Together with the reciprocity of A it becomes clear that the matrix A has $n - 1$ degrees of freedom only. The decisions taken for the first supradiagonal $a_{1,2}, a_{2,3}, \dots, a_{n-1,n}$ define all other matrix elements by consistency and reciprocity. It is easy to see that all column vectors in the matrix are a multiple of the first column; thus, such matrices have exactly one principal right eigenvector, namely the one column with the least eigenvalue. After normalization by setting the eigenvalue $\lambda = 1$, for the principal eigenvector x holds $Ax = x$. This eigenvector can be interpreted as a prioritization decision metrics, since regardless how many times A is applied to x, it always yields the same result, balancing all side effects optimally.

3-1.4 APPLYING AHP TO REAL-WORLD PROBLEMS

In the real world, if people compare stones, or business alternatives, they cannot rely on accurate measurements. Typically, decisions rely on incomplete data, partial knowledge, and expert judgments.

Value-based paired comparison yields different results when applied repeatedly to the same decision problem, whereas AHP smoothers human errors by approximating the principal eigenvector and thus eliminating the measurements and decision errors. This theoretical background makes it the most successful decision method in today's business environment.

The spreadsheet shown in Figure 3-2 deals with the same alternatives as Figure 3-1; however, for the comparisons, we must enter ratios. As proposed by Saaty, we replace the zero values in Figure 3-1 by $1/9$ and the corresponding ones by 9, and limit our ratio scale to $1/9, 1/7, 1/5, 1/3, 1, 3, 5, 7, 9$. The intermediary values in Figure 3-1 become now 5, respectively $1/5$. The ranking no longer changes, when replacing the possibly wrong equality in importance between A7 and A5 by a more consistent 9.

Figure 3-2: Decision Metrics with AHP; calculations not shown

AHP Priorities

Requirements	A1	A2	A3	A4	A5	A6	A7	Weight	Ranking	Profile
A1	1	1	1	1/9	1	9	1/9	6%	5	0.14
A2	1	1	9	1	1/9	9	1/9	9%	4	0.20
A3	1	1/9	1	1/9	1/9	9	1/9	5%	7	0.10
A4	9	1	9	1	9	1/5	1/9	22%	2	0.48
A5	1	9	9	1/9	1	9	1	20%	3	0.43
A6	1/9	1/9	1/9	5	1/9	1	1/9	6%	6	0.12
A7	9	9	9	9	1	9	1	33%	1	0.71

Convergence Gap:	**0.02**	< 0.1: Solution found
Schurr Radius:	**0.06**	If value < 0.1,
Convergence Range:	0.10	judgements are consistent,
Convergence Limit:	0.20	else judgements are doubtful

It is obvious that such matrices cannot always be consistent, as for instance a multiple of 1/7 and 5 yields 5/7, which is not part of Saaty's scale. Reducing available values supports making decisions faster on all freely selectable matrix elements in the upper right matrix triangle. This affects $(n-1)*(n-2)/2$ elements. By forcing decision makers to restrict themselves to only a few choices and allowing corrective decisions while comparing all alternatives, overall decision quality and accuracy increases. The profile column in Figure 3-2 shows a normalized eigenvector, not the weights in percent.

3-2 THE DETAILS BEHIND THE AHP

Sound decision metrics are often not available when selecting the single best solution. In fact, if one knows which ways to earn € 20'000 or € 200'000, there is no decision needed – most people go for the second choice without having problems with decisions. However, when analyzing the symptoms of a sick person, as a doctor, or of a sick economy, as a business decision-maker, no absolute scales are available, even when based on reliable data.

Saaty says in (Saaty, 2003) that "The Analytic Hierarchy Process is rigorously concerned with the scaling problem and what sort of numbers to use, and

Figure 3-3: Saaty's Ratio Scale

9	Overruling importance
8	
7	Much higher importance
6	
5	Clearly higher importance
4	
3	Somewhat higher importance
2	
1	Equal importance
1/2	
1/3	Somewhat smaller importance
1/4	
1/5	Clearly smaller importance
1/6	
1/7	Much smaller importance
1/8	
1/9	No importance at all

how to correctly combine the priorities resulting from them". He uses relative scales rather than ordinals, and derives the measurements from it. This makes AHP powerful. It is not necessary to establish a measurement theory first; it is simply comparison between objects. Alternatively, as Saaty puts it: Measurement results from comparing relative scales.

3-2.1 VECTOR DISTANCE MEASUREMENTS

Since the measured objects are points in a multidimensional vector space, the distance between them is measurable as a linear vector space norm. Traditional paired comparison, but also traditional Quality Functional Deployment, takes not Euclidian norm – $\mathcal{L}_2$ – but rather the absolute distance norm named $\mathcal{L}_1$ as a distance between vector components. The absolute distance metric $\mathcal{L}_1$ does not allow for using statistical methods (ISO 16355-1:2015, 2015), contrary to $\mathcal{L}_2$. $\mathcal{L}_1$ does not provide a ratio scale for rating and ranking, as Saaty pointed out (Saaty & Alexander, 1989). See also section 3.4 *Profiles and Weights* in *Chapter 2: Transfer Functions*.

The consequences of this change from the $\mathcal{L}_1$ to the $\mathcal{L}_2$ distance measurement method are important for AHP. It makes the power of both linear algebra and its statistical interpretation available to decision-making. With the $\mathcal{L}_2$ distance measurement method, the decision-space receives the structures of a statistical events space – which in turn is a linear vector space whose dimensions are the number of free degrees that exist in decision-making – and all the power of the Six Sigma practice becomes available to decision-making.

3-2.2 CALCULATING THE EIGENVECTOR

The vector distance between two decisions visualizes their aberration. Two decisions may be almost identical if the components of the decisions differ only slightly, or decisions may become opposite it the vectors in the vector space point at very different directions.

Total consistency of AHP matrices is the exception in business; inconsistencies such as shown in Figure 3-1 are unavoidable. The question is, how big such inconsistencies are, and whether they have any impact on decisions based on AHP. The answer is *Sensitivity Analysis*, a technique well known in Six Sigma (Bana e Costa & Vansnick, 2008).

In the case of an ideal consistent paired comparison matrix such as Figure 3-2, finding the unknown eigenvector x is easy. As before, let $a_{i,j}$ be the coefficients of the reciprocal matrix A such that $a_{i,j} = {}^1\!/a_{j,i}$, $1 \le i,j \le n$ holds. The normalized solution vector $x = \langle x_1, x_2, \dots, x_n \rangle$ is the result of applying the transpose matrix A^T to x component-wise:

$$n * \xi_i = \sum_{j=1..n} \alpha_{i,j} * \xi_j = \sum_{j=1..n} \frac{\xi_i}{\xi_j} * \xi_j \tag{3-6}$$

since $\alpha_{i,j} = \xi_i/\xi_j$. Thus, every column of A is a multiple of the solution vector x and therefore, finding x is only a matter of normalization.

The most popular method for calculating the eigenvector by normalization is the "*Annihilator*" method. In short, the method intersects a vector space with its algebraic *Dual Space*. The algebraic dual space exits for all vector spaces. Intuitively, the dual space reverses cause and effect. Since cause and effect coincide for decision problems such as addressed by AHP, it is necessary to come to the same decisive conclusion from both sides. The calculation of the AHP priority profile indeed reflects that idea. By normalization, all rows become equal in theory, namely to the principal eigenvector, as explained in the previous chapter; the inequalities that exist between rows reflect the inconsistencies in the decisions itself. For details; see e.g., (Bourbaki, 1989) or any other textbook about Linear Algebra.

Figure 3-4: Spreadsheet calculated

	A1	A2	A3	A4	A5	A6	A7
A1	1	1	1	1/9	1	9	1/9
A2	1	1	9	1	1/9	9	1/9
A3	1	1/9	1	1/9	1/9	9	1/9
A4	9	1	9	1	9	1/5	1/9
A5	1	9	9	1/9	1	9	1
A6	1/9	1/9	1/9	5	1/9	1	1/9
A7	9	9	9	9	1	9	1
	22.1	21.2	38.1	16.3	12.3	46.2	2.6

0.06	0.09	0.05	0.22	0.20	0.06	❹		Weight	1/Weight	
0.05	0.05	0.03	0.01	0.08	0.19	0.04	0.44	6%	15.74	0.97	7%
0.05	0.05	0.24	0.06	0.01	0.19	0.04	0.64	9%	10.99	1.36	9%
0.05	0.01	0.03	0.01	0.01	0.19	0.04	0.33	5%	21.16	0.71	5%
0.41	0.05	0.24	0.06	0.73	0.00	0.04	1.53	22%	4.58	3.13	21%
0.05	0.42	0.24	0.01	0.08	0.19	0.39	1.38	20%	5.07	2.36	16%
0.01	0.01	0.00	0.31	0.01	0.02	0.04	0.39	6%	17.79	1.23	8%
0.41	0.42	0.24	0.55	0.08	0.19	0.39	2.29	33%	3.06	4.8	33%
0.10	0.12	0.07	0.16	0.22	0.13	0.10	7	0.46	14.57		

Enter ratios <1 for less important, >1 for more important.
Use ratio scale 1/9; ... 1/2; 1; 2; ... 9 only
The direction is: Row compared to Column.
If row entry is less important than column entry, enter <1.

Convergence Gap:	0.02	< 0.1: Solution found
Schurr Radius:	0.06	If value < 0.1,
Convergence Range:	0.10	judgements are consistent,
Convergence Limit:	0.20	else judgements are doubt

The Annihilator algorithm goes as follows:

Step 1: Normalize the components $a_{i,j}$ by their total component sum by columns;

Step 2: Calculate sum of rows r_j; this yields a vector $r = \langle r_1, r, ..., r_j \rangle$;

Step 3: Again, normalize the vector r by its total component sum;

Step 4: Normalize the vector r by the maximum component, yielding the solution candidate x';

Step 5: Calculate $x = Ax'$;

Step 6: The distance $\|x - x'\|$ indicates how near to the principal eigenvector the two solutions are.

The distance $\|x - x'\|$ is the *Convergence Gap*. The Annihilator method does not always yield a solution; however, since the principal eigenvector always exists for reciprocal matrices, it is sufficiently often yielding a good-enough result.

3-2.3 CASCADING AHP

The big advantage of AHP comes when comparing more than just a few alternatives. Paired comparison with n alternatives means assessing $(n - 1)(n - 2)/2$ pairwise comparisons. This is much work and its correctness is difficult to guarantee, not mentioning consistency.

AHP allows for a hierarchy of decisions. You can group the alternatives into affinity groups, compare only the few elements per group among each other and then compare the groups. The ratios used for the groups directly affect the comparisons for the group elements by multiplication. The total numbers of comparisons sharply decrease when doing AHP with ratio scales, as compared with paired comparisons with value-based additive indexes.

Figure 3-5: Sample Decision Hierarchy for Evaluating the Best School

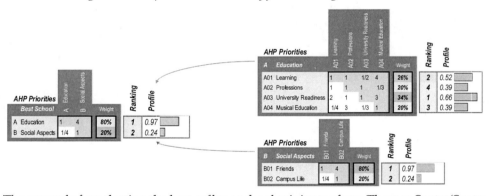

The example for selecting the best college school originates from Thomas Saaty (Saaty, 2003); however, the values presented here rather might reflect European preferences. The question is what kind of school education best fits the career opportunities of a child. In this example (Figure 3-5), using a decision hierarchy reduces the number of paired comparisons from 15 down to 5.

In the first step, the decision makers compare two categories 1) educational and 2) social aspects. Because the educational aspects such as learning facilities and equipment such as laboratories, number of professions that can be reached, and school rating (again!) and say that they prefer education three times more than the social aspects.

This is an individual preference and is not a generalized rule for everybody. In the second step, the decision maker rates and ranks both second level decision categories.

The advantage of the vector-space calculations is that subhierarchy results easily combine for the total ranking. Since all profile vectors are normalized to unit length; i.e., $\|x\| = 1$, they can be vector-multiplied yielding another profile vector on the higher hierarchy level.

Figure 3-6: Top Ranking Criteria for Evaluating the Best School

Top Targets
Best School

		Top Targets	Attributes		Weight	Profile	
A Education	A01	Learning	Ease of learning	Trained to learn	21%	0.48	
	A02	Professions	Wide range of professions	Find an excellent start	16%	0.36	
	A03	University Readiness	Will successfully conduct studies	Well prepared	27%	0.61	
	A04	Musical Education	Learns to perform	High precision	16%	0.36	
B Social Aspects	B01	Friends	Makes friends for life	Good insider relationship	16%	0.36	
	B02	Campus Life	Socialize	Behavior	4%	0.09	

The combination of the two sublevel decisions yield the overall ranking across the hierarchy, see Figure 3-6 that also shows the hierarchy. Such a hierarchical decision process breaks complex decision tree down to manageable levels; its power is obvious. AHP is widely in use for medicinal diagnostics, strategic decisions, for uncovering Voice of the Customer (VoC) and in many more areas where complex decision are needed. There is one caveat: AHP works only among comparable criteria; apparently, much misuse originates from comparing oranges with apples. This is not possible, not even with AHP.

3-2.4 THE SCHURR RADIUS

Saaty has proposed a method of calculating a consistency index for AHP decision matrices (Saaty, 1990). It should measure the degree of consensus among decision makers. There is also much work done on reconciliation of decision matrices from different decision teams trying the same decision.

In this book, consistency is not discussed since it is probably irrelevant. Looking at AHP matrices as special transfer functions that collect measurements carrying important measurement errors is more beneficial. The *Schurr Radius* is a concept with more potential.

The goal of the consistency dispute is to measure gut feeling. It is difficult to conceive how mathematical methods can improve expert decisions. Some recent research has shown that gut feeling is better than algorithmic methods in finding the best business decision (Gigerenzer, 2007). As we have seen before, in section 3-1.1, some algorithmic methods in use today are faulty. In contrary, transfer functions seem to model what humans experience as "gut feeling" and thus are at least as good.

This observation leads to proposing another metric. Measuring the variations in the bundle of row – or column – vectors in the AHP matrix yields a measure for the variations in decisions. In three dimensions $(n = 3)$, its visualization looks convincing; or any dimensions, the variation is reflected by the median distance between the n column vectors and the eigenvector found. In case of $n = 3$, the convergence gap is visualized in Figure 3-7, according an idea presented in (Schurr, 2011).

The calculation of eigenvectors is easily possible with any industry standard linear algebra package; see the appendices of this book. Thus, eigenvector theory validates the choice of cost drivers but cannot ascertain their correct label and meaning. It enhanced their calculation by removing inconsistencies in the cost profiles, improving quality of cost driver profiling.

Figure 3-7: Small and Large Schurr Radius for Three Dimensions

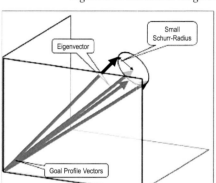
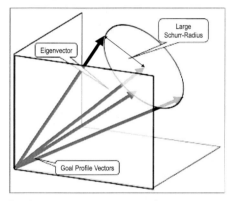

Let $\Delta_1, \Delta_2, \dots, \Delta_n$ be these differences, namely the convergence gaps between eigenvector and the column vectors in the AHP matrix. Then the formula resembles the standard deviation σ:

$$Schurr\ \ Radius = \sqrt{\frac{\sum_{i=1..n}\Delta_i^2}{n-1}} \qquad (3\text{-}7)$$

The variability index (3-7) is called *Schurr Radius*. It measures the total of variations among the individual decision vectors as selected by the decision makers in the AHP process.

3-3 AHP PROJECT MANAGEMENT APPLICATIONS

In many project management areas, AHP can be utilized to crystalize the best possible solution for a given project management problem.

3-3.1 EARLY PROJECT COST ESTIMATION

In the proposal stage of a project, it is difficult to know requirements and possible solution approaches sufficiently well to base estimates on an undisputable work breakdown structure. Software suppliers thus base their estimates on functional size measurements and experts' opinions; see *Chapter 6: Functional Sizing* and *Chapter 12: Effort Estimation for ICT Projects.* Since cost factors are known but difficult to measure, AHP is the method of choice for understanding the relative impact of each cost factor.

Santillo combines COSMIC and AHP to better understand early project estimation based on functional size (Santillo, 2011). For this, he first uses the early & quick COSMIC function point counting technique. This technique provides practitioners with an early and quick forecast of the functional size, based on the hierarchical system representation of the system under scrutiny, for preliminary technical and managerial decisions at early stages of the development cycle. Obviously, this raises the question whether AHP speeds up measurements of such hierarchical systems. For instance, it should be possible measuring only a part of the hierarchy, and use AHP to compare the different hierarchy branches. Santillo says, "When we document a software structure, we usually name the root as the application level and then we go down to defining single nodes, each one with a name that is logically correlated to the functions included; we reach the leaf level when we don't think it is useful to proceed to a further decomposition." At least the AHP reduces the uncertainty inherent to each size measurement applied to some leaf; nevertheless, the validity of the approach for practical purposes is still under investigation.

3-3.2 PROJECT PORTFOLIO MANAGEMENT

In his famous book, Bhushan (Bhushan & Rai, 2004) uses the AHP for ranking projects and experts using AHP. Criteria such as project similarity, complexity factors, functionality, technology, process, and both business and application domain proficiency allow classifying and evaluating a project portfolio. Furthermore, ranking experts concerning their business domain expertise, estimation experience, development platform and process exposure weights their effort estimations when initiating a software project. AHP thus assigns more weight to senior experts' opinion avoiding the traps of averaging.

3-3.3 PRODUCT MANAGEMENT

AHP is useful even when no customers are at hand that can be asked for their preferences. Thus, it is the method of choice for determining a goal profile for product development. Section 3-4 presents *The Kitchen Knife Case* as an example for successful product management with AHP.

3-4 THE KITCHEN KNIFE CASE

Refflinghaus (Refflinghaus, et al., 2014) used a kitchen knife project to teach product development. With AHP, the customer's needs became apparent; although the customer base involved into the AHP consisted of a small focus group only, the results are quite convincing. The question was what product features affect the willingness to buy some specific kitchen knife most, if branding is not a decisive issue. Customers simply compare a few kitchen knife offerings, for instance in the department's store household alley. While inspecting closely the offered samples, they can judge about things like haptics or blade characteristics.

3-4.1 THE KITCHEN KNIFE PRODUCT FEATURES AND CHARACTERISTICS

The author selected the following 43 product features as candidates for affecting overall product success; see Table 3-8.

Table 3-8: The Kitchen Knife Product Features

A	Product Characteristics	B	Material Properties
A01	Material of the Grip	B01	Dishwater Staunchness
A02	Weight	B02	Blade's Strength at Rupture
A03	Quality of Workmanship	B03	Life Span
A04	Grip haptics	B04	Material Fatigue
A05	Effort needed for Cutting	B05	Blade Abrasiveness
A06	Stability when Cutting	B06	Stability
A07	Bulge for Fingers	B07	Blade Hardness
A08	Versatility	B08	Blade Grinding
		B09	Sensitivity
		B10	Blade Material

C	Constitutive Criteria	D	Service Portfolio
C01	Appearance	D01	Commercial Presentation
C02	Coloring	D02	Guarantee Period
C03	Innovation	D03	Guarantee Scope
C04	Blade Engraving	D04	Money-back Guarantee
C05	Packaging Design	D05	Free Sharpening
C06	Appropriation Symbol	D06	Customer List Enrollment
		D07	Extensibility of the Knife Set

E	Delivery Scope	F	Quality Characteristics
E01	Manual	F01	Certificates
E02	Blade Protection	F02	Promoters
E03	Custody Slot	F03	Brand Name
		F04	Recommendations
		F05	Originality
		F06	Exclusivity
		F07	Test Reports
		F08	Sales Channel
		F09	Price/Benefits Ratio

3-4.2 THE ANALYTIC HIERARCHY PROCESS FOR THE KITCHEN KNIFE

Refflinghaus (Refflinghaus, et al., 2014) defined the top hierarchy of AHP for the kitchen knife product by deciding the relative weights between the following six feature groups:

A. Product Characteristics;
B. Material Properties;
C. Constitutive Criteria;
D. Service Portfolio;
E. Delivery Scope;
F. Quality Characteristics.

Each of the feature groups has a few topics that may add to the overall buyer perception.

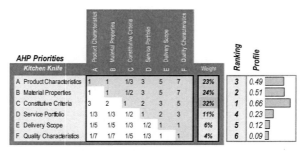

AHP Priorities Kitchen Knife	A Product Characteristics	B Material Properties	C Constitutive Criteria	D Service Portfolio	E Delivery Scope	F Quality Characteristics	Weight	Ranking	Profile	
A Product Characteristics	1	1	1/3	3	5	7	23%	3	0.49	
B Material Properties	1	1	1/2	3	5	7	24%	2	0.51	
C Constitutive Criteria	3	2	1	2	3	5	32%	1	0.66	
D Service Portfolio	1/3	1/3	1/2	1	2	3	11%	4	0.23	
E Delivery Scope	1/5	1/5	1/3	1/2	1	1	6%	5	0.12	
F Quality Characteristics	1/7	1/7	1/5	1/3	1	1	4%	6	0.09	

The top levels are first due for determining relative priority; see Figure 3-9.

From the six feature group characteristics, Figure 3-10, The material used for the knife is a priori probably the most susceptible as key criterion customers look for when buying a kitchen knife. However, this is an erroneous assumption.

Figure 3-11, and Figure 3-12 reveal the three top most groups with their AHP matrix. The other three AHP decision matrices affect the overall hierarchy to a lesser extent.

Figure 3-10: AHP for Kitchen Knife Product Characteristics

AHP Priorities A Product Characteristics	A01 Material of the Grip	A02 Weight	A03 Quality of Workmanship	A04 Grip haptics	A05 Effort needed for Cutting	A06 Stability when Cutting	A07 Bulge for Fingers	A08 Versatility	Weight	Ranking	Profile
A01 Material of the Grip	1	1	1/2	1	1/5	1/9	1/3	1/7	4%	8	0.09
A02 Weight	1	1	2	1/5	1/3	1/7	1/2	1/2	5%	7	0.12
A03 Quality of Workmanship	2	1/2	1	1/2	2	1/7	1	1	9%	6	0.21
A04 Grip haptics	1	5	2	1	1	1/5	2	1	12%	3	0.28
A05 Effort needed for Cutting	5	3	1/2	1	1	1	3	2	17%	2	0.39
A06 Stability when Cutting	9	7	7	5	1	1	3	5	34%	1	0.78
A07 Bulge for Fingers	3	2	1	1/2	1/3	1/3	1	3	10%	4	0.23
A08 Versatility	7	2	1	1	1/2	1/5	1/3	1	9%	5	0.21

The material used for the knife is a priori probably the most susceptible as key criterion customers look for when buying a kitchen knife. However, this is an erroneous assumption.

Figure 3-11: AHP for the Kitchen Knife Material Properties

B	Material Properties	B01 Dishwater Staunchness	B02 Blade's Strength at Rupture	B03 Life Span	B04 Material Fatigue	B05 Blade Abrasiveness	B06 Stability	B07 Blade Hardness	B08 Blade Grinding	B09 Sensitivity	B10 Blade Material	Weight	Ranking	Profile
B01	Dishwater Staunchness	1	1	1/2	1/2	1/7	1/5	1/6	2	1/4	1	4%	10	0.12
B02	Blade's Strength at Rupture	1	1	2	2	1/2	1	1/2	2	1	1	8%	5	0.23
B03	Life Span	2	1/2	1	1/3	1/7	1/6	1/5	5	1/5	1	5%	8	0.15
B04	Material Fatigue	2	1/2	3	1	1/9	1/7	1/7	3	1/6	1	5%	9	0.14
B05	Blade Abrasiveness	7	2	7	9	1	1	2	7	1	3	21%	1	0.59
B06	Stability	5	1	6	7	1	1	2	3	1	2	17%	2	0.46
B07	Blade Hardness	6	2	5	7	1/2	1/2	1	1	2	1	14%	3	0.39
B08	Blade Grinding	1/2	1/2	1/5	1/3	1/7	1/3	1	1	2	1	6%	7	0.17
B09	Sensitivity	4	1	5	6	1	1	1/2	1/2	1	1	12%	4	0.33
B10	Blade Material	1	1	1	1	1/3	1/2	1	1	1	1	7%	6	0.19

AHP Priorities

Figure 3-12: AHP for the Kitchen Knife Constitutive Criteria

AHP Priorities

C	Constitutive Criteria	C01 Appearance	C02 Coloring	C03 Innovation	C04 Emblem Blade Engraving	C05 Packaging Design	C06 Appropriation Symbol	Weight	Ranking	Profile
C01	Appearance	1	3	1/3	1/5	1	1/5	7%	4	0.15
C02	Coloring	1/3	1	1/2	1/7	2	1/6	6%	5	0.12
C03	Innovation	3	2	1	1/2	5	1	19%	3	0.38
C04	Emblem Blade Engraving	5	7	2	1	9	1	34%	1	0.69
C05	Packaging Design	1	1/2	1/5	1/9	1	1/7	4%	6	0.08
C06	Appropriation Symbol	5	6	1	1	7	1	29%	2	0.58

Combining the local profiles with the top profile shown in Figure 3-9 yields a somewhat surprising result.

Figure 3-13: Top Eight Business Drivers for the Kitchen Knife – sorted

Top Business Drivers
Kitchen Knife

Top Business Drivers	Attributes		Weight	Profile	
C04 Emblem Blade Engraving	Noble	Knight	21%	0.56	
C06 Appropriation Symbol	Recognition		18%	0.47	
A06 Stability when Cutting	Trust	Ease of Use	15%	0.41	
C03 Innovation	Prestige		12%	0.31	
B05 Blade Abrasiveness	Sharp	Cut	10%	0.27	
B06 Stability	Fitting	Comfort	8%	0.21	
A05 Effort needed for Cutting	Strong		8%	0.20	
E03 Custody Slot	Cushy		8%	0.20	

8 Top Requirements Selected

The noble knight attributes of the blade engraving seem to dominate all other characteristics when shown in the department shop. Appropriation symbol and stability when cutting come next; however, the dominant feature is not hardware but rather the trust created when linking the knife's brand with some noble ancestry – obviously experienced in knifes, swords and the like.

3-4.3 TOP TOPICS PROFILE

Selecting eight of the 43 features reflects the 7 ± 2 rule of Miller (Miller, 1956). This rule tells marketers and communication experts that their customers cannot handle more than 7 ± 2 features when making up their mind on something. We selected eight because the three last business drivers proved to be almost equal.

The eight top business drivers appear in the ordering inherited by their hierarchical position; this yields a profile that might become a goal profile in some product deployment. Ordering of profile topics does not matter. Both Figure 3-13 and Figure 3-14 are equally valid.

Figure 3-14: Top Eight Business Driver's Profile for the Kitchen Knife

Top Business Drivers
Kitchen Knife

	Top Business Drivers	Attributes		Weight	Profile	
A Product Characteristics	A05 Effort needed for Cutting	Strong		8%	0.20	
	A06 Stability when Cutting	Trust	Ease of Use	15%	0.41	
B Material Properties	B05 Blade Abrasiveness	Sharp	Cut	10%	0.27	
	B06 Stability	Fitting	Comfort	8%	0.21	
C Constitutive Criteria	C03 Innovation	Prestige		12%	0.31	
	C04 Emblem Blade Engraving	Noble	Knight	21%	0.56	
	C06 Appropriation Symbol	Recognition		18%	0.47	
E Delivery Scope	E03 Custody Slot	Cushy		8%	0.20	

3-4.4 LEARNINGS FROM THIS SAMPLE CASE

It is obvious that no product development team ever will admit that all their artisanship is less important for success than the simple and relatively cheap engraving of an emblem, most probably a knight's arms, on the blade of the knife. Marketing people probably will understand it better, knowing that buyers often have no time and expertise to investigate into more relevant product features.

The case shows that what seems obvious to some people, sometimes is not true. The example strengthens the claim of Thomas L. Saaty that AHP helps reconciling different views, avoiding conflicts. Nevertheless, since AHP relies on hierarchies, it also grades hierarchies. This means that if the AHP team moves some topic to an unimportant part of the hierarchy – in the case above those in the three less important feature groups – it has almost no chance to pop up high in the overall rating. Thus, the AHP is only as good as the hierarchy is. Moreover, if the feature groups have unequal size, as with the E: Delivery Scope in the kitchen knife example, its topics get a higher local weight than those in large feature groups such as F: Quality Characteristics. This too affects their global ranking.

However, such considerations are somewhat elusive – an AHP is still much better than any other, statistically useless decision method, because it considers the crossover effects among different topic areas. This is what some people call "gut feeling" (Gigerenzer, 2007).

3-5 PRODUCTIVITY IMPACT FACTOR DETERMINATION BY AHP

Estimating the overall effort and overall cost of a software development project is a most challenging and critical task. Beni et al. (Beni, et al., 2011), Heemstra and Kusters[5], and Morgenshtern, Raz and Dvir (Morgenshtern, et al., 2007) list project Performance Impact Factors (PIF) which (might) affect the effort and cost estimation of the software development project, see also Frohnhoff (Frohnhoff, 2009) in the context of the *Use Case Points* estimation methodology.

Since 2007, a working group of the Italian Software Metrics Association GUFPI-ISMA (Beni, et al., 2011) has collected existing approaches to understand cost factors in software projects. They combined the *General System Characteristics* of older IFPUG standards (IFPUG Counting Practice Committee, 2010) with the parameters used by IFPUG (Hill, 2010) to classify benchmark data, the cost drivers used in COCOMO (Boehm & et.al., 2000), and with the benchmarking parameters collected by the International Software Benchmarking Standards Group (ISBSG). Table 3-15 lists the resulting *Productivity Impact Factors* (PIF).

It is difficult to measure most of these 27 factors directly; neither measurement procedures nor measuring units are readily available, for most of the factors. Moreover, combining these factors with functional size, the primary cost driver when developing software, is far from straightforward – although in *Chapter 12: Effort Estimation for ICT Projects*, it is shown how to accomplish that.

Table 3-15: GUFPI-ISMA Productivity Impact Factors

Personal		Process		Product		Technology	
H1	Domain Knowhow	P1	Organization Maturity	S1	Product Size	T1	Programming Language
H2	Personnel Capability	P2	Schedule Constraints	S2	Product Architecture	T2	Development Tools
H3	Technology Knowledge	P3	Requirement Completeness	S3	Product Complexity	T3	Technical Environment
H4	Team Turnover	P4	Reuse	S4	Other Product Properties	T4	Technology Change
H5	Management Capability	P5	Project Type	S5	Required Documentation	T5	Technical Constraints

Personal		Process		Product		Technology
H6	Team Size	P6	Methodology	S6	System Integration	
		P7	Stakeholder Cohesion	S7	Required Reusability	
		P8	Project/Program Integration			
		P9	Project Logistics			

3-5.1 PRODUCTIVITY IMPACT FACTORS AS ANALYTICAL HIERARCHY

Santillo uses AHP for early and quick counts based on the ISO/IEC 19761 COSMIC standard (Santillo, 2011); thus, it seems very reasonable to use AHP also for measuring PIF. The result of measurement will not be absolute numbers, rather a profile as usual used when working with transfer functions. Beni et al. (Beni, et al., 2011) define four key groups or classes of PIF; see overview in Table 3-15. These groups define the hierarchy; the resulting profile describes the impact on productivity for each of the PIF.

Each of the fours group splits into several subgroups; each subgroup is detailed by some terms and phrases. This structure enables the determination and assessment of those PIF, which are most relevant to a project under consideration. In addition, this structure can easily be extended by adding new project specific impact factors provided by e.g., Frohnhoff (Frohnhoff, 2009).

3-5.2 THE PIF GROUP PERSONAL, OR HUMAN FACTORS

This class assembles personnel-related characteristics of the team involved in the project or process of developing or maintaining software, and consists of six subgroups.

3-5.2.1 H1: DOMAIN KNOWLEDGE

This subgroup reflects the level of knowledge of the subject application treated by the working group. Typical terms and phrases are: precedentedness, staff application knowledge, experience of working in the field, experience on the system, knowledge of the application context, experience of the application domain, availability of people who have worked on similar projects.

3-5.2.2 H2: PERSONNEL CAPABILITY

This subgroup indicates the ability or experience of the project team members with respect to skills of analysis, to skills of programming, analysis, communication and cooperation in terms of self-reliance and effectiveness. Typical terms and phrases are: analyst capability, programmer capability, staff tool skills, staff analysis skills, ability of analysts, ability of programmers, inexperience of human resources, possibility sharing experience and expertise, togetherness of resources, communicability, cooperation.

3-5.2.3 H3: Technology Knowledge

This subgroup specifies the knowledge of working tools applied in the considered project such as requirements engineering tools, programming languages tools and configuration management, and the knowledge of hardware and software environment of the project. Typical terms and phrases are: application experience, platform experience, language and tool experience, experience of the project team, staff tool skills, technical expertise, known/new technology platform, knowledge of the tools and hardware/software platform, experience of working tools.

3-5.2.4 H4: Team Stability (vs. Turnover)

This subgroup concerns the availability and level of staff fluctuation in the project team (including the percentage of allocation of resources, which are working exclusively for the project). Typical terms and phrases are: team cohesion, personnel continuity, staff availability, project team, availability of resources, experienced and stable team resp. inexperienced and/or unstable team, shared use of human resources, turnover team (discontinuity of use), exclusive human resources, number of available resources.

3-5.2.5 H5: Management Capability

This subgroup indicates the ability or experience of the project manager per application of methods and project management techniques adopted in leading the project. Typical terms and phrases are: capacity of the project leader, experienced resp. inexperienced project manager, project management, experience of conducting the project, personal motivation, possibilities for reuse or parallelize resources shared with other projects, better organization and management of relations between the working groups, organization within the group clear and well defined, recognition of group work, investments to improve the knowledge of the elements of the working group.

3-5.2.6 H6: Team Size

This subgroup relates to the number of the project team members.

3-5.3 The PIF Group Process

This class collects the inherent characteristics of the work process adopted and the project management approach used.

3-5.3.1 P1: Process Maturity

This subgroup reflects the level of effectiveness of the use, consolidation and/or standardization of methods, techniques and practices in the development and management of the project. Typical terms and phrases are: development flexibility, architecture/risk resolution, documentation match to life-cycle needs, standards use, relative complexity of the project, approach to project management, maturity of the development process, tracking requirements, flexibility of process, standardization of production processes, definition and clarity of responsibilities, planning the control of the production process, tracking deadlines.

3-5.3.2 P2: Schedule Constraints

This subgroup describes time restrictions per considered project related to stringent requirements. Typical terms and phrases are: required development schedule, duration, presence of binding deadlines (milestones).

3-5.3.3 P3: Requirements Completeness

This subgroup indicates the level of comprehensibility and stability of the requirements between project start and subsequent steps (includes the quality and variability of the requirements). Typical terms and phrases are: stability of user requirements, requirements volatility, technological changes during construction, stable vs. unstable requirements or specifications, quality requirements, incomplete requirements, quantity and quality of available documentation, clarity and definition of objectives.

3-5.3.4 P4: Reuse

This subgroup specifies the implementation of portions or phrases of the project or product by utilizing already existing or implemented portions or phrases of projects or products. This subgroup does not solely focus on software (source) code. Typical terms and phrases are: percentage of reusable code, technical or functional reuse of existing software.

3-5.3.5 P5: Project Type

This subgroup relates to the type of a project under consideration and indicates whether the project is a development from scratch, a re-development, a maintenance or customization project, and includes the type of the implementation approach such as unique, distributed or incremental. Typical terms and phrases are: new development or enhancement, release strategy, project structure.

3-5.3.6 P6: Methodology

This subgroup shows which methods of analysis and design and programming are applied in the project. Typical terms and phrases are: methods and tools used, methods of analysis and design, programming methods.

3-5.3.7 P7: Stakeholder Cohesion

This subgroup depicts the level of communication and cooperation, and the (technical) information exchange between the various stakeholders and actors involved in the project. Typical terms and phrases are: customer participation, user participation, good vs. low customer commitment, division among stakeholders, plurality of clients, agreement of clients.

3-5.3.8 P8: Integration of Other Projects and/or Programs

This subgroup shows the correlation or integration with other projects or programs. Typical terms and phrases are: integration resp. correlation with other projects.

3-5.3.9 P9: Project Logistics

This subgroup relates to the project logistics and the geographical distribution of the project team members. Typical terms and phrases are: multi-site development, where developed, software developer, better working environment, and knowledge is not centralized.

3-5.4 The PIF Group Product Characteristics

This class collects the inherent characteristics of the software product under development or maintenance and consists of seven subgroups.

3-5.4.1 S1: Product Size

This subgroup contains the features related to the size of the software product. Typical terms and phrases are: database size, functional dimension, and functional size.

3-5.4.2 S2: Product Architecture

This subgroup concerns the features related to the architecture of the software product such as the number of sites from which it is possible to use the product, or how to distribute the computational logic of the product. Typical terms and phrases are: execution time constraint, distributed data processing, data communications, multiplicity of sites, requirements of communications resp. networking, operating system architecture, single or multiple site delivery, and number of distinct target architectures.

3-5.4.3 S3: Product Complexity

This subgroup describes the characteristics, which determine the complexity of application of the product such as the complexity of algorithms, complex structure of the data to be processed. Typical terms and phrases are: complex processing, (software's) logical complexity, and higher level of testing due to system criticality.

3-5.4.4 S4: "Other" Product Properties

This subgroup relates to the characteristics that are different from those given in S2 and S3 such as security characteristics, ease of operations, and type of interface and/or interaction with the user. Typical terms and phrases are: required software reliability, frequency of transactions, interactive data input, efficiency for end-user, update interactive, ease of installation, easy to change, ease of operations management, security/privacy/controllability, efficiency requirements, application type, quality requirements, performance, facilitation of changes, types of applications, degree of innovation, system performance requirements, database management systems, quality/reliability/usability request, performance characteristics, quantity of sites/users/interfaces, constraints on processing times, and quality of product.

3-5.4.5 S5: Required (Product) Documentation

This subgroup concerns documentation features to be issued, and whether the documentation can be complete and extensive. Typical terms and phrases are: the need for user training, installation requirements, level of deliverable documentation, and documentation required by the project.

3-5.4.6 S6: (Other) System(s) Integration

This subgroup contains features that define the product integration with other software systems. Typical terms and phrases are: requirements from other applications, direct use of the application by a third party, interfacing with other systems, integration with other systems, and complexity of the application context.

3-5.4.7 S7: Required (Product) Reusability

This subgroup reflects the consideration that when writing the code, the reuse aspect must be considered. Typical terms and phrases are: required reusability, usability in other applications, and replication software functionally identical or similar on different platforms.

3-5.5 The PIF Group Technology

This class collects the inherent characteristics of the technology used or required for the development or enhancement of the software and consists of five subgroups.

3-5.5.1 T1: PROGRAMMING LANGUAGE

This subgroup concerns the type of the programming language. Typical terms and phrases are: programming language used and development languages.

3-5.5.2 T2: DEVELOPMENT TOOLS

This subgroup describes all the tools such as the framework, DBMS, testing tools, etc., which affect the implementation of the product under consideration. Typical terms and phrases are: use of software tools, level of technical environment (e.g. languages), tools use minimal tools, editors, tool for testing and debugging, tool for automatic testing, tool for analysis of source code, using advanced tools, using dedicated tools such as database management system and reporting system, availability of utilities that can be parametrized, availability of utilities operating system and platform, availability of the most efficient tools.

3-5.5.3 T3: TECHNICAL ENVIRONMENT

This subgroup concerns characteristics of reliability and stability, respectively, and the industrialization of the environment (HW/SW). Typical terms and phrases are: development environment adequacy, tool for configuration management, and quality of technical development (hardware/software).

3-5.5.4 T4: TECHNOLOGICAL CHANGE

This subgroup describes the eventuality that the development team must modify or replace a significant infrastructure component during the development of the product. Typical terms and phrases are: platform volatility and technological changes during duration.

3-5.5.5 T5: TECHNOLOGICAL CONSTRAINTS

This subgroup contains limitations and restriction to be respected during the development. Typical terms and phrases are: main storage constraints, intensive use of configuration, hardware type, number of DBMS, operating systems, technical factors and technology platform, number of different target architectures.

3-5.6 THE AHP PROCESS FOR THE PIF

The measurement subject is a project, in our case, an ICT project. Different projects, especially in different environments, will yield different profiles. The following example of a PIF measurement assesses on a technology-driven project, named "Project X".

3-5.6.1 THE TOP LEVEL

On the top level of Figure 3-16, we distinguish the relative weights of the four PIF groups *Personal, Process, Process*, and *Technology*.

Figure 3-16: Top Level AHP Matrix for Project X

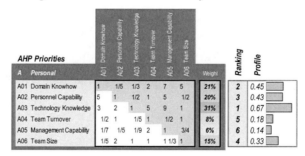

The abbreviations H, P, S, and T used in Table 3-15 refer consecutively to the matrices named A, B, C, and D, in Figure 3-16.

3-5.6.2 THE FOUR LOWER LEVEL AHPs

All pairwise comparisons are subject to expert's judgment only. They will differ for other projects. Since knowledge about technology is key to personal characteristics in the project X, it wins in Figure 3-17.

Figure 3-17: AHP Decision Matrix for Personal PIF

The project process relies to a high degree on requirements completeness – something that is hard to achieve in most ICT projects; however, for a technology-driven project it wins in weight against other process characteristics; see Figure 3-18.

Figure 3-18: AHP Decision Matrix for Process PIF

AHP Priorities

B Process	B01 Organization Maturity	B02 Schedule Constraints	B03 Requirement Completeness	B04 Reuse	B05 Project Type	B06 Methodology	B07 Stakeholder Cohesion	B08 Project/Program Integration	B09 Project Logistics	Weight	Ranking	Profile
B01 Organization Maturity	1	1/7	1/9	1	1	2	2	3	1/3	7%	5	0.18
B02 Schedule Constraints	7	1	1/5	3	2	3	3	4	1/2	16%	2	0.39
B03 Requirement Completeness	9	5	1	7	5	9	1	2	1	29%	1	0.72
B04 Reuse	1	1/3	1/7	1	1	2	1/2	5	1/3	7%	6	0.16
B05 Project Type	1	1/2	1/5	1	1	1	1/2	2	1	6%	7	0.15
B06 Methodology	1/2	1/3	1/9	1/2	1	1	1/2	3	1	6%	8	0.14
B07 Stakeholder Cohesion	1/2	1/3	1	2	2	2	1	9	2	14%	3	0.36
B08 Project/Program Integration	1/3	1/4	1/2	1/5	1/2	1/3	1/9	1	1	4%	9	0.11
B09 Project Logistics	3	2	1	3	1	1	1/2	1	1	12%	4	0.29

As expected for a technology project, product complexity is most important and has highest impact on productivity (Figure 3-19).

Figure 3-19: AHP Decision Matrix for Product PIF

AHP Priorities

C Product	C01 Product Size	C02 Product Architecture	C03 Product Complexity	C04 Other Product Characteristics	C05 Required Documentation	C06 System Integration	C07 Required Reusability	Weight	Ranking	Profile
C01 Product Size	1	1/2	1/2	9	2	4	1	15%	3	0.32
C02 Product Architecture	2	1	1	9	5	7	2	27%	2	0.57
C03 Product Complexity	2	1	1	9	9	5	5	33%	1	0.69
C04 Other Product Characteristics	1/9	1/9	1/9	1	1	1	1/5	3%	7	0.07
C05 Required Documentation	1/2	1/5	1/9	1	1	3	1/2	6%	5	0.12
C06 System Integration	1/4	1/7	1/5	1	1/3	1	1/5	4%	6	0.07
C07 Required Reusability	1	1/2	1/5	5	2	5	1	13%	4	0.27

Ongoing change in technology affects productivity; see Figure 3-20.

Figure 3-20: AHP Decision Matrix for Technology PIF

AHP Priorities

D Technology	D01 Programming Language	D02 Development Tools	D03 Technical Environment	D04 Technology Change	D05 Technical Constraints	Weight	Ranking	Profile
D01 Programming Language	1	1/2	1/3	1/5	2	9%	4	0.16
D02 Development Tools	2	1	1/2	1/3	5	17%	3	0.31
D03 Technical Environment	3	2	1	1/2	3	24%	2	0.43
D04 Technology Change	5	3	2	1	9	46%	1	0.83
D05 Technical Constraints	1/2	1/5	1/3	1/9	1	5%	5	0.09

3-5.6.3 THE PIF PROFILE

Combining these AHP decision matrices yields a profile for the 27 PIF. The overall winner is again *D04: Technology Change*, followed by *D03: Technical Environment* and *A03: Technology Knowledge*, from the Personal PIF group.

Figure 3-21: The Final Productivity Impact Profile for Project X

Top Targets
PIFs for Project X

Top Targets		Attributes			Weight	Profile	
A Personal	A01 Domain Knowhow	Insufficient	Learning	Environment	5%	0.19	
	A02 Personnel Capability	Business	Competition	Users	5%	0.18	
	A03 Technology Knowledge	Scarce Resource	Evolving		7%	0.28	
	A04 Team Turnover	Personality	Stress		2%	0.07	
	A05 Management Capability	Internal	Public	Product	1%	0.06	
	A06 Team Size	Differences	Politics		3%	0.14	
B Process	B01 Organization Maturity	Environment	Heritage		2%	0.07	
	B02 Schedule Constraints	Always	Too	Short	4%	0.15	
	B03 Requirement Completeness	Agile	Knowledge	Acquisition	7%	0.27	
	B04 Reuse	Quality	Capability	Maturity	2%	0.06	
	B05 Project Type	Well documented			1%	0.06	
	B06 Methodology	Agile			1%	0.05	
	B07 Stakeholder Cohesion	Too big	Just Right	Too few	3%	0.13	
	B08 Project/Program Integration	Impact	Stand-alone		1%	0.04	
	B09 Project Logistics	Distributed	Organized		3%	0.11	
C Product	C01 Product Size	Components			3%	0.11	
	C02 Product Architecture	Complex	Adaptable		5%	0.20	
	C03 Product Complexity	Even higher	Adapting	Flexible	6%	0.24	
	C04 Other Product Characteristics	Open			1%	0.02	
	C05 Required Documentation	Audience			1%	0.04	
	C06 System Integration	Environment	Middleware	Cloud	1%	0.03	
	C07 Required Reusability	Important	Paid for		2%	0.09	
D Technology	D01 Programming Language	Appropriate	Framework		3%	0.12	
	D02 Development Tools	Evolving			6%	0.23	
	D03 Technical Environment	Changing	Inflationary	Challenging	8%	0.32	
	D04 Technology Change	Fast	Paradigm	Change	16%	0.62	
	D05 Technical Constraints	Stability	Incidents		2%	0.07	

The resulting profile (Figure 3-21) is specific to our technology-driven project X. Other ICT projects likely will have different PIF profiles and therefore different cost drivers. Doing such an AHP assessment is almost free when compared to other costs incurred when doing project estimation. In *Chapter 12: Effort Estimation for ICT Projects*, we show how to use such a PIF profile for predicting project cost.

3-6 CONCLUSION

AHP is a simple and easy to use tool for supporting decision-making. It can be implemented in a simple spreadsheet; however, professional software is available that not only supports decision-making but also manages decision-taking, tracking resources and intermediate results influencing the final decision. Thomas Saaty introduced the powerful concept of eigenvectors to decision-making; since then eigenvectors had a huge effect as the world leader in search, Google, uses the PageRank algorithm to propose best hits, see e.g., (Kressner, 2005), (Langville & Meyer., 2006), (Gallardo, 2007).

The major limitation of AHP is that it only compares similar topics. It does not compare well between alternatives of a different kind, and it does not analyze cause-effect relationships.

CHAPTER 4: QUALITY FUNCTION DEPLOYMENT

Quality Function Deployment (QFD) developed at the beginning of 1970s in Japan is the method of choice when developing a product or a service, which relates to the needs, delights and expectations of customers, or in other words, which relates to the Voice of the Customer (VoC).

QFD helps to identify and to prioritize the most attractive product or service attributes or qualities defined by the customer, so that the final product or service conforms exactly to what the customer demands or wants. One prominent benefit when utilizing QFD in product or service development is, that the development cycle time can significantly be decreased.

QFD is part of the Six Sigma roster. It is pivotal to Design for Six Sigma.

4-1 QUALITY FUNCTION DEPLOYMENT IN A NUTSHELL

Quality Function Deployment (QFD) is possibly one of the most successful but least known methods for product design. It aims at products based on customer's needs. Successful industries such as automotive or smartphones use it extensively; however, these industries do not relate much about the method because it relates to business secrets.

However, QFD is an open methodology, and helps in developing countries to make their growing industry competitive for the world market. QFD became famous when it did this job for Japan 40 years ago. It is not surprising that interest in Turkey, Iran, China, India, to name just a few upcoming industrialized countries, is overwhelming. For traditional industrial nations, the belief that an existing product design remains best for all times makes QFD look unnecessary – a belief that sometimes is appropriate, unless paradigms change suddenly.

QFD is part of the Six Sigma roster; in particular, it is pivotal to *Design for Six Sigma* (DfSS), first mentioned in *Chapter 1: Lean Six Sigma, section 1-1.1* .

The QFD method relies on detecting customer's needs by listening to customer's voice. Most of these needs remain hidden because customers, when asked, usually mix up solutions with needs. For instance, the need of moving physically between different places required different technical solutions, after the advent of motorized cars and with the future ability to construct self-controlled vehicles. However, over all times and technologies, the need to move remains the same. Thus, detecting and understanding customer's needs is much more than just listening to customer's voice. Voices alone might be misleading.

The choice of technical solutions that best fit customer's need is the crucial step in QFD. Some solutions are unnecessary because they do not contribute to any need; some are excessively solving problems that the customer did not have before using the product. In Six Sigma terms, these technical solutions are *Controls*. Not all controls need to exhibit technology. For instance, branding and other soft factors that come with a product might be part of the controls that all together cover customer's needs.

4-2 EARLY ADOPTION OF QUALITY FUNCTION DEPLOYMENT

Possibly the earliest application of transfer functions outside of physics and signal theory for analyzing controls in a process is QFD. QFD usually starts with measuring customer's needs, or other applicable strategic goal topics, by means of set of techniques called *Voice of the Customer* (VoC). From VoC, the *Voice of the Engineers* (VoE) is derived, namely the solution controls that most closely matches the VoC profile. These techniques and methods control the desired system response.

QFD originated in shipyard's industries in Japan, where high complexity was characterized by dozens of mostly conflicting customer's needs and hundreds of technical parameters requiring adjustment to meet customer's needs. In QFD, ships are systems that implements a freight or passenger carrying process. Key point was visualizing complexity and stir discussions when inserting data into matrix cells. Finally, this method established consensus among the many disciplines involved.

4-2.1 THE ORIGINS OF QUALITY FUNCTION DEPLOYMENT

Today, QFD is the method of choice when designing and manufacturing cars, planes or other highly complex engineered device. The big advantage of QFD is that all decisions rely on measurements, be it for the strategic goals measured among customer segments by marketing, or for the impact technical features and solutions have per matrix cell. This is the key benefit of QFD, and Prof. Akao demonstrated in a verbal communication that a simple QFD – with much less rows and columns than in shipyards – would have been able to predict correctly the risks that finally caused the Fukushima burndown of nuclear power plants.

Because many of the couplings between controls and responses proved to have contrary effects on each other – e.g., stabilizing the system could be achieved by using heavier material, contradicting the customer's needs for a lightweight solution – teams added a triangle roof to the matrix, indicating whether there is some favorable or unfavorable cross-effect. Therefore, QFD became synonymous with the "House of Qualities" – namely the qualities of VoC versus VoE with the roof indicating contradicting needs.

4-2.2 WHERE QFD IS NOT USED DESPITE ITS USEFULNESS

QFD is a powerful remedy against the inherited shortcomings of humans, namely to rely on human herd instinct, trusting power struggle more than logic or reasoning.

This book targets not only at people who want to understand how to manage complexity, but also at managers willing to scrutinize their inclination towards personal power and influencing. It should help them to unleash the power of structured thinking against unstructured beliefs. The reason why at times QFD is so successful is the generic Six Sigma measurement matrix, the transfer function; uncovering coupling cells based in data that traditionally remained hidden, leveraging them from a matter of influencing minds to a factual discussion. It is hard to negate something you clearly can see there. The empty cell in the matrix might simply wait for measurements. That made QFD in shipyards successful.

On the other hand, some people do not like matrices. Their perception simply frightens them. They prefer to look for unanimity based on unsolicited belief that, e.g., "a Tsunami higher than 5 meters cannot occur", even if recurrence to historical data would have demonstrated the contrary.

4-2.3 IMPLICATIONS OF MIXING UP CAUSE AND EFFECT ON SOCIETY

In many political discussions, even experts do not rely on measurement data but feel an emotional need to seek conformance to public opinion, mixing cause and effect. Because people belief in experts, they take their statements for granted. In turn, the experts are reluctant to state the truth because such statements would harm their reputation, career, and future income. This occurs in many such public topics such as climate change and the question, which kind of plants should provide the electricity needed by a growing population that becomes more and more dependent from ICT technology.

4-2.4 QFD IN REQUIREMENTS ELICITATION

There is often much confusion about requirements. The "What's" and the "How's" are freely mixed if requirements are collected without the QFD discipline. This is typical for many projects: not only the needs of the users and the customer is part of the requirements; also, requirements postulate certain technical means. Naturally, retaining a technical strategy can be extremely important, for instance to avoid a mixture of different technologies providing the same functionality. However, in such cases, not the choice of technology is the customer's requirement. The need is to continue adhering to a technology choice that coexists with previous investments. This stirs creativity in defining suitable solution controls rather than blocking innovation by keeping to some chosen strategy. The difference between the two approaches can be substantial.

4-2.5 QFD FOR PREDICTIONS

In many cases, QFD experts guess the couplings in the matrix cells. This is typical if the system under question is under development and not yet built, and predictions govern optimum design decisions.

Such expert judgments are kind of measurements. This change of attitude probably makes the difference between traditional and modern QFD. Traditionally, the team used symbols representing strong, medium, or weak coupling. The advantage of this was that the visual perception helped the team to judge whether the matrix represents the transfer function between controls and responses. The disadvantage clearly is that it was very difficult, and needed long experience with QFD, to detect shortcoming or inconsistencies in QFD matrices. Experienced QFD moderators used esthetics – the look of the matrix – for detecting inconsistencies. Some inconsistencies are easy to detect – e.g., empty columns, indicating that this control has no affect at all, or empty rows, indicating a lack of support for some of the expected responses. In addition, empty areas indicate a lack of coherence within either the response topics or the solution controls.

4-3 QUALITY FUNCTION DEPLOYMENT BASICS

Asking the right question is already half the answer. QFD can provide a choice of solution controls to all kind of endeavors; however, the choice of controls is never a uniquely given decision. This kind of thinking is what makes QFD so powerful for complex undertakings.

4-3.1 FROM VOICE OF THE ENGINEER TO VOICE OF THE CUSTOMER

Quality Function Deployment (QFD) ensures that the VoE (*Voice of the Engineer*) delivers what the VoC (*Voice of the Customer*) expects (Akao, 1990). This is not common practice; many people try it the other way around. The VoC is the response that the system, or process, must deliver, and the VoE are the controls. If a product fails to match the VoC requirements, it is very likely that it becomes a failure, despite excellent engineering or high quality in manufacturing or delivery.

It is always easier inventing new contributions to VoE, or improving existing ones, than detecting new aspects of the VoC topics. However, customers decide based on their needs and expectations, and both are always evolving. The challenge for anybody looking for success in the marketplace is to uncover hidden, or new, customer's needs – ideally those that not even the customers knew about before.

Once uncovered, the engineers adapt their solution approach to the new or changed needs. QFD provides a framework to identify the need for new solution approaches among the solution topics; however, neither uncovering hidden VoC nor identifying new VoE works in an automated way.

Transfer functions model the relationship between VoE and VoC. Thus, the QFD discipline is about studying and measuring the transfer function. Not understanding the difference between VoC and VoE, expected response and controls, is one of the most common sources of failures in product management and service improvement. It means for instance that solutions are mistaken as customer's needs, making it easy for competition to perform better.

4-3.2 BLITZ OR NO BLITZ?

During the last years, there was a tendency to shorten the QFD process and restrict it to understanding customer needs. It seems more important saving (customer's) time than understanding relationships between process controls and process response.

The so-called Blitz QFD® (Zultner, 1995) is a shortened process focusing on the *Customer Value Table* rather than on the transfer function. It aims at understanding customer needs from its values. This approach proved extremely valuable when writing proposals. Thomas Saaty's *Analytic Hierarchical Process* (AHP) also analyzes customer needs, see *Chapter 3: What is AHP?* prioritizes them with ratio scales, and assesses for consistency. This seems much quicker and more reliable than assessing the Voice of the Engineers, and then trying to understand how it transforms into something that meets Voice of the Customer's requirements.

However, the Blitz QFD® does not allow uncovering hidden customer's needs from other sources than value table, contrary to the Six Sigma approach. Six Sigma *Defines, Measures, Analyzes, Improves,* and *Controls* (DMAIC) relationships between process controls and process responses. QFD is about transfer functions that describe how to transform controls used by engineers into responses expected by customers. If there are more than one response characteristics, and more than one control affects the response, the QFD matrix is the natural choice for representing the transfer function.

Doing QFD without matrices is nonsense; however, QFD with bad mathematics is also nonsense, as very concisely pointed out by Mazur (Mazur, 2014).

The transfer function is used to understand the signal received from the Voice of the Customer and allows to detect not only more controls but also uncover hidden customer's needs by means of analyzing inconsistencies. The convergence gap is a simple measure indicating the need to research for different controls, and if competition proposes different solution approaches, then the *New Lanchester Strategy* (Taoka, 1997) allows detecting customer's needs not addressed by current offerings. Because QFD is not limited to a single house of quality, it is possible to establish a *Deming Chain* (Fehlmann, 2003) that combines transfer functions between different topics, again uncovering hidden needs and overlooked trends. We explain Deming chains later in section 4-9 and in more detail in *Chapter 13: Dynamic Sampling of Topic Areas*.

4-4 ELICITING THE VOICE OF THE CUSTOMER

Voice of the Customer (VoC) is the start on all QFD projects. Goal of the VoC approach is to understand the needs of customers, user, or more general, targeted market. Customers are not always readily available and find time to let their voice hear; especially in the case of new product development, customers are mostly a vision for the future.

We present a choice of available methods. The most popular methods are surveys and questionnaire; the most effective method is *Going to the Gemba*.

4-4.1 WORKSHOP WITH CUSTOMERS OR SENSING GROUPS

If real customers are not available, a popular but not very reliable workaround is working with sensing groups.

Figure 4-1: Sample Voice of the Customer by Voting

	Customer's Needs Topics	Attributes		VoC Input	Weight	Profile	
Y.a Group 1	y1 Target 1	Attribute 1.1	Attribute 1.2	16	10%	0.25	
	y2 Target 2	Attribute 2.1		34	22%	0.52	
	y3 Target 3	Attribute 3.1	Attribute 3.2	38	25%	0.59	
	y4 Target 4	Attribute 4.1	Attribute 4.2	27	18%	0.42	
Y.b Group 2	y5 Target 5	Attribute 5.1		21	14%	0.32	
	y6 Target 6	Attribute 6.1	Attribute 6.2	6	4%	0.09	
	y7 Target 7	Attribute 7.4	Attribute 7.2	12	8%	0.19	

Experts sit together and play the roles of users and customers. Sometimes it works quite well, if the sensing groups can understand customer experience; sometimes the approach fails because sensing groups follow group mechanisms rather than the real customer or user. If everything else fails, a sensing group is still better than starting a project without even trying to understand the customer's needs profile. A missing goal profile is the most common reason why projects fail.

Workshop teams, customers or sensing groups, create a profile by voting. Workshops start with brainstorming for topics that could qualify; then, they cluster these topics. Now every member gets a certain number of points that she or he can assign all to one, or distribute among, topics selected. These votes constitute the VoC input in Figure 4-1.

The result is a goal profile, from which the workshop team can remove certain topics, because they got less than a minimum number of votes. This works similar as with AHP. The rules for normalization of the VoC input are those introduced in *Chapter 2: Transfer Functions*.

Let c_i be the number of votes per topic, $i = 1..n$, n being the total number of topics, then normalization of the vector $c = \langle c_1, ... c_n \rangle$ yields the profile $x = \langle x_1, ... x_n \rangle$ as follows:

$$x_i = \frac{c_i}{\|c\|} = c_i \bigg/ \sqrt{\sum_{i=1}^{n} c_i^2}, \ \ for \ \ i = 1, ..., n \tag{4-1}$$

This vector $x = \langle x_1, ... x_n \rangle$ is a profile for the VoC input; see section 3.2.4 about normalization of vectors in *Chapter 2: Transfer Functions*.

4-4.2 ANALYTIC HIERARCHY PROCESS (AHP)

The AHP presented in *Chapter 3: What is AHP?* captures VoC together with customers or a sensing group. If the number of available topics or alternatives exceed a certain minimum number, usually twelve, then AHP is always the better choice than voting.

Figure 4-2: Sample AHP as an Alternative to Voting

AHP Priorities	y1 Target 1	y2 Target 2	y3 Target 3	y4 Target 4	y5 Target 5	y6 Target 6	y7 Target 7			
Customer's Needs								Weight	Ranking	Profile
y1 Target 1	1	1/5	1/6	1/2	1/2	2	1	7%	6	0.16
y2 Target 2	5	1	1/2	2	2	5	2	23%	2	0.55
y3 Target 3	6	2	1	1/2	2	6	2	26%	1	0.61
y4 Target 4	2	1/2	2	1	1	3	1	17%	3	0.39
y5 Target 5	2	1/2	1/2	1	1	2	1	12%	4	0.27
y6 Target 6	1/2	1/5	1/6	1/3	1/2	1	1	5%	7	0.13
y7 Target 7	1	1/2	1/2	1	1	1	1	10%	5	0.24

Convergence Gap:	0.00	< 0.1: Solution found
Schurr Radius:	0.03	If value < 0.1,
Convergence Range:	0.10	judgements are consistent,
Convergence Limit:	0.20	else judgements are doubtful

An AHP done with actual customers is probably the best available method for prioritizing customer's needs, and the method of choice when working with sensing groups.

Its only disadvantage is that people sometimes find it difficult to understand how eigenvectors work, and thus prefer more direct but less reliable methods such as voting or the unreliable pairwise comparison as shown in Figure 3-1.

Normalization of profile is the same as in equation (4-1). Figure 4-2 shows an AHP matrix that compares with Figure 4-1, sharing the same pairwise decisions, however with a different ranking. AHP has the advantage that it handles well even a significant amount of inconsistency. With voting, inconsistency is never apparent. This sample comes with a Schurr Radius of 0.03, indicative for only small variations in individual ratings.

4-4.3 TRADITIONAL QUESTIONNAIRES AND SURVEYS

Traditional VoC analysis uses survey questionnaires, among future customers, beta users, or sensing groups. Today, questionnaires, especially over the web, have become so abundant that their relevance becomes more and more questionable. Who responds to a 4-page 80-question survey? Moreover, questionnaires often reflect the viewpoint of the supplier rather than customer's voice; thus, the answers are biased anyway.

For determining the value for the customer, so-called Kano questionnaires ask for quantification of both importance and satisfaction with certain features or characteristics of a product. For surveys, especially customer satisfaction surveys, *Net Promoter®️ Score* (NPS) surveys (see *Section 4-4.7 Net Promoter®️ Score*) have replaced traditional questionnaires to a large extend. Especially for understanding customer's values, NPS surveys provide significant better results than traditional questionnaires, even when utilizing the Kano approach to ask for both importance and satisfaction.

On the other hand, if used with care, questionnaires may still give valuable insight to answer questions such as how much people are ready to spend.

4-4.4 THE KANO METHOD

The *Kano Model* (Kano, et al., 1984) provides a very simple, timesaving and effective method. The model distinguishes three kind of attributes: *Delighters, One-to-one Requirements*, and *Expectations*. Position these attributes in a square comparing degree of customer satisfaction to degree of technical excellence needed when implementing it. The geometric locus of all delighters is a hyperbolic curve in the diagram; similar for the expectations but with inverted prefix.

We use the hyperbolic curve of the delighters decreasing its steepness towards zero for the bottom line and derive the customer priority out of it. In Figure 4-3, we see the three attributes, together with was QFD considers "Sales Points. The higher left a feature positions itself, the higher is its value as a "sales point"; thus, something that can persuade a prospect to buy something. The deeper right – technically complex but without any impact on customer satisfaction – a feature positions itself, the less it adds value for the customer.

Figure 4-3: The Kano Model with its Weight Categories

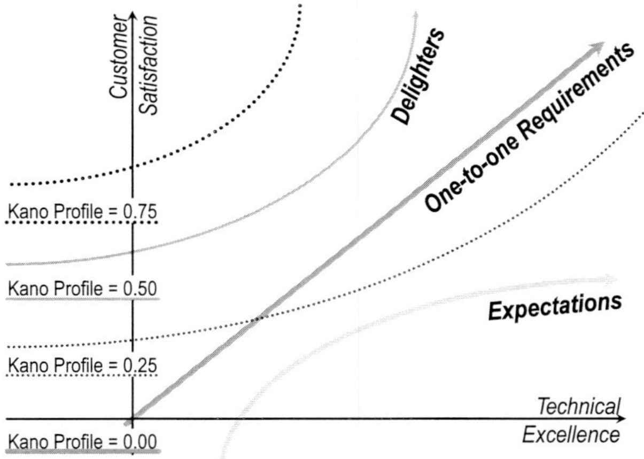

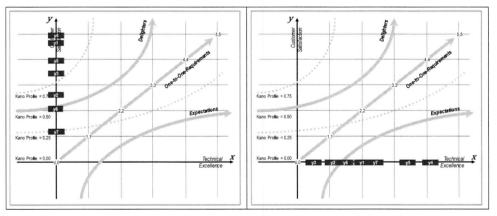

Thus, the Kano model distinguishes "must-be" features situated in the lower right from "delighters" up left. The hyperbolic curves shown in Figure 4-3 hint at their profile value. The curves shown correspond to a profile that is flat; i.e., all topics on these curves have equal weight.

Figure 4-4: Step 1 and Step 2 of Kano Workshop

The Kano profile value grows from downright up to upper-left, and this is easy to explain to a workshop team.

Cheap ways to satisfy the customer always have precedence over excellent but expensive solutions that do not matter much to the customer. A Kano workshop starts as follows: Brainstorm and cluster relevant feature attributes as before. The customer representative starts the analysis by recording and ordering the feature attributes along the vertical customer satisfaction axis. The technical people independently judge the technical excellence needed to achieve on the horizontal axis.

The position on the axis tells how important, respectively how difficult an attribute is. The team can use cards for positioning them at some quadrant, or even with more precision, at a specific place.

Figure 4-5: Step 3 of Kano Workshop

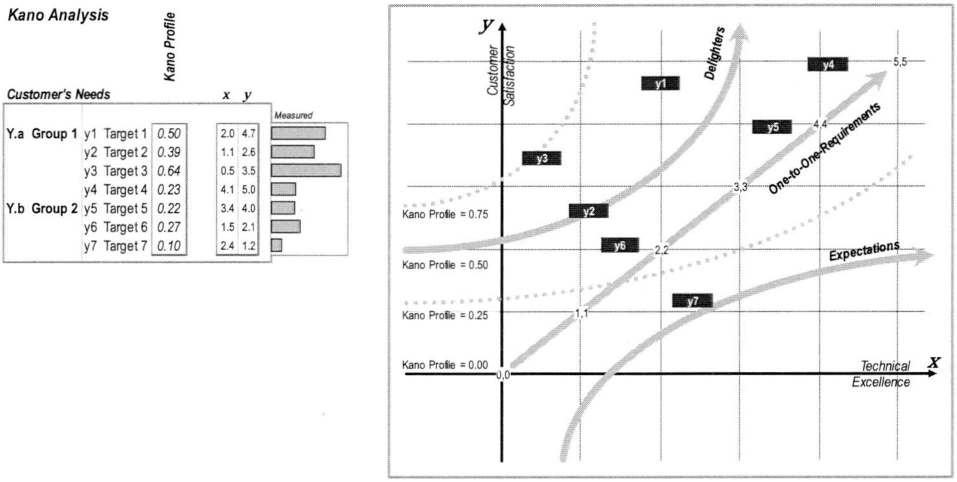

An important point of the technique is that both teams cannot judge two different features as equally important, or difficult. They must decide what comes first and what comes next. This somewhat limits the applicability of the Kano technique. As already pointed out, it works best for analyzing customer's preferences and decision criteria, the "Sales Points".

In Figure 4-5, the *y7: Target 7* almost disappears from the profile, because it is an expectation that seems not all easy to ascertain in the product. This position tells that the feature is not helping the customer with its buying decision; however, it might still be a technically important feature and require significant attention.

4-4.5 KANO FOR SAFETY AND OTHER QUALITY CHARACTERISTICS

Sometimes, teams repeat the Kano workshop with different aims, for instance aiming at security relevance, or safety, instead of customer satisfaction. The results of all the Kano workshops combined might give a better target for product design than concentrating on a single aim.

Thanks to the normalization property, different profiles simply combine, sometimes with different weights. The Kano method has the advantage to visualize the target profile at an early stage, and without spending too much time and effort on it. A Kano workshop requires 10 minutes only and is an excellent communication tool.

The Kano model is adaptable to other topic areas than customer's satisfaction. For instance, it could position features about safety requirements. Some features are more relevant to safety than others are; e.g., car speed or motor power. Both are delighting customers but relevant to safety.

Figure 4-6: Kano Analysis for Safety Relevance

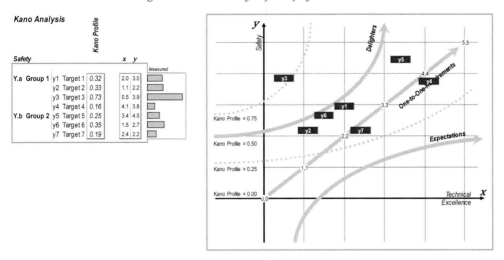

The value profile for safety is likely to look somewhat different. Assuming that the selection of topics *y1: Target 1* to *y7: Target 7* affect both customer satisfaction and safety, a similar analysis might yield different profiles; see Figure 4-6. The x-axis still holds for technical excellence required to implement some features, and thus the x-coordinates in Figure 4-5 and Figure 4-6 are identical; the safety relevance has some distinct alterations against customer satisfaction.

The Kano method is most versatile because of the reconciliation of different views. Kano complements the AHP tool for decision support; often both are used, and results combined.

4-4.6 GOING TO THE GEMBA (現場)

Gemba – 現場 in Japanese – means going to the real place where the essential things happen. In lean manufacturing, the idea of gemba is that the problems are visible, and the best improvement ideas will come from "going to the gemba". The gemba walk is an activity that takes management to the front lines to look for waste and opportunities for practical shop floor improvement, see Womack (Womack, 2013). For lean software development, the gemba walk means that the development team visits future users at their workplace. Observing, talking with them, looking for waste and opportunities for process improvement are activities that should happen before developing any software. Such gemba walks fit well to agile methodology.

Moreover, QFD techniques exist how to go to other gemba than the shop floor. The most popular are verbatim analysis in all kind of documents a customer produces, even if non-related to the specific project topic area, like monitoring social media.

Such gemba techniques define priorities. With priorities, Thomas Saaty (Saaty, 1990) has shown that it is necessary to handle priorities as vectors; i.e., normalized to their vector length, not to its maximum coefficient as old QFD did, and use ratio scales, not symbols or symbolic values. Vectors are additive and normalized vector profiles are comparable, thus they qualify as goal profiles for transfer functions.

Quantifying gemba visits is not straightforward. Depending on the kind of observations, it is possible to measure time, number of clicks, number of failed attempts, or number of complaints to understand the value for the customer. Sometimes, gemba visits furnish topics for AHP or Kano prioritization rather than directly a profile. However, if the observable events are measurable, a priority profile results from normalization using equation (4-1), see section 4-4.1 .

4-4.7 NET PROMOTER® SCORE

Fred Reichheld, Bain & Company, and Satmetrix Systems, Inc. have introduced and trademarked *Net Promoter® Score* (NPS) as a measurement method for customer loyalty (Reichheld, 2007). The NPS method is part of the Six Sigma toolbox for understanding Voice of the Customer; see Fehlmann and Kranich on how to use transfer functions to analyze an NPS survey (Fehlmann & Kranich, 2012-3). Together with transfer functions, NPS is an excellent, reliable and relatively cheap tool. *Chapter 5: Voice of the Customer by Net Promoter®* of this book introduces the method and explains the transfer functions needed.

4-4.8 SOCIAL MEDIA METRICS (NPS 2.0)

The same method can analyze big data collected from social media. Clustered verbatim statements graded by NPS and compared with the major cluster keywords, relate the score with the potential reasons for giving importance and being satisfied with product characteristics using transfer functions, see section 5-5.3 in *Chapter 5: Voice of the Customer by Net Promoter®*.

4-4.9 NEW LANCHESTER STRATEGY

Like NPS, the *New Lanchester Strategy* uses transfer functions to analyze root causes for observed market share profiles. The root causes constitute the value system of the buyers in some product market. *Chapter 11: Application to Product Management* explains how to use New Lanchester Strategy for understanding VoC. Contrary to NPS, New Lanchester Strategy is part of the QFD tool box since its inception in the Nineties (Taoka, 1997).

4-5 TRANSFER FUNCTIONS IN A QFD CONTEXT

Quality Function Deployment (QFD) is a method for analyzing customer's needs and creating solutions that best deliver value for the customer. The QFD relies on matrices of the form $A \in \mathbb{R}^{m \times n}$ with $m \leq n$. The cell entries constitute the relationship strength among the solution controls and the response. Hence, the results of the previous per multiple response linear transfer functions apply to the QFD context. In this section, we show a sample QFD. QFD benefits from applying multiple response linear transfer functions, eigenvalue theory and dual space linear forms as introduced in *Chapter 2: Transfer Functions*. Because QFD matrices represent multiple response transfer functions, the theory is applicable and qualifies QFD analysis. The new proposed international standard (ISO 16355-1:2015, 2015) paves the way.

Mathematics cannot replace experts solving a problem. Eigenvector theory is primarily about numbers representing the solution characteristics proposed by experts. However, the theory predicts the outcome of the solutions found with a QFD and this is the contribution of Six Sigma to managing complexity.

4-5.1 COMBINING VARIOUS VOICES OF CUSTOMERS

Thanks to the *£2* normalization property, combining several VoC analysis activities is straightforward. Figure 4-7 demonstrates the combination of four profiles with different weights.

Such combinations add trust to VoC analysis, if done with equal precision and care. It is possible adding extra weight to certain VoC methods. For instance, in Figure 4-7, the QFD team weighted safety twice as much as the other profiles.

Figure 4-7: Combining Several Voices of the Customer Methods

Targets
Combined Profile for 4 VoC Input

Topics		Attributes	QfdVocPriority 1	QfdKanoStep3 1	QfdSafety 2	QfdAhpPriority 1	Σ	Weight	Profile	Combined Profile
Y.a Group 1	y1 Target 1	Attribute 1.1 Attribute 1.2 Attribute 1.3	0.25	0.50	0.80	0.39	0.25	10%	0.25	
	y2 Target 2	Attribute 2.1	0.52	0.39	0.84	0.32	0.52	22%	0.52	
	y3 Target 3	Attribute 3.1 Attribute 3.2	0.59	0.64	1.05	0.63	0.59	25%	0.59	
	y4 Target 4	Attribute 4.1 Attribute 4.2 Attribute 4.3	0.42	0.10	0.47	0.21	0.42	18%	0.42	
Y.b Group 2	y5 Target 5	Attribute 5.1	0.32	0.22	0.64	0.36	0.32	14%	0.32	
	y6 Target 6	Attribute 6.1 Attribute 6.2	0.09	0.27	0.88	0.36	0.09	4%	0.09	
	y7 Target 7	Attribute 7.4 Attribute 7.2 Attribute 7.3	0.19	0.10	0.47	0.21	0.19	8%	0.19	

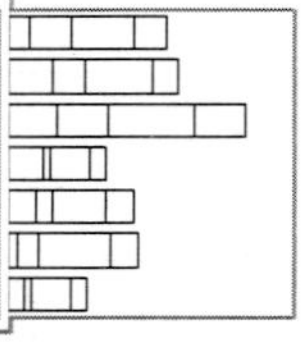

4-5.2 TRANSFER FUNCTIONS FOR CAUSE-EFFECT RELATIONSHIPS

Hu and Antony (Hu & Antony, 2007) observed, how transfer functions map process controls onto responses, for instance, design decision into product features in Design for Six Sigma (DFSS), or Voice of the Engineer (VoE) onto the Voice of the Customer (VoC) in QFD. The response is known, is observable and measurable, whilst the choice of controls is uncertain, and the optimum control profile sought for designing a product or service is right. This is strong evidence that the derived solution profile for the controls is a valid indirect measurement.

Thus, transfer functions uncover the cause of some observable effect, just like in astronomy transfer functions uncover otherwise unobservable exoplanets by analyzing its effects on the luminosity of its star.

4-5.3 THE SPRING CANON EXAMPLE REVISITED

Recall the spring canon example described in *section 1-1.2 : A Simple Six Sigma Case*. Assume the target coordinates ⟨3.80, 8.90, 2.50⟩ are the (metric) coordinates for the spring canon target. These coordinates constitute a goal profile in the sense that they specify the target for the spring canon with sufficient precision.

Normalizing the target coordinates to a vector of length one yields the goal profile calculated in Figure 4-8:

Figure 4-8: Goal Profile of Target Coordinates

Target Topics	Attributes			Weight	Profile	
Pos-x Coord x	Meter	North	4	25%	0.38	
Pos-y Coord y	Meter	East	9	59%	0.89	
Pos-z Coord z	Meter	Height	3	16%	0.25	

The transfer function Figure 4-10 below calculates the four control dimensions in Figure 4-9 as follows:

Figure 4-9: Critical Target Control Topics as Control Dimensions

Critical to Precision

	Critical to Precision Topics	Attributes		Weight	Profile	
				Priority		
SC-1 Canon	SC-1.1 Angle	Angle	North	25%	0.48	
	SC-1.2 Spring Force	Spring	Compression	27%	0.52	
SC-2 Optical	SC-2.1 Accuracy Notch Sight	Notch	Precision	12%	0.23	
	SC-2.2 Accuracy Info Transfer	Reading	Precision	35%	0.67	

All target controls are measurable in some suitable metrical system.

When adjusting, say, the angle, it is necessary to look at all three coordinates $\langle x, y, z \rangle$, not only the x coordinate.

Figure 4-10: Parameters for Hitting the Target

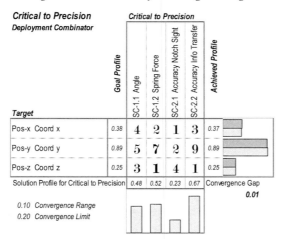

The matrix in Figure 4-10 describes the working of the controls, and their impact on hitting the target. The rows of the matrix simply represent the three coordinates $\langle x, y, z \rangle$ of the target, based on some suitable room coordinate system. The columns represent the four control dimensions, angle, spring force, accuracy of the notch sight, and accuracy of information transfer. In this case, information transfer refers to how well the optical systems allows reading the coordinates needed for the target controls.

The system adjusting the spring canon for hitting the target has the generic form shown in Figure 4-10. The cell numbers represent the coordinates that ensure the spring canon hits the target in some suitable measurement scale. The yellow profile shows the achieved position, and the convergence gap is the deviation between hit and target.

To make the system run and hit targets effectively, one must convert the generic coupling factors in the cells to some real parameters in the spring cannon; this is mainly a matter of dimensional adjustments. The solution profile represents the relative importance of each factor for the accuracy of our spring cannon. In turn, the convergence gap is a prediction of how accurate our spring cannon hits the target.

4-5.4 APPLYING TRANSFER FUNCTIONS TO QFD

The most popular transfer function utilized in a QFD context is the *House of Qualities* (HoQ). Its more popular name is 'House of Quality', although more than just one quality is considered. Customer's quality topics connect with the engineers' quality perceptions. Hence, two views on quality met that the QFD method reconciles.

4-5.4.1 VOICE OF THE CUSTOMER

In a HoQ, one of the qualities is the response expected by customers. This is the response profile y expected to be sufficiently close to the target profile τ_y. The response profile y characterizes the qualities customers expect from a product or service. These are called *Customer's Needs*; in product management, *Market's Needs,* or *Business Drivers* by Denney, (Denney, 2005).

We discuss techniques for uncovering the profiles of customer's needs in *Chapter 3: What is AHP?* and *Chapter 5: Voice of the Customer by Net Promoter®.*

4-5.4.2 PREDICTING THE TRANSFER FUNCTION

QFD uses workshop techniques to assess the transfer function. Traditionally, a matrix is filled with 0, 1, 3, and 9, indicating how much a specific control quality contributes to some response quality. Ratio scales are compulsory according the ISO/IEC standard (ISO 16355-1:2015, 2015), following AHP (Saaty & Özdemir, 2003). In QFD, things are different as the roof of the matrix is used to record contradictions, and some practitioners use negative values in the matrix coefficient. If the corresponding matrix AA^{T} remains positive definite, negative values in the matrix cause no problems. If this condition no longer holds, there is no longer a principal eigenvector of AA^{T}, see *Appendix B-1 Basic Definitions and Results.*

The benefit of using transfer functions with solutions to the eigenvalue problem is apparent. When groups of expert people try to identify the QFD matrix, they use several techniques to check their matrix for this benefit, including visual balance of the distribution of strong, medium, and weak correlations. A solution to the eigenvalue problem for AA^T validates their work.

To measure a transfer function sometimes is quite easy. For instance, when developing software, the cells in the house of quality usually combine user stories as controls with customer's needs. The correlation values of the cells define the impact that the corresponding user story has on customer's needs. These cells have measurable correlation values. Measuring with their functional size works well; see *Chapter 6: Functional Sizing*.

4-6 TRANSFER FUNCTIONS AS QFD MATRICES

In *Chapter 2: Transfer Functions*, it is shown that transfer functions can be represented as matrices. A central element of the QFD approach is establishing a matrix for mapping the control profile – or profile of solution characteristics – onto the Voice of the Customer profile. Of course, QFD matrices are transfer functions, although many QFD transfer functions have not been measured by Design of Experiment or other Six Sigma measurement techniques but are the result of QFD workshops.

QFD workshops represent a technique measuring expert's opinions about how controls affect the response. The difficult step is how to measure the strength of relationships between controls and processes for each control/response pair – for each cell in the matrix. In QFD workshop, experts predict and estimate the most likely measurement result per cell. QFD workshops held with experienced people replace the Design of Experience (DoE) measurements in production, held when doing Design for Six Sigma. If these expert judgments can be compared to measurements later, cell by cell, when executing the process, it is even better. Thus, matrices in QFD are the container for both process predictions and measurements, and yield much greater value than just doing AHP, or Blitz QFD®. The question is whether this justifies the additional effort.

4-6.1 BRIDGING THE COMMUNICATION GAP

Usually, engineers have different view than marketers, or top management, and the real difficulty is not only finding the solution but also communicating it to the various stakeholders. This is soft, fuzzy data.

Consider the example of some customer Call-in Service. This service process' response consists of answers to customer's inquiries concerning anything that relates to products or services of the sponsor's organization.

Assume, the Call-in Service needs to become more brilliant and better meet the customer's needs profile (the response **y**) in aspects such as friendliness, response time, and accuracy of the response.

The improvement project might include software, and adaptation of ICT support; however, the project is not limited to ICT. The Kano analysis (Figure 4-11) requires a short workshop uniting actual customers, or sales-oriented people, with the technical team. It will yield a customer's needs profile, weighting friendliness, response time, and accuracy of the response against each other, according the method as explained above (section 4-4.4).

Figure 4-11: Sample Kano Analysis for Using Call-in Service

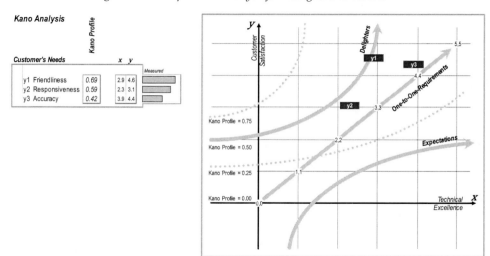

The method yields a profile for customer's priorities for an improved Call-in Service. In this case, no must-be quality or expectations went into the profile. This is fine in a product improvement project where the existing Call-in Service product already addresses these topics somehow. Figure 4-12 shows the result of the Kano VoC analysis.

Figure 4-12: Business Drivers reflecting Customer's Needs in Using Call-in Service

Customer's Needs Topics	Attributes		Weight	Profile	
y1 Friendliness	Remains cool	Always friendly	40%	0.69	
y2 Responsiveness	Understands the problem	Finds a way to solve	35%	0.59	
y3 Accuracy	Complete information	Compelling	25%	0.42	

Thus, *y1: Friendliness* tops *y3: Accuracy* by a factor of 0.69/0.42 = 1.7. If the Call-in Service provider wants to improve its operations, by doing some investments, the question is how much to spend in each of the controls times response cells of the matrix. As controls for the process (the x), we might have training of the Call-in Service employees, performance of the ICT equipment, salary system, and work environment.

Every item in a matrix cell needs communications and costs effort. The total of the cells' content is proportional to total cost. However, benefit comes from the response only. When doing highly complex decisions, matrices are the preferred visualization method for looking at the interdependencies is compulsory.

Figure 4-13: Call-in Service QFD Matrix: Aligning the Budget based on Measurable Contributions

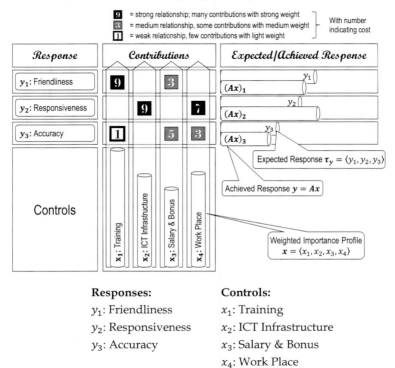

Responses:
y_1: Friendliness
y_2: Responsiveness
y_3: Accuracy

Controls:
x_1: Training
x_2: ICT Infrastructure
x_3: Salary & Bonus
x_4: Work Place

The cells of the matrix (Figure 4-13) contain the work items, as for instance cell $\langle x_1, y_1 \rangle$: *Train Call-in Service staff how to be friendly at the phone* with value 9, or $\langle x_2, y_2 \rangle$: *Invest into faster ICT equipment*, also with value 9.

Each cell has impact, indicating how much it contributes to the respective response in the row. Traditionally, QFD teams recorded impact as symbols having values 1–3–9 assigned; however, modern QFD understands these are ratio scale values. Any other ratio scale will work as well. It depends upon the accuracy of data and availability of predictions and measurement methods what kind of ratio scale to choose; however, a ratio scale is mandatory, as the QFD community learned from Saaty's AHP (Saaty & Peniwati, 2008).

In Figure 4-13, the first challenge is to measure the transfer function; i.e., the matrix. Traditional QFD takes a scale 1-3-9 for low – medium – high relationship, depending upon the contribution achieved in that cell from the controls to the expected response. This scale is a ratio scale only if assuming 'high' is three times 'medium'; and 'medium' is three times 'low'. Although this sounds like familiar lore, better measurements are possible in this sample case, because it deals with financial impact on process improvement for the call-in service. In the financial manager's view, spending one Euro on training is equivalent to spending one Euro on ICT; thus, if the transfer function is measured as cost per cell, depending upon contributions, it is a ratio scale indeed and not a three-step scale. Instead, the transfer function consists of the relative spending per matrix cell, shown in Figure 4-14.

4-6.2 TRADITIONAL QFD

The interesting part is how financial managers deal with such measurement. The traditional approach to distributing the budget among the functions involved into Call-in Service improvement is summing up the columns of the matrix, weighted by customer's needs. This is the traditional evaluation method for QFD matrices; see Figure 4-14.

Figure 4-14: Traditional QFD Matrix Evaluation

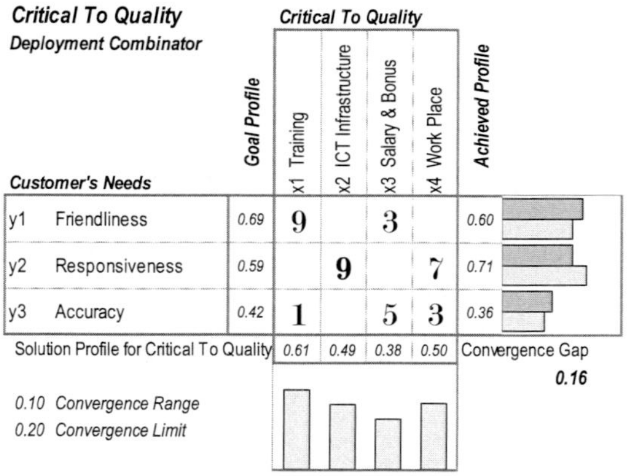

Critical To Quality Deployment Combinator	Goal Profile	x1 Training	x2 ICT Infrastructure	x3 Salary & Bonus	x4 Work Place	Achieved Profile	
Customer's Needs							
y1 Friendliness	0.69	9		3		0.60	
y2 Responsiveness	0.59		9		7	0.71	
y3 Accuracy	0.42	1		5	3	0.36	
Solution Profile for Critical To Quality		0.61	0.49	0.38	0.50	Convergence Gap	
						0.16	

0.10 Convergence Range
0.20 Convergence Limit

Investing into training is most effective, followed by ICT Infrastructure und Work Place improvements. This is the original meaning of (*Quality*) *Function Deployment* (QFD) in Japanese; function refers to organizational unit. Note that this traditional calculation of the proposed solution profile applies the transposed matrix A^T to the expected response y, representing the weight of the respective customer's need. This yields $x = A^\mathsf{T}y$ as the proposed effective solution, shown below the expected solution in both Figure 4-13 and Figure 4-14. Since the expected response y is not necessarily an eigenvector to AA^T, the quality of the proposed solution is doubtful.

Indeed, looking at the convergence gap $\|y - AA^Ty\|$, the result looks doubtful. Smart financial managers will not believe it, they prefer calculating the effectiveness of spending money it the other way around using $A^Ty = x$, where $x = \langle x_1, x_2, x_3, x_4\rangle = \langle 0.61, 0.48. 0.39, 0.49\rangle$. and they will find the convergence gap widely open.

The transfer function A, applied to A^Ty, maps controls to response, thus summing up the impacts per row, weighted by the investment money spent (i.e., the measured control profile), and yields the total effective impact on the customer's needs; i.e., the total money spent on meeting customer's need. The response profile describes the effectiveness of the investment made; it results from the transfer function. In fact, the achieved solution reverts the relative effects on the first two expected responses; in Figure 4-14, the *y1: Friendliness* and the *y2: Responsiveness*. Obviously, $y = \langle y_1, y_2, y_3\rangle$ is no eigenvector of AA^T; the cause-effect system is unstable, and our smart financial manager will probably not approve our proposed solution, for reasons that are good but difficult to comprehend.

The achieved profile calculates as follows from $A = (a_{i,j}), i = 1 \dots 4, j = 1..3$:

$$\langle x_1, x_2, x_3, x_4\rangle = A^Ty = \langle \sum_{i=1}^{3} a_{i,1}y_1, \sum_{i=1}^{3} a_{i,2}y_i, \sum_{i=1}^{3} a_{i,3}y_i, \sum_{i=1}^{3} a_{i,4}y_i\rangle \tag{4-2}$$

$$y' = Ax = AA^Ty = \langle \sum_{j=1}^{4} a_{1,j}x_j, \sum_{j=1}^{4} a_{2,j}x_j, \sum_{j=1}^{4} a_{3,j}x_j\rangle \tag{4-3}$$

The difference between the expected profile y and the achieved profile y' is the convergence gap, as defined in *Chapter 2: Transfer Functions*.

$$\|y - y'\| = \left\| \langle y_1 - \sum_{i=1}^{4} a_{i,1}x_i, \ y_2 - \sum_{i=1}^{4} a_{i,2}x_i, y_3 - \sum_{i=1}^{4} a_{i,3}x_i\rangle \right\|$$
$$= \sqrt{\sum_{j=1}^{3}\left(y_j - \sum_{i=1}^{4} a_{i,j}x_i\right)^2} = \sqrt{\sum_{j=1}^{m}(y_j - y'_j)^2} \tag{4-4}$$

Traditionally, the length of unit vector in QFD was set to 5, not 1, see (Fehlmann, 2005-2). Thus, the convergence gap would stay between 0 and 5 instead of between 0 and 1. Such kind of traditional normalization only adds unnecessary complexity when comparing or combining profiles.

In Figure 4-14, the financial manager can see that too much effort is spent either in cell $\langle x_2, y_2 \rangle$, or cell $\langle x_4, y_2 \rangle$; the achieved solution bar (0.71) exceeds the expected solution bar (0.59) in the second row by 120%. Smart financial managers regularly keep an eye on such convergence gaps to make sure investments are effective, with the gap as small as possible. This simple validation technique is useful for any QFD as well. A small gap tells the QFD team that the matrix controls (the x) cause the expected responses (the y), making them predictable. Hence, the convergence gap is a quality measure for QFD matrices. It allows the QFD method dealing with soft, fuzzy transfer functions, such as found in cosmology, and in Google search algorithms.

Calculating the convergence gap is utterly simple; every spreadsheet can accomplish it. Communicating the QFD matrix to people and stakeholders is straightforward: strong relationship corresponds to active work – explaining issues, providing training, spending money, or ordering tools for improving process performance. Management can set up their communication plans by means of a QFD matrix, and project managers their project plan. The QFD matrix even works for estimation of projects; see *Chapter 12: Effort Estimation for ICT Projects*.

4-6.3 CONSISTENCY

The problem with the traditional method is its inconsistency. The convergence gap in Figure 4-14 is 0.16, meaning that the vector length difference of the two unit vectors y and $Ay' = AA^\mathsf{T}y$ is around 16%. This proposed solution vector points quite into some other direction. Following the strategy proposed with the solution profile calculated with the traditional QFD calculation method would result in the wrong direction.

The interesting question is what happens when we calculate the eigenvector y_E of AA^T and derive the solution profile $x_E = A^\mathsf{T}y_E$. The solution profile x_E are the eigencontrols of the matrix A; see section 2-4.2 in *Chapter 2: Transfer Functions*. It is far from obvious that spending the same amount of money for $\langle x_1, y_1 \rangle$ as for $\langle x_2, y_2 \rangle$ yields an unbalanced solution but it does. Figure 4-15 demonstrates that the transfer function's goal profile $y = \langle 0.69, 0.58, 0.42 \rangle$ is not an eigenvector of AA^T shown in Figure 4-16.

The eigencontrols $x_E = \langle 0.12, 0.75, 0.13, 0.64 \rangle$ of the matrix in Figure 4-15 are quite different from the weighted proposed solution $x' = \langle 0.61, 0.48, 0.39, 0.49 \rangle$ in Figure 4-14, and both yield an insufficient convergence gap. This is an indication that the transfer function A is not consistent with the goal profile y. The usual threshold is 10%, corresponding to a convergence gap of 0.1. The QFD transfer function of Figure 4-14 seems biased towards x2: *ICT Infrastructure*, as can be seen when calculating the eigenvector calculation in Figure 4-16.

Figure 4-15: ICT-biased Solution Approach

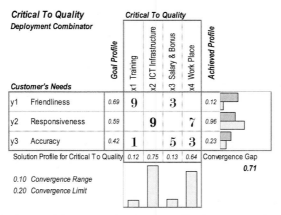

Figure 4-16: Calculation of the Principal Eigenvector for the ICT-biased Transfer Function

A :

9	0	3	0
0	9	0	7
1	0	5	3

AA ^{T} :

90	0	24
0	130	21
24	21	35

Jacobi Iterative Method for Finding Eigenvalues:

98	0	0
0	135	0
0	0	22

Eigenvectors:

0.94	**0.12**	-0.33
-0.19	**0.96**	-0.18
0.30	**0.23**	0.93

From the matrix in Figure 4-15, it is obvious that the x2: *ICT Infrastructure* control just gets too much weight, compared to the other controls. It happens as often in such cases; outbalanced investments devaluate all others. Eigenvectors uncover such inconsistencies. Then, the selection of controls needs change and improvement. The principal eigenvector in Figure 4-16 is the second from the three candidates.

4-6.4 MODERN QFD

Adapting the transfer function more precisely to the problem yields the final solution. It saves four cell contributions out of a total of 37 (11%) by adapting the transfer function such that the principal eigenvector moves to the first position in Figure 4-18, and *y* almost becomes an eigenvector of the modified transfer function *A* in Figure 4-17.

This sample demonstrates that the eigenvector property of *A,* with respect to *y*, adds quality assessment of a measured transfer function. This Call-in Service sample is a service undertaking and the transfer function reflects the amount of planned or actual investment into the process – thus, it can be adapted to customer's needs. A small modification of the transfer function *A* solves the problem; see Figure 4-18.

Figure 4-17: Final Solution Approach

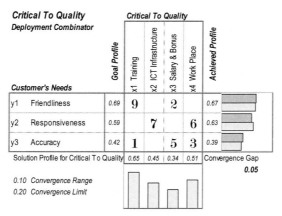

Critical To Quality Deployment Combinator		Goal Profile	x1 Training	x2 ICT Infrastructure	x3 Salary & Bonus	x4 Work Place	Achieved Profile	
Customer's Needs								
y1	Friendliness	0.69	9		2		0.67	
y2	Responsiveness	0.59		7		6	0.63	
y3	Accuracy	0.42	1		5	3	0.39	
Solution Profile for Critical To Quality		0.65	0.45	0.34	0.51	Convergence Gap		

0.05

0.10 Convergence Range
0.20 Convergence Limit

Figure 4-18: Calculation of the Principal Eigenvector for the Final Transfer Function

A :

9	0	2	0
0	7	0	6
1	0	5	3

AA^T :

85	0	19
0	85	18
19	18	35

Jacobi Iterative Method for Finding Eigenvalues:

96	0	0
0	85	0
0	0	24

Eigenvectors:

0.67	-0.69	-0.29
0.63	0.73	-0.27
0.39	0.00	0.92

AHP or a Blitz QFD® does not provide such decision support, neither for doing financial investments into processes, nor when managing stakeholder communication. If management must communicate its investment decisions, the QFD matrix provides explanations to teams and stakeholders. AHP, or Blitz alone, do not.

By strictly keeping to the 1-3-9 scale, the finance manager's QFD approach would have failed, lacking necessary precision. As Saaty (Saaty, 1990) remarked, the scale for the correlation values must be a ratio scale.

4-6.5 CONFIDENCE

For the measurement in the matrix **A**, it would be helpful to have an indicator for the presumed measurement errors. When changing one of the matrix cells, how much does it affect the outcome of the process?

In statistics, the *Confidence Interval* answers such questions, for instance for the 95 percentiles. This is the probability that any further independent measurement falls within a ±5% range around the true value. To calculate such probabilities, a large enough measurement sample must be available.

In this case, there are no repeatable measurements samples – usually we cannot repeat QFD workshops several times. However, the assumption that whatever score selected in a QFD workshop or other kind of measurement in the range 0…9 can differ by 10%; i.e., one score higher or lower, might yield an estimate. An estimate for the 95 percentile is obtainable by calculating the convergence gap to the eigencontrols, provided the team decisions are normally distributed and not biased to a measurable extend.

This confidence interval suits more to management needs than the convergence gap, since gut feeling for zero usually is not as good as for 100%. However, convergence gap and confidence interval are two very different metrics. The convergence gap indicates the difference between the eigencontrols and the goal profile; the confidence interval depends from the transfer function only. Consequently, confidence intervals for small $n \times m$ matrices are relatively low, increasing with the number of measured cells. This is because the number of measurement point increases, balancing out measurement errors. However, when calculating the confidence intervals for the 4×3 matrices shown in Figure 4-13 to Figure 4-17, you cannot get high confidence if measuring only a small number of cells in a sparse matrix.

4-6.6 Identifying the Right Controls

QFD is about building transfer functions that model product design or development. It is not possible to calculate the eigenvector for AA^T and derive the needed controls out of it; the vector difference between goal response profile and next-best eigenvector is only indicative for the measurement error. To calculate the convergence gap it is necessary to find a set of controls, and try to build the transfer function to get the goal response as good as possible. This requires sound domain expertise, and makes the QFD process so special. QFD is not an automatic problem solver.

However, even if convergence gap closes and measurement error is small, this does not guarantee the best possible solution. By adding or replacing better controls, the overall solution quality improves. It must be investigated whether AA^T hints where to improve, or if there is nothing than gut feeling of highly skilled experts that helps identifying the matrix cells needing adjustment. Negative values in QFD matrices up to certain amount are harmless. The condition is that AA^T remains positive definite even if some of the matrix elements in the transfer function A are negative. If this is not the case, eventually there are no eigenvectors near the expected goal profile and the problem is intractable with the chosen controls.

An interesting question is thinking about strategies to identify minimum solutions, considering that the more controls the worse; and costs increase with the number of non-empty matrix cells. In fact, one strategy is minimizing the number of controls; however, this also increases the need for exact measurements among the few matrix entries left. Alternatively, the other way around, any measurement error, or expert slippage, can yield disastrous results.

4-7 DESIGNING SOLUTIONS FOR BUSINESS DRIVERS

We demonstrate an example from agile software development. QFD workshop haptic normally is part of agile software development practices. When agile teams collect story cards and put them on a pin wall for sprint planning, matching *User Stories* to *Business Drivers*, following (Denney, 2005), and use story points for weighting story cards, they typically construct a QFD transfer function that is measurable.

4-7.1 THE SOFTWARE DEVELOPMENT EXAMPLE

When developing software, visualization of work items communicates to the team where it stands, what problems and blockers they encountered are, and what they plan to do next. One of the well-known agile practices is Scrum (Schwaber & Beedle, 2002) that prescribes such practices in the daily *Stand-Up Meeting*. The development team writes work items on *Story Cards*. Story cards relate to some *User Story* during a *Sprint*. In this book, *Chapter 8: Lean & Agile Software Development* elaborates this practice further and puts it into a wider context.

The narrower context of this chapter refers to an aspect of story card, namely that they typically relate to effort. Scrum uses story points to assign a number representing its effort, and they write that number to the card. Story points are not a measure, such as Function Points, and usually do not exactly relate to something like hours needed for completing that story card. They are indicative for the relative work effort felt appropriate by the team, in comparison to some "standard candle" used for calibration.

The example is a project taken from customer communication management: assume a large postal mailing organization that needs to print monthly bill statements, mail stamp the envelope and manage geographic indications for large mailings. Addresses arrive in a large unordered file – "Source File" – that needs reordering such that the mails fit into trucks that serve specific geographic areas. Printing mails in the right order allows packaging them immediately for sending them to destinations. Other facets for this sample case such as preparation of content, collection of transactional data and added marketing material is not within the scope of this project.

4-7.2 IDENTIFYING STORY CARDS

Agile methodologies rely on user stories that constitute the elements of the *Product Backlog*. They represent what the sponsor wants to get out of the software development process. Since user stories do not fit well into the time slots allocated for sprints, and since often user stories require quite different things to implement, usually user stories will split into *Story Cards*. Some of these story cards will not add any measurable functionality or even in code because the address quality issues such as performance or reliability.

4-7.3 THE SEVEN USER STORIES

The seven user stories describe the project "SendMail". There is only one application within this project's scope, namely an application for mail stamp and geo indication for large mailings.

4-7.3.1 USER STORY 1: ANALYSIS OF THE SOURCE FILE

As a dispatcher, I want the software read and analyze the source file containing the addresses and the individual message content. I want to know the file size, the number of data lines, the maximum line length and the number of data fields for files in ASCII format, such that this information is available for cleaning the source file in *User Story 2* and persistently recorded into some internal store, called *Print Job Data*.

4-7.3.2 USER STORY 2: CLEANING THE SOURCE FILE

As a dispatcher, I want the data records handed over in the source file checked for the following:

➢ Follows the postcode the exact format for the destination country?
➢ Does the "City" field contains only the city and not the postcode?
➢ Does the "Item weight" field contain positive values without any decimal places?
➢ Line length for ASCII file is constant (reference is the first line)?
➢ Number of fields for ASCII file is constant (reference is the first line)?

Such that all erroneous data records are identified as non-usable and written into some output report *Undeliverable Items*, deleted from the internal store *Print Job Data* such that the data supplier has the possibility of correcting these data and saving postage costs for undeliverable items. It is acceptable that after cleaning the source file some addresses remain that do not really exist.

4-7.3.3 USER STORY 3: GEOCODING

As a dispatcher, I want to check all addresses for validity, existence and deliverability based on *GeoData* (geographical data) from postal services such that the system processes valid and deliverable data records only in the following steps. As an outcome of this step, report rejected addresses as *Addresses Not Found* and earmarked in *Print Job Data* as non-deliverable.

4-7.3.4 USER STORY 4: PRE-SORTING DATA RECORDS

As a dispatcher, I want all data records belonging to the job sorted in accordance with rules defined by the postal service, reflecting trucking routes and modes. These rules originate from an external service *Trucking Rules* maintained by another application.

4-7.3.5 USER STORY 5: BUNDLE AND PACKING-UNIT FORMATION

As a dispatcher, I want all packing units for one destination to have approximately the same dimensions and weights, then sort output data for Euro-pallets first, then for disposable pallets and finally the rest put into boxes.

4-7.3.6 USER STORY 6: MAIL STAMPING

As dispatcher, I now want to stamp all valid mailings in *Print Job Data* according the mail stamp values set forth in by *Mail Tax*, the trucking cost parameters according *Trucking Rules*, so that mail stamp values are added to *Print Job Data* and I got an *Accounting Report*.

4-7.3.7 USER STORY 7: COMPILING THE PRODUCTION FILE

As dispatcher, I finally want the print jobs bundled, sorted and prepared for transportation, together with the *Datamatrix* ("the mail stamp") to go into print production, called the *Production File*, such that the *Print Job Monitor* reports the printing process, and successfully printed jobs are marked as printed in *Print Job Data*. In addition, the trucker's *Shipment Request* must indicate where to drive and deliver pallets. Boxes must become available on the respective output channels; e.g., as a mobile app for the truckers.

4-7.4 DEFINING CUSTOMER'S NEEDS

Assume the customer has three needs that he wants to fulfil with implementing the *SendMail* project: *y1: Optimize Cost*; *y2: High Reliability*; and *y3: Interoperability*.

Figure 4-19: SendMail Customer's Needs

Customer's Needs Topics		Attributes			
Cost	y1 Optimize Cost	Reduce Waste	Simplify processes	End-to-end optimization	
Growth	y2 High Reliability	Tracking Log	Recovery	Monitoring	
	y3 Interoperability	Hardware	Software	Industry Standards	Multi-Platform

The AHP process yield a normalized goal profile of ⟨0.85, 0.47, 0.26⟩, see Figure 4-20.

Figure 4-20: AHP for Priority Determination

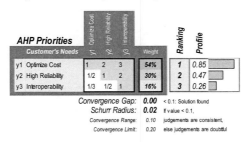

The team used the recommended practice to validate the AHP finding by the Kano method. Comparing the AHP profile with the corresponding profile originating from the Kano diagram achieves this.

Figure 4-21: Kano Analysis for SendMail

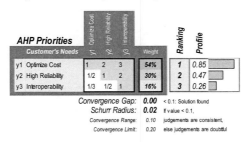

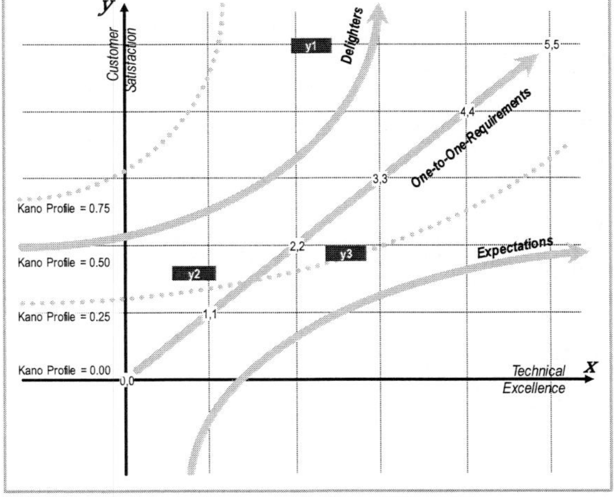

To do this, the QFD team tries to position the three customer needs topics in the Kano diagram such that their profile closely matches that of the AHP. If that is possible, and if the positions in the Kano diagram make sense, it confirms the findings from the

AHP. If in contrary the Kano analysis would yield a different goal profile for the needs of the customer, there is reason to revise both the AHP priorities and the Kano analysis in Figure 4-21 until both methods match; i.e., the convergence gap – here 0.01 – goes towards zero.

Figure 4-21 compares the AHP profile with the Kano approach; this is a common practice for getting confidence with AHP decision matrices. Although a profile, be it from AHP or else, does not uniquely define a Kano diagram, since the exact positions of the profile components on the hyperbolic curves remain undefined; if the same criteria are used as with Kano, the AHP and the Kano profile must be similar at least; i.e., yielding the same ranking.

With only three topics, the validation is not as strong as with more complex value profiles; nevertheless, Figure 4-21 demonstrates that Kano and AHP might yield the same profiles under favorable conditions up to a convergence gap of 0.01.

4-7.5 MAPPING USER STORIES TO CUSTOMER'S NEEDS

Figure 4-22: Pin Wall Mapping Story Cards that meet Customer's Needs

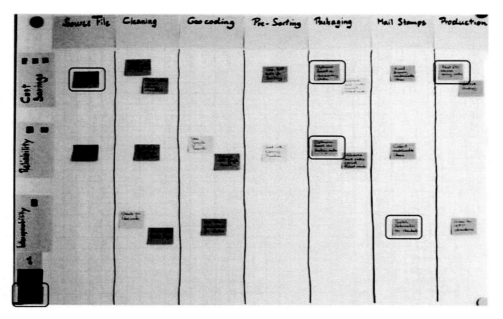

The software development team uses the customer need's prioritization for the matrix as shown in Figure 4-22, putting story cards that record the work related to customer's needs in the respective cells. The color represents the amount of story points; i.e., the effort needed to do that work as defined by the development team. The darker the more effort needed, those squared most (red: 6 points; green: 3 points; light blue: 1

point). The matrix might not record all work. There might be preparatory work such as setting up a database, not related to any of the needs mentioned but required for the basic functionality needed.

The pin wall matrix is a QFD matrix for planning purposes. Figure 4-23 below shows its evaluation as a transfer function.

Figure 4-23: Evaluating the Pin Wall QFD Matrix for Planned Effort

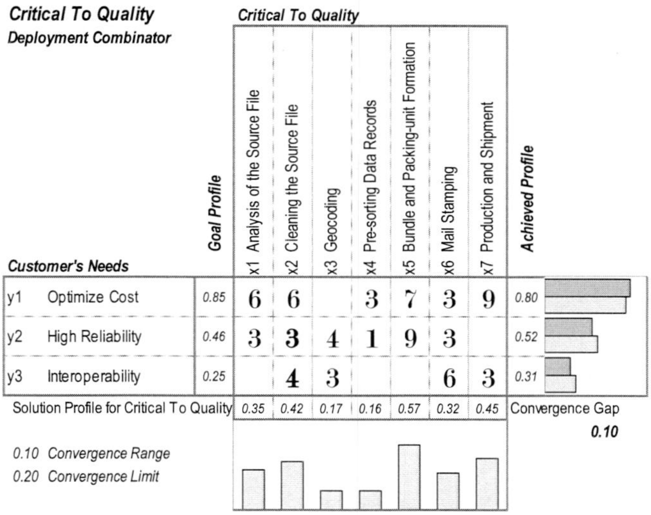

Critical To Quality Deployment Combinator	Goal Profile	x1 Analysis of the Source File	x2 Cleaning the Source File	x3 Geocoding	x4 Pre-sorting Data Records	x5 Bundle and Packing-unit Formation	x6 Mail Stamping	x7 Production and Shipment	Achieved Profile
Customer's Needs									
y1 Optimize Cost	0.85	6	6		3	7	3	9	0.80
y2 High Reliability	0.46	3	3	4	1	9	3		0.52
y3 Interoperability	0.25		4	3			6	3	0.31
Solution Profile for Critical To Quality		0.35	0.42	0.17	0.16	0.57	0.32	0.45	Convergence Gap 0.10

0.10 Convergence Range
0.20 Convergence Limit

In addition, some functionality might not be specific to some need, and consequently the matrix does not record it. Nevertheless, the matrix proves to the team – and to the sponsor – that they have understood customer's needs and are going to spend effort to it. Moreover, it is not just "some effort", the effort is focused towards what matters most for the customer. Cost savings are kept more important that reliability and interoperability when finishing the software.

4-7.6 EVALUATING THE QFD

Obviously, Figure 4-22 shows a transfer function mapping work effort to fulfilling customer's needs. It is straightforward to evaluate, using the story points provided from the team during planning poker as input. The planning matrix shows a suitable distribution of effort according customer's needs.

The planning matrix shows a suitable distribution of effort according customer's needs. The numbers indicate the relative effort; in whatever "dimension" the team uses for effort estimation within sprints. No restrictions apply; fractions are useful as well, if numbers are positive. In the above example, the team used Excel and documented the content of the stickers in the Excel cells as comments. Better tooling exists for QFD if needed.

4-7.7 MONITORING ACHIEVEMENTS

After completion of the sprints, or at some intermediate *Retrospective*, the actual achievement – plus estimate to complete – monitors and gauges efforts most effectively towards customer's needs. There might be some slight adjustments. In the case shown in Figure 4-24, these adjustments led to improving the convergence gap.

Figure 4-24: Achieved Effort

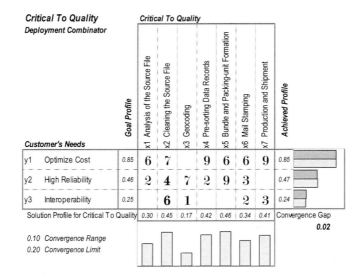

Again, the measurements in the matric cells can be anything – time reports, days spent, or any other measure reflecting the impact of the team's effort on customer's needs. The team should agree on what they like to measure and keep to that decision such that the units used show how the actual matrix differs from the planned matrix.

It is obvious that when using such a tool with the team, people adjust their efforts towards meeting the goals set by the customer. Is some way, the team uses QFD as a daily means of monitoring their achievements and setting priorities accordingly.

Note that the matrix above might represent only part of the effort. Some effort might be needed that is not contributing to the three customer's needs set as goals, as those goals do not reflect functionality, for instance. Usually, *Functional User Requirements* (FUR) exist and cause even more work to do than just topping with non-functional requirements. *Chapter 6: Functional Sizing* and *Chapter 12: Effort Estimation for ICT Projects* contain more on functional and non-functional requirements.

4-8 CONDUCTING A QFD WORKSHOP

If correlation items are not that easily identifiable and measurable as with story cards in agile software development, workshops are conducted to identify the strengths of correlation. The participants act as measured entities; the higher their expertise the better measurement results. It is therefore crucial to have both experts for the VoC as experts for the VoE, together with experts who understand in detail how the VoE-controls contribute to meeting the expectations set by the VoC.

4-8.1 SELECT THE TEAM

An ideal team consists of

- A moderator with the necessary skills;
- One or two VoC-experts (e.g., sales representative or account managers);
- One or two VoE-experts that understand what controls are available;
- One or two communication experts who understand who VoE contributes to VoC; e.g., marketing experts.

Finding the communication experts with domain expertise in both areas is usually most difficult. However, their contribution is crucial for getting the transfer function right. Usually, engineers and sales people know that correlations exist, but they are not obliged to identify exactly, how.

On the other hand, a VoC is not limited to just asking existing customers, or prospects. It is a challenge to find people having the vision that allows them expressing their needs. They tend to propose solutions, not needs, limiting the ability of engineers to identify optimum solutions.

4-8.2 DO A VOICE OF THE CUSTOMER FIRST

The VoC process is the first step to a successful QFD. The technique of choice is verbatim analysis. It consists of collecting customer's voice to understand what values they esteem. Counting the number of occurrence of certain repeated notions or topics, helps identify the weight of such values. Verbatim analysis is a clustering technique. The result is a list of *Topics*.

Not all topic values are equal. Certain values weight more than others. This is crucial information, since these value profiles constitute system responses that allow selecting suitable controls. As an example, a survey among customers or prospects allows identifying customer's needs if analyzed for customer segments without asking them directly. Why not ask customers directly? Because usually, customers do not know; and if they know, they will not tell. This situation typically occurs within a Business-to-Consumer (B2C) setting; however sometimes even in a Business-to-Business (B2B) environment, customers tend conservatively to keep to what they are used to and are not akin to innovation.

Thus, the primary concern when doing a VoC is how to avoid being stalled in common customer's habits and finding topics of interest to the customer.

4-8.3 ALWAYS DO A KANO ANALYSIS

Once the QFD team selected some topics, the Kano model shall analyze their value for the customer. To this purpose, the team places topics on a two-dimensional area representing *Technical Excellence* on the x–axis and *Customer Satisfaction* on the y–axis. Ideally, sales people select the y–position and technical experts select the x–position. He higher customer satisfaction is, and the less technical excellence is needed, the more important a topic is in the needs profile. These points constitute hyperbolic curves, called *Sales Points*, because they can help selling a product or idea to others.

A Kano diagram is a measurement tool suited for transferring human visual perception in something like value for the customer. The hyperbolic functions are quite hypothetic and measurements are not very precise. Nevertheless, for identifying the topics of interest for the customer, a Kano diagram is a useful workshop tool. As a minimum, it helps understanding what story cards provide value to the customer.

4-8.4 BRAINSTORMING THE CONTROLS

How to materialize such value? This means finding the controls to the QFD matrix, expressed as user stories. This is the most difficult step because innovation and creativity is required. Not all user stories qualify as a control; there must be some correlation to one or more of the goal topics, the business drivers, that makes them effective.

Also, if there is no impact to any of the goal topics, rather equal impact to all of them, it means that this control is not specific to customer's needs and therefore probably not needed. The controls are not a comprehensive list of technical requirements; in Six Sigma, these are selected requirements called *Critical for Quality* (CtQ). It might be necessary to consider other technical needs that are not critical for quality; however, they provide basic functionality, security, or another feature that is not immediately visible to the customer.

4-8.5 ELIMINATING CONTROLS

Controls can also be redundant in the sense that they provide the same kind of impact. As linear vector, they are linearly dependent. You can eliminate such controls without affecting the system and its convergence gap.

Do not eliminate and consolidate controls automatically. Sometimes, in a QFD workshop, it is important to discuss what consolidation means; doing so might bring additional insights to the QFD team.

4-8.6 CONTROLLING THE CONVERGENCE GAP

The convergence gap indicates how well our controls are capable to affect the result. A large convergence gap indicates, the project does not meet the stated goals – either because the goals are wrong or the controls are not appropriate.

Doing a QFD without calculating the convergence gap is fraudulent (Fehlmann, 2003).

4-8.7 DOCUMENT FINDINGS

Finally, the team must discuss and document all findings in a QFD matrix. The reasoning for assigning a value to a cell is especially valuable. If no other data than experts' knowledge is available, agreeing on the weight of contribution per cell can take a significant amount of time. For some people, getting agreement at an early stage in the project might look like a waste of time. Consequently, they question the usefulness of QFD matrices. While this might be a valid concern for fuzzy correlations – such as, for instance, how do people experience the haptic of a surface – in many cases, the correlation is measurable and discussing in the team and finally documenting the weight of a cell is highly rewarding.

The cell content of the QFD matrix is the knowledge repository for a process, a system or an organization, whatever topics affects QFD outcome. QFDs often last longer, sometimes even longer as the product life cycle that it initially described. QFDs are long-lasting investments and should be documented and maintained accordingly. QFD documentation, especially if regularly reviewed, is an asset in the organization's memory.

4-9 COMPREHENSIVE QFD – THE DEMING CHAIN

Akao introduced QFD not as the single house of qualities but as a chain of matrices. He called it *Comprehensive QFD*, also known as *QFD in the Large* (Akao, 1990). He drafted extensive Deming chains in this famous book on QFD. Because Deming mentioned value chains of cause-effect coupling first, we call them *Deming Chain* of matrices. Deming described the value chain in manufacturing processes subject to Six Sigma for process control and optimization (Deming, 1986).

4-9.1 DEPENDENT PROCESSES

Each QFD matrix represents some organizational process that uses the responses of previous processes as controls. Deployment means for Akao (Akao, 1990), starting with the final response targeted versus customers, and then adjusting all previous processes such that their response profiles match the optimum control profiles of the upstream processes. The solution control profiles become the goals of the upstream process QFDs.

Figure 4-25: A Deming Chain Connecting Two Processes

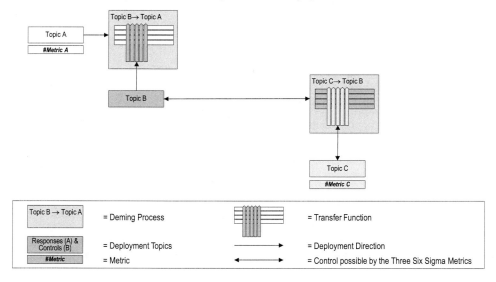

This process repeats as many times as needed. Figure 4-25 represents a simple Deming chain with two processes, one deploying topic A into topic B, the other one deploying topic C into topic B. The final goal topic is topic A – possibly customer's needs – while topic B depends from the responses of the second process controlled by topic C.

4-9.2 COMPREHENSIVE QFD

The *American Supplier Institute* (ASI) made a special variant of the Deming chain popular: the four-house chain for simple hardware manufacturing deploying customer's needs into solution characteristics. Those in turn deploy into solution design, and finally into production process parameters and vendor selection, modelling a full production process. This Deming chain caused much confusion (Rahman & Mitchell, 1994) – Bob King repeatedly apologized for it – but is still used for teaching QFD to Six Sigma Black Belts.

In manufacturing practice, and even more in software development or any other modern Six Sigma application area, the four-house chain (Figure 4-26) of the ASI is unusable – except for training purposes when explaining the concept of a Deming chain.

Figure 4-26: The Four-House QFD Deployment for Simple Hardware – Part of 1990 ASI Training

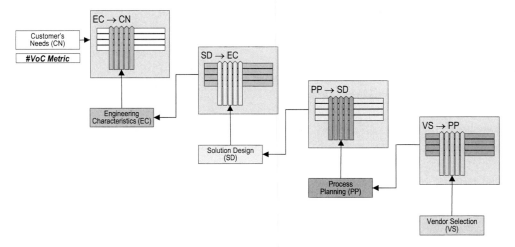

The international standard ISO 16355 (ISO 16355-1:2015, 2015) describes the current state of the art on comprehensive QFD. It expands the system of QFD across the entire organization, encompassing all levels of business functions from top management strategy, project portfolio management, system level design, R&D, detailed design, testing, down to day-to-day operations, ongoing support, and product retirement, making up for far more than four houses.

The other less famous QFD training model for service industries of the ASI was much better adapted to practice but surprisingly had significantly less impact, although used for teaching in the QFD Master Courses at the same time. Clausing (Clausing, 1994) enhanced Comprehensive QFD for Total Quality Development and Concurrent Engineering significantly. About ten years later, the German QFD Institute defined a best practice approach for Software QFD, identifying the deployments needed to develop new software (Herzwurm & Schockert, 2006). However, there exists no final rule determining which matrices to deploy. In this book, we will present a number of comprehensive QFDs in the form of the Deming chain, and ending with a theoretical exposure how such combination work for highly complex undertakings (*Chapter 13: Dynamic Sampling of Topic Areas*).

There is one caveat when using Deming chains. The combination of matrices works only in the *£2* norm; the *£1* norm, which earlier QFD experts traditionally used, introduces glitches that might endanger the QFD process. Maybe this is the reason why comprehensive QFD in the Nineties did not perform as expected, especially not in the West. In Japan, QFD experts developed the capability to "guess" the convergence gap just by looking at some matrix. They can tell whether the matrix is "balanced"; i.e., has a small convergence gap, or not. Western experts might benefit even more from the handy convergence gap metrics.

4-10 CONCLUSION

QFD is a very powerful method, largely used for product development. The ISO 16355 standard (ISO 16355-1:2015, 2015), developed by an international team from US, Europe and Japan, has given many new impulses to the method; previous misconceptions being sorted out and better aligned to the popular AHP decision method. The eigenvector method of weighting the solution is ultimately based on Saaty's invention that he made for AHP, see Saaty (Saaty, 1990).

In future, QFD will be no longer be exclusively used in product development, because its nature as a transfer function allows using the method for other problems as well, for instance, for software testing treated in Chapter 10. Moreover, with the spread of the Internet of Things (IoT), QFD will evolve from fixed-size matrices to variable-size rule sets, as explained in *Chapter 13: Dynamic Sampling of Topic Areas*.

CHAPTER 5: VOICE OF THE CUSTOMER BY NET PROMOTER®

Not only hardware production and software development are part of Deming Chains. Social networks and marketing also add value. Consequently, defects in social networking, bad rumors or missing referrals can destroy market share. Therefore, social status must be measured and variations controlled with Lean Six Sigma.

Marketing surveys based on the one-question approach such as Net Promoter® Score (NPS) have become popular for understanding customer's experience with respect to various touch-points.

The one-question approach avoids the pitfalls of multi-paged questionnaires. However, two issues arise:

(i)   Does such a one-question survey allow collecting specific information what drives customer business?

(ii) Are such surveys suitable to reveal the Voice of Customer (VoC), as needed for Lean Six Sigma?

5-1 INTRODUCTION

The last few years have seen a flooding of surveys and questionnaires; some of them intended to orientate an organization towards customer's needs; most indeed for other marketing purposes. You cannot stay in a hotel overnight without getting a questionnaire, you will receive an invite to share your experience on every travel platform you use; even your telecom provider or commuter railway asks you whether you found helpful and professional people in their stores or services, and whether you will recommend it.

Because interactive surveys via the web have become so cheap, marketers use them to collect all kind of information from customers or prospects. However, are surveys appropriate for collecting customer's voice? If so, which ones? Customers know that collectors of information usually know whom they ask, thanks to their CRM system and new sparkling information sources such as social media. So, will they answer honestly, even if they depend from that service?

Even worse, some surveys ask for employees' behavior characteristics, naturally without considering circumstances. Salary and bonuses might depend from such answers.

So, what will people answer when they realize that their answer affects the salary payment of their waiter, or any other business contact?

Sometime, surveys are created that contains dozens of questions; sometimes, silly questions are asked such as "How skilled were our consultants?" if I hire them exactly for the skills I personally have not? How should I rate them?

All this is neither lean, nor is it suitable to collect Voice of the Customer (VoC). Questionnaire-based surveys reflect the suppliers' view on products and services, and responses at best express the customer's consent. That is not VoC!

5-1.1 THE GEMBA WAY FOR GETTING FEEDBACK

Lean Six Sigma better looks at gemba ways to get feedback. *Gemba*, (現場 *genba*), is a Japanese term meaning *"the real place"*. Gemba refers to the place of value-creation: the factory floor, the sales point or where the service provider interacts directly with the customer (Imai, 1997) and (Yang & El-Haik, 2009).

Mazur uses this term in Quality Function Deployment (QFD) to denote the customer's place of business or lifestyle (Mazur & Bylund, 2009). To act customer-driven, one must go to the customer's gemba to understand his use of product or service, using all one's senses to gather and process data.

However, gemba visits for services provided are not always possible and not always are physical visits able to assess customer's experience with such services. For software, usability tests have been widely accepted as a kind of gemba visits, although testers usually conduct their tests in lab environments. Rather than at value-creation, these tests target for better human interaction design.

On the other hand, opportunities for gemba in services exist: helpdesk tickets and feedback from support interventions. Helpdesk tickets describe an unintended use of some product or service, and thus is a treasure of information for the supplier of the software or service. Nevertheless, organizations ignore this wealth of information because analyzing support data is difficult and seems unattractive.

5-1.2 NET PROMOTER® SCORE

Reichheld, Bain & Company, and Satmetrix Systems, Inc. have introduced and trademarked *Net Promoter® Score* (NPS) as a measurement method for customer loyalty (Reichheld, 2007). NPS measures the likeliness of customers to recommend a product or service privately to friends or relatives. Likeliness is recorded on a scale zero to ten (or as a probability between 0% and 100%, in steps of 10%). Moreover, the customer is asked for the reasons why she or he selected such grading, see (Fehlmann, 2011-3).

The NPS itself is the difference between *Promoters'* share (the percentage of those responding with a 9 or 10) from *Detractors'* share (those responding with 0 to 6). Those responding with a 7 or 8 are called *Passives.*

Figure 5-1: Net Promoter Score

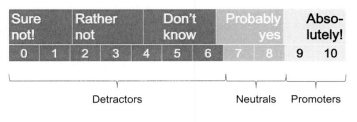

Net Promoter Score = %*Promoters* – %*Detractors*

➡ **NPS = 40%:**   40% more Promoters than Detractors – Everything's fine!

➡ **NPS = 0%:**   Equally many Promoters than Detractors – Something's wrong

➡ **NPS < 0%:**   Less Promoters than Detractors – All hands off board!

Clearly, NPS is a kind of attractiveness metric; however, it relates to companies and organizations rather than to product characteristics. The metric considers only the two best scores as promoters; thus, models customer loyalty well, and predicts future business. We have developed an approach using transfer functions that analyzes the underlying product characteristics. Based on the known customer profiles of the respondents, it yields goal profiles for business drivers, see (Fehlmann & Kranich, 2013-1), and (Fehlmann & Kranich, 2012-3).

5-1.3 THE NET PROMOTER® SURVEY

When Reichheld introduced the *Ultimate Question* approach for surveys (Reichheld, 2007), it was a big step away from the supplier view. By asking one question on how likely the customer is to recommend the supplier towards friends and relatives (not business partners) by means of a scoring scale from 0 (= not at all likely) to 10 (= extremely likely) he brought the customer view, his or her emotions and feelings, into surveys. This is the gemba approach.

Additionally, the customer should comment in his own words why he selected that score. These comments are the source treasure for analyzing VoC. The Net Promoter Score (NPS) itself is not a linear measurement – the respondents scoring with 9 and 10 are called *Promoters*, those scoring between 0 to 6 are *Detractors*, and the 7 and 8 *Passives*; the NPS is calculated as the difference between promoter percentage and detractor percentage of the total sample. That metric has been much disputed (Schneider, et al., 2008). Nevertheless, Reichheld claims NPS is predicative for future business growth, and since the NPS approach is not limited to measuring and statistical analysis but to acting upon the feedback received by the customer, he is most probably right (Owen & Brooks, 2009). The predicative effect is not due to some statistical reason but for the closed-loop work, that is also part of the NPS approach. Companies use the NPS to eliminate detractors, convert passives into promoters, and keep promoters by product and process improvement, and this in turn is instrumental to business growth.

Figure 5-2: NPS Ultimate Question Score Overview

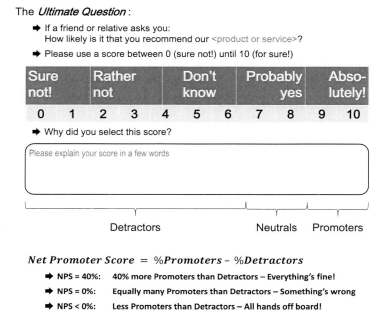

The **Ultimate Question** :

➡ If a friend or relative asks you:
How likely is it that you recommend our <product or service>?

➡ Please use a score between 0 (sure not!) until 10 (for sure!)

Sure not!	Rather not	Don't know	Probably yes	Absolutely!
0 1	2 3	4 5 6	7 8	9 10

➡ Why did you select this score?

Please explain your score in a few words

Detractors   Neutrals   Promoters

Net Promoter Score = *%Promoters* – *%Detractors*

➡ NPS = 40%:   40% more Promoters than Detractors – Everything's fine!
➡ NPS = 0%:   Equally many Promoters than Detractors – Something's wrong
➡ NPS < 0%:   Less Promoters than Detractors – All hands off board!

5-1.4 NPS – A LEAN EXTENSION TO THE SIX SIGMA TOOLBOX

The aim of this chapter is to explain how to analyze customer's comments, together with the NPS score and previous knowledge about the customer segment. It uses standard Six Sigma techniques; see e.g., (Creveling, et al., 2003). However, we will not discuss statistical or metronomic significance of NPS, which for instance Schneider disputes (Schneider, et al., 2008).

NPS in the product development lifecycle allows for a continuous monitoring of variations in the importance for the customer, and of customer's satisfaction, with the major existing or missing features of a product, listening closely to customer's voice.

5-2 WOM Economics

Word of Mouths (WOM) is a major driver for business. People highly estimate experiences made by others and try to copy them if those people have anything like a paragon standing against them. However, WOM is hard to measure directly, because people find it difficult to remember what they said and relate to it later. Asking them will not yield reliable answers.

However, it is not difficult to analyze the effects that WOM has, and identify the parameters that determine its effect.

5-2.1 The Loyalty Effect

Following (Reichheld, 2001), promoters' recommendations have effect on buying decision, while detractors' warning delay or inhibit them. There is reason to believe that recommendations on average have less impact than warnings by detractors.

Thus, the interesting parameters for WOM are:

- The frequency promoters recommending a product
- The frequency detractors warning against buying a product
- How many recommendations overcome warnings received against a product?

Assume that the *Net Customer Value* (NCV) – the net profit made with a customer after cost of sales, delivery and support – is known for promoters: NCV_{Pr}, neutrals: NCV_{Av}, and detractors: NCV_{Dr}. Every promoter issues R_P positive referral per period (e.g., year), every detractor makes R_N negative referrals during the same period.

To persuade a new customer to buy, let us assume that he or she needs R_T positive referrals. However, once a customer gets one negative referral, it needs R_S positive referrals to overcome this obstacle. Obviously, $R_S > R_T$, usually even significantly.

$$V_{Pr} = NCV_{Pr} + \frac{R_P}{R_T} \ NCV_{Av} \tag{5-1}$$

Equation (5-1) says that every promoter, in addition to provide net revenue NCV_{Pr}, also adds more revenue by providing referrals that add a new (average) customer.

$$V_{Dr} = NCV_{Dr} - \frac{R_S R_N}{R_T} \ NCV_{Av} \tag{5-2}$$

Equation (5-2) in turn says that its net value depreciates by providing negative referrals and thus blocking revenue from new additional customers. Equations (5-1) and (5-2) yield the total value in terms of net profit – that usually is negative for detractors because they relate their bad experience to other potential customers, damaging net profit more than what they add to the overall profitability by their already low contributions in new sales and support revenue.

$$Pr * V_{Dr} + Dr * V_{Dr} = 0 \tag{5-3}$$

Equation (5-3) identifies the equilibrium between positive and negative effects.

This yields formula (5-4) that shows the simple relationship between net values of promoters and detractors.

$$\frac{Pr}{Dr} = -\frac{V_{Dr}}{V_{Dr}} \tag{5-4}$$

You need P_r/D_r more promoters than detractors to outweigh detrimental effects by negative referrals. Finally, you must take care of the total number of customers – i.e., the neutrals – before translating ratio (5-5) into a threshold *NPS* value.

$$NPS = \frac{Pr - Dr}{\#Respondents} \tag{5-5}$$

This threshold enables future growth for a company or a product.

5-2.2 A SAMPLE CASE

Figure 5-3: Sample WOM Economics for a B2B Business Case

Word-of-Mouth Economics (WOM)

Test Sheet

Part 1: Net Customer Value (NCV) - New Customer 2010

Discount Factor: 5%

Expected Profit per Year for Average Customer

		2015		2016		2017		2018		2019		2020		2021		2022		2023		2024		2025
Total Revenue	€	200'000	€	50'000	€	50'000	€	50'000	€	50'000	€	90'000	€	65'000	€	65'000	€	65'000	€	65'000	€	65'000
Total Cost	€	180'000	€	6'000	€	9'000	€	9'000	€	9'000	€	39'000	€	7'800	€	11'700	€	11'700	€	11'700	€	11'700
Net Profit	€	20'000	€	44'000	€	41'000	€	41'000	€	41'000	€	51'000	€	57'200	€	53'300	€	53'300	€	53'300	€	53'300
Present Value	€	15'671	€	32'833	€	29'138	€	27'750	€	26'429	€	31'310	€	33'444	€	29'679	€	28'266	€	26'920	€	25'638

Net Present Value € 307'078

Part 2: Word-of-Mouth Economics

Average positive referrals per Promoter:	2.8		Promoter NCV	120%	€ 368'494
Average negative referrals per Detractor:	0.7		Average NCV		€ 307'078
One negative referral needs	7	or more positive counterweights	Detractor NCV	35%	€ 107'477
It takes	3	positive referrals to persuade prospect			

Value of Promoter:	€ 657'058	(Promoter NCV + 2.8/3*Average NCV)	**Business Growth Threshold**
Value of Detractor:	€ -396'811	(Detractor NCV - 7*0.7/3*Average NCV)	*Promoter* 60% (how many more promoters we need to outweigh detractors)
Value of turning			Assume Neutrals 40% (of total number of customers)
Detractor -> Promoter:	€ 1'053'869		*NPS* 36% (that's the NPS Score needed to enable business growth)

In the sample case above, the company is in the B2B sector supplying long-lived investment goods to other businesses. The net customer value in this case calculates over an investment period of five years that typically includes a mid-term re-purchase needed to keep up with changing business needs, or keep the product alive and running. The revenue in the years between originate from maintenance and support, or cloud rental cost.

The average positive referrals for promoters and detractors refer to an unpublished survey from Germany. All other data are assumptions.

The NPS threshold needed to maintain business growth is quite high. The ballpoint calculation suggests 36% – assuming that 40% of the customers behave as neutrals – and if there are less neutrals, the ratio equation (5-5) between promoters and detractors shows that NPS must become even higher.

The sample in Figure 5-3 shows that monitoring NPS is vital especially in the B2B area. In B2C it is similar; however, it follows different patterns because end customers can switch much easier and faster, and time runs at a tremendous pace.

5-2.3 SAMPLE NPS SCORES IN THE ICT AREA

A report from Satmetrix Systems, Inc. (Figure 5-4), from 2007 shows typical NPS rates and trends for the computer industry, see (Satmetrix Systems, Inc., 2007). Although data are from 2003 to 2007, the trend against the makers of traditional laptops can already be seen; Apple reaches the top with an NPS crossing 80%, only followed by Sony with 60% but dropping, and the other manufactures, some of them already gone by now, show a clear trend from 50% downwards. Thus, the advent of tablets and the big change in computing paradigm from offices to mobile that started in 2008 appeared in NPS at least from 2005 onwards.

Figure 5-4: Apple's NPS Trends 2003 – 2007 – with permission from Satmetrics, Inc.

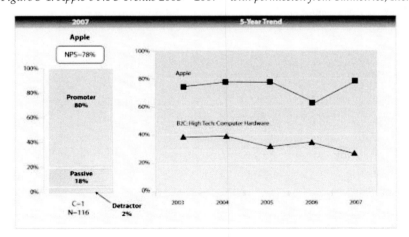

Consequently, an NPS of 40% does not protect against missing new trends.

Even more important than the high NPS compared to competition is that average high-tech customers in the USA – almost 80% of them promoters – spent $2'344 on average compared to $1'615 for general computer hardware consumers—a gain of 45% that directly added to Apple's revenue stream.

5-2.4 THE IMPACT OF NPS TO CORPORATE STRATEGY

It is obvious that such data greatly facilitates decisions, how much to spend for marketing, research, and product innovation. Not only because the revenue stream is here; the strategy, as perceived by customers, promises safe payback in the future. NPS could drive strategic decisions. However, in that case it is necessary understanding what exactly are the reasons that cause customers to recommend, and to spend their money on these products.

To do this, we use transfer functions. They behave as analytical instruments that allow identifying the root causes for the high – or low – NPS observed.

5-2.5 TRANSFER FUNCTIONS FOR UNDERSTANDING CAUSE-EFFECT

If root causes in gemba visits hide behind the observed effects; e.g., the VoC behind a support call or complaint, then the method of choice to uncover such causes are transfer functions. As explained in *Chapter 2: Transfer Functions*, they explain the root causes for NPS responses when finding correct business drivers for the observed NPS. The convergence gap quantifies how well these business drivers explain the observed NPS response. A small convergence gap indicates evidence for a cause-effect relationship between the selected business drivers and the observed NPS response; a large gap warns against the selected business drivers be taken as explanation for the observed NPS response.

5-2.6 FINDING THE RIGHT CONTROLS

In NPS surveys, controls are best determined by clustering customer responses. Such techniques exist that even work for big data and allow identifying to which notions or keywords customers most often refer. If the convergence gap indicates that the initial NPS analysis did not use the right controls, repeating the analysis with better categories for the business drivers might help.

A natural question arises: how many controls explain the expected response? Most often, the controls need more dimensions than the observed response profile. Therefore, $m \leq n$ holds, if n is the dimension of the solution profile. Otherwise, some of the responses would be interdependent and thus redundant; the dimensions of control and response profiles behave like degrees of freedom in mathematical statistics.

In practice, control profiles map to some real domain controls that allow guaranteeing the observed response accurately enough. This works will as well for VoC analysis.

5-3 VOICE OF THE CUSTOMER ANALYSIS

Generally, VoC consists of words and statements. The most popular and widely used analysis method for such VoC is *Verbatim Analysis*. The basic technique of verbatim analysis is to apply Six Sigma's affinity diagram developing procedure to VoC.

5-3.1 BUILDING TRANSFER FUNCTIONS FROM VOC

Building a transfer function from VoC is hard to automate, because of the variety of natural language and the difficulty of recognizing semantics. Most customer verbatim require a context for understanding. Nevertheless, for surveys or support tickets that amount to a few hundred samples, manual analysis is feasible. Since there is not much of higher value than a customer verbatim, it pays off. It is no waste.

The verbatim analysis is done by identifying and counting keywords in the responses that relate to some suspected customer's need, or business driver. This yields a matrix, mapping keywords in the responses onto customer segments, which must be known.

The matrix, acting as a transfer function, maps keyword frequency into both importance and satisfaction of the known customer segments. Thus, two transfer functions are constructed that allow comparing the NPS measured with the NPS how it should come out from the answers.

Apparently, the verbatim analysis is a customer voice measurement method validated by customers' own words.

5-3.2 BUSINESS DRIVERS

Transfer functions origin from analysis of signals and systems such as Fourier transforms, see Girod, Rabenstein, and Stenger (Girod, et al., 2001). The principle of customer verbatim analysis is like signal theory, called *Frequency Analysis*. Instead of counting wave functions, the number of references of certain words or notions defines the frequency. The customer verbatim categorize into references for business drivers. It is advisable to take as few business drivers as possibly needed for explaining the observed behavior in the response. Most domain experts know business drivers quite well, and the customers itself supply keywords that allow detecting true business drivers. Analyzing social media and community discussions might help as well.

The frequency of references collected and matched to the various customer segments defines a multi-response transfer function that maps frequency of verbatim to segment responses. The frequency yields a profile for the x-axis of the matrix. The response profile we get from the NPS score by segments.

5-3.3 MEASUREMENT ERRORS

The observed response profile y as well as the transfer function A (the result of verbatim analysis) is subject to measurement errors. Respondents may not understand the NPS scale and skew their score. Availability of context information and verbal skills affect the verbatim analysis; language sometimes is fuzzy and people do not express them correctly. Written statements often contain syntax flaws and the semantics are not clear. Humans knowing the context may understand better, but for an analysis tool, it might become rather complicated. However, the convergence gap yields a test for such frequency analysis. It yields the total measurement error of both the response measurement y and the transfer function A.

5-3.4 ADJUSTING THE ANALYSIS

The convergence gap might identify mistakes done in the verbatim analysis. In verbatim analysis, most often, the fuzzy meaning may refer to several different interpretations, and the preferred way is such that the analysis result matches the overall observed response; i.e., the measurement error decreases.

This principle allows for defining algorithms that do the verbatim analysis automatically. Not unlike multiple linear regressions, searching for minimal measurement error yields an improved transfer function A and therefore an improved τ as well from the frequency analysis. With that improved transfer function A, the control profile $x = A^\mathsf{T} y$ can be calculated and compared to its convergence gap.

5-3.5 BENEFITS

NPS surveys can deliver more information than traditional questionnaire-based surveys by not reflecting the supplier view but by stimulating customer's voice through an emotional channel.

The Six Sigma community understands verbatim analysis well enough and tools are available to extend such analysis to much larger response samples than only a few hundred. NPS surveys are popular, well-received by customers, transport an important message to the respondent, namely how important it is to recommend his or her preferred supplier, and allow for better gemba in service areas.

The predictive value of NPS lies in the actions and follow-ups, in personal contacts with respondents, in process improvement where needed, in better meeting business requirements of customers.

5-3.6 LIMITATIONS

If importance and satisfaction differ too much, or if NPS is not applicable because it is negative, the eigenvector validation method does not work as nice. However, this is easy to test by checking the convergence gap of the two transfer functions mapping the VoC onto importance and customer satisfaction, respectively, and the responses itself are still helpful, even if they cannot explain the observed NPS. The fact that responses might not be as interesting and lead to a better understanding of customers.

Verbatim analysis for large survey samples requires a sophisticated tool set for automatic analysis of the responses. Such tools are available from Big Data analytics. Additionally, the eigenvector solution profile calculation method also softens mistakes made in the analysis.

5-4 A SAMPLE NPS SURVEY

The following section exemplifies the NPS approach described in this chapter in a practical context.

5-4.1 SAMPLE NPS PROFILE

We assume a travel company collected feedback using an NPS survey for their helpdesk service helping in case of service disruption or cancellations; see the normalized NPS profile in Table 5-5.

Table 5-5: Sample NPS Profile According Customer Segments

Profile				Overall NPS: 18%	
Customer Segments	**Attributes**			**NPS Profile**	**NPS**
CS1 Young Traveler	Listens to music	Chats with peers	May want attention	0.56	20%
CS2 Business Traveler	Wants to work	Phone calls	Needs a table	0.47	17%
CS3 Holiday Traveler	Enjoys landscape	Lot of luggage	With children	0.56	20%
CS4 Elderly Leisure	Smalls groups	Wants it quiet	Need toilets	0.40	14%

NPS scores are not continuous satisfaction ratio scales. The eleven selectable scores between 0 and 10 collapse into three groups: promoters, passives, and detractors. Only promoters and detractors are explicitly included in the calculation. When both customer segments *Young* and *Holiday Travelers* score with Chats with peers, a significant difference in the number of passives still might exist. Passives moving into promoter or detractor status affect the overall score.

Nevertheless, we use the NPS scores as a ratio scale, based on the evidence suggested by its inventor, Fred Reichheld, keeping in mind that this assumption is not scientifically proved, as most of the statistical material used remains proprietary (Reichheld, 2001). The reason for this is that the experiences made with that assumption were always positive, and that a scientific proof for its validity is unnecessary. For our purpose, it is sufficient to investigate which controls we need to impact to increase the NPS score. Whether a higher NPS leads to higher sales and to increased customer loyalty is not within our scope.

5-4.2 COUNT OF VERBATIM

The sample NPS analysis uses five business drivers for verbatim analysis, shown in Table 5-6. The business drivers reflect the customer's viewpoint. The customer voice analyst can guess these business drivers, or find them by clustering survey responses from the customers. People working with customers; e.g., sales folk or helpdesk supporters, know the suspects well.

Table 5-6: Suspected Business Drivers for Verbatim Analysis in the Sample Case

	Topics	Attributes		
a) **Helpdesk**	BD1 Responsiveness	*Fast Response*	*Few Selections*	*Known Issue*
	BD2 Be Compelling	*Able to resolve issue*	*Committed to help*	*Knowledgeable*
	BD3 Friendliness	*Keeps calm*	*Unagitated*	*Cool*
b) **Software**	BD4 Personalization	*Knows who's calling*	*Knows travel plans*	*Knows delays*
	BD5 Competence	*Can figure out solution*	*According my preferences*	

Analyzing the customer verbatim now means counting references made to each of the five suspected business drivers in one of the responses. It can be explicit or by some predefined keyword that refers to it; e.g., "know me", "preferred", "my preference" for *BD4: Personalization*, or implicit by context.

The reference can be positive or negative; see Figure 5-7. Counting all references means assessing how important the topic is for the sample. Satisfaction counts positive minus negative.

Positive and negative references are indicative for satisfaction with the respective business driver. The passives also influence frequency analysis. Since passives cannot affect the overall NPS, the weight of the verbatim count decreases with the number of passives. This is the *Attenuation Factor*. This factor is a linear decrement of the response frequency r_S by the percentage of neutrals n_S/t_S, where the total of respondents from customer segment S is t_S and the number of passives is n_S.

NPS surveys do not measure customer satisfaction but something else: the willingness of customers to let the supplier's business grow by recommending it. This is doable even if satisfaction is not perfect; there are many reasons to recommend solutions that still have unsatisfactory aspects – most probably expecting that the supplier will address those weaknesses in future releases. Promoters are likely to tell so in their verbatim comments.

Figure 5-7: Verbatim Analysis - Extract

Customer Response

Customer Segments Name	Customer Segments	Score	Comment	Satisfaction BD1	BD2	BD3	BD4	BD5	Action Needed
Business Traveler	CS2	10	Prominent service, good cost/perfomance ratio		1	1		1	No
Young Traveler	CS1	9	As a young person, I haven't that much experience but I like it!	1					No
Business Traveler	CS2	10	Travelling means productive time; they know us and help us when needed		1		1	1	No
Holiday Traveler	CS3	3	I don't like being treatened like a leisure traveller when I'm using that service with my fami	-1	-1		-1		**Yes**
Elderly Leisure	CS4	8	Not sure whether they understand that I'm no longer that flexible and easy-moving				-1	1	No
Elderly Leisure	CS4	2	Waiting, no help; should this be my vacations?	-1	-1		-1		**Yes**
Business Traveler	CS2	7	As a Business Traveler, I love to know when I'll probably arrive. Thus a weak good score	-1			-1	1	No
Business Traveler	CS2	9	Very diligent, responsive service!			3			No
Young Traveler	CS1	10	Party, party!			2			No
Young Traveler	CS1	9	They pointed me to excellent saving offers: EURO-Rail Ticket		1		1		No
Business Traveler	CS2	7	While traveling you can work and arrive in good shape	1	1	1			No

Importance: 6 6 7 6 4
Satisfaction: 0 2 7 -2 4

(Responsiveness, Be Compelling, Friendliness, Personalization, Competence)

Importance: 1
Satisfaction: -1

Equation (5-6) shows the effective weight w_S.

$$w_S = r_S * \left(1 - {}^{n_S}/_{t_S}\right) \qquad (5\text{-}6)$$

In fact, responses obtained in an NPS survey help much more than traditional satisfaction surveys to uncover weaknesses and detect issues that otherwise remain unknown to the supplier.

5-4.3 SAMPLE VOC ANALYSIS

Conducting frequency counts results in two transfer functions: one explains the observed NPS score from the importance; see Figure 5-8, the other explains a different response profile by looking at satisfaction. The latter sometimes exhibits negative scores on the profile; see Figure 5-9. If the squared matrix with its transpose remains positive definite and the convergence gap remains close enough to zero, the result is still significant.

Not both transfer functions are always capable of explaining the observed score. Usually, a positive NPS better explains importance than satisfaction does, especially since a positive definite transfer function cannot explain negative NPS. Satisfaction analysis is possible if the satisfaction transfer function's eigenvector matrix $AA^\top$ remains positive definite. Thus, with satisfaction below a certain level, this VoC–Analysis might be inappropriate for the purpose. A convergence gap of 0.29 is just at the limit.

From comparing Figure 5-8 with Figure 5-9, it is obvious that there are no problems with *BD3: Friendliness*, but with *BD4: Personalization*. In terms of importance, it tops the other business drivers; however, satisfaction is even negative. Opportunities for improvement, however, of lower importance, can also be seen in *BD1: Responsiveness*, and in *BD2: Be Compelling*.

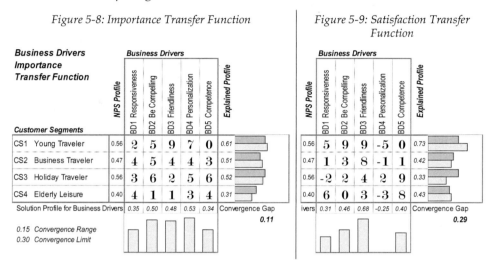

Figure 5-8: Importance Transfer Function | *Figure 5-9: Satisfaction Transfer Function*

5-4.4 FINDINGS FROM THIS ANALYSIS

In our case, there is dissatisfaction with the ability to recognize frequent travelers personally from their web login or mobile number. For the respondents, this is considered quite important, yielding clear advice where to invest into service improvement.

Usually, satisfaction explains the observed NPS less good than importance – it could hint at the willingness of respondents that the sample company can improve, and better meet the important business drivers.

5-4.5 COMBINING IMPORTANCE AND SATISFACTION INTO A PROFILE

Neither importance nor satisfaction alone provides a valid profile for the business drivers. This is because NPS asks for reasons to recommend; if something is not a reason to recommend it will not appear in a positive comment. It still might in a negative; thus, the satisfaction profile affects the total weight of business drivers.

There is no general rule how to combine the importance profile with the satisfaction profile. Probably the best approach is using the *Satisfaction Gap*. The satisfaction gap weights negative statements exponentially; it thus stretches the importance profile in case of dissatisfaction. If customers are dissatisfied with an unimportant topic, the satisfaction gap remains small and does not affect the profile.

Figure 5-10: Combination of Importance and Satisfaction Profile

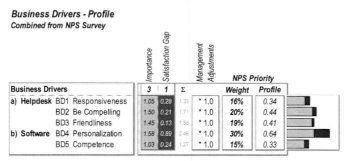

Business Drivers - Profile
Combined from NPS Survey

Business Drivers			Importance 3	Satisfaction Gap 1	Σ	Management Adjustments	NPS Priority Weight	Profile
a) Helpdesk	BD1	Responsiveness	1.05	0.28	1.33	* 1.0	16%	0.34
	BD2	Be Compelling	1.50	0.21	1.71	* 1.0	20%	0.44
	BD3	Friendliness	1.45	0.13	1.58	* 1.0	19%	0.41
b) Software	BD4	Personalization	1.58	0.89	2.46	* 1.0	30%	0.64
	BD5	Competence	1.03	0.24	1.27	* 1.0	15%	0.33

The higher importance is, the more impact has dissatisfaction and applying equation (5-7) reflects this in the combined profile in Figure 5-10 as follows:

$$Combined \ \ Profile = x_I + e^{-ax_S} \qquad (5-7)$$

Negative satisfaction adds more weight to the importance profile x_I by turning negative satisfaction x_S into additional profile weight for business drivers. The normalization parameter a adds nothing to the combined profile for the highest satisfaction profile component, thus equation (5-7) corrects missing importance for missing business drivers in a product or service. The satisfaction profile must affect importance only in case of overwhelming dissatisfaction.

The equation (5-7) adds weight to those business drivers that leave customers dissatisfied. In this case, the driver *BD4: Personalization* gets the additional considerable weight, see Figure 5-10; the others only as much as customer's dissatisfaction requires. The equation (5-7) is empirically derived and not based on any theoretical background.

The resulting business driver combined profile shows the priorities needed for future products to become successful. Importance alone explains why the customer recommends, but non-importance does not mean that such topics are not relevant to customers, especially not if they show dissatisfaction. Thus, interpreting NPS results always requires sound considerations of customers' experiences.

5-4.6 COMPARING NPS SCORES WITH KANO ANALYSIS

With the sample survey before, if we compare the survey analysis results regarding importance with the Kano method, we see as a possible interpretation the picture shown in Figure 5-11. With a convergence gap of 0.10, Kano and NPS almost converge.

Figure 5-11: Kano Analysis compared to Combined Profile with Graphical Kano Analysis Tool

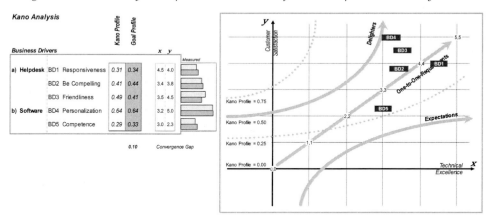

As explained in *Chapter 4: Quality Function Deployment*, the profile only defines the locus of the Kano profile curves, not the exact position on it. Each placement is unique only up to the geometric locus with identical profiles. The exact position depends from the relation between customer satisfaction level and technical excellence. Usually, developers know technical excellence best, resolving ambiguities; sales representatives can better forecast customer satisfaction. In Figure 5-10, it can be seen that *BD4: Personalization* tops in Kano again; the others are more or less in the area of *One-to-One Requirements*; i.e., expected to be as good as possible, with high potential to delight the customer if present. This is especially true for *BD3: Friendliness*. The noteworthy exception is *BD5: Competence*, that goes as an expectation.

Kano is a means for quality control an NPS survey as well, as false assumptions become immediately visible, and results crosscheck for consistency with delighters and expectations. However, the NPS values only indicate the profile and not the position where on the diagonal the business driver appears. Thus, conducting a Kano analysis provides additional value to an NPS survey analysis. In the sample shown in Figure 5-11, it seems that the Kano analysis gave additional weight to *BD3: Friendliness* compared to the NPS survey.

To some extent, Kano might adapt findings to product strategy, as it identifies those drivers that add particularly much to customer satisfaction by delighting them, even if customers do not mention them as reasons for recommend.

5-5 APPLYING NPS

NPS is applicable in many settings and has become an integral part of the VoC toolbox in Six Sigma. In some way, it is the gemba for Internet times.

5-5.1 NPS AS A DEMING CHAIN

Figure 5-12: NPS as a Deming Chain

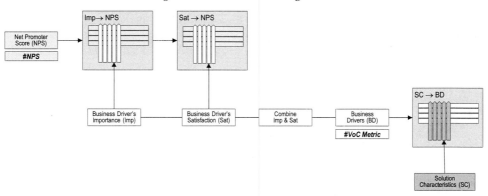

For NPS, both importance and satisfaction contribute to the business driver's profile. This can be understood as a Deming chain; transforming value from the business drivers up to customer's willingness to recommend. Figure 5-12 shows – additionally to the combination of importance with the satisfaction gap, as seen in section 5-4.5 – how the business driver profile deploys into a technical solution for the product or service.

Thus, NPS is the method of choice for interactively develop and improve products are services to meet the changing needs of customers.

5-5.2 NPS FOR CONTINUOUS PROJECT MONITORING

A similar application area for NPS are projects. Projects, not only ICT projects, often fail because the stakeholders' views start diverging during the project. An excellent practice is to conduct NPS surveys every day. Smartphone apps might be the means to do it; after each day with the project, the team members and all involved stakeholder anonymously respond to the "Would you recommend" – question regarding whether they believe the project will reach its stated goals. Comments why they recommend, or not, are extremely helpful to early detect weak or critical points that the project team eventually can address well before viewpoints start diverging. Even without such comments, monitoring the trend is extremely valuable for the project team. Such monitoring is possible at almost no cost.

5-5.3 NPS with Social Media

The main difference when using social media data as a base for NPS analysis is that there is no personal contact with the respondents. This makes the closed loop difficult and correctly identifying the customer segment. Moreover, sometimes it is not possible to get scores from big data; you rather get statements that need interpretation for whether it represents a promoter, a passive, or a detractor. The usual way of applying transfer function might not work as expected.

However, these hurdles mastered, today's *Big Data* tools allow finding business drivers and perform verbatim analysis automatically. It is a typical big data problem – although you cannot trust data on the individual level, the trends they uncover might still be most significant. Many big companies today monitor social media for detecting trends and avoiding shit-storms that threaten reputation, and thus profitable business. A good strategy is combining such monitoring with conducting actual NPS surveys among its known customer base. Interdependencies are likely to exist and if so, they might be quite insightful.

5-5.4 Predictive Analysis

The area of *Predictive Analytics* has just begun; for instance, Siegel presents in his book (Siegel, 2013) a variety of possible applications. One are of application that matters most for managing complexity is getting customer's preferences right out of social media, and adapting to changed preferences fast. The NPS method is just a forerunner of it.

Predictive analysis uses the same methods as presented here for analyzing customer's voice by verbatim frequency analysis of blogs, tweet and other statements in social media. Intelligent algorithms allow to assess how positive or negative such statements are meant, although in the single case the failure rate, for instance in recognizing cynics, is probably high. Nevertheless, averaging over big data samples, predictive analysis can predict relevant events and decisions in the future, based on transfer function.

The mathematical techniques presented in this book do not cover the problem of solving large matrices; however, apart from the different mathematical toolset, the same principles apply.

5-6 Conclusion

The Net Promoter® Score (NPS) is much more than the marketing tool as promoted by Reichheld (Reichheld, 2007). As shown in this chapter, NPS combined with transfer functions provides the opportunity to discover the meaning of customer voice. This technique is like the big data mining techniques, nowadays widely popularized and much discussed inside the big data community. The mathematics behind NPS is as simple as Google's secret, which is explained by, e.g., Gallardo (Gallardo, 2007); basically, it means applying linear algebra, and eigenvector theory to large data sets.

The problem with NPS in the context of this book is, that originally this is not meant as a ratio scale. Hence, NPS must be interpolated (done silently in this chapter), to be used in transfer functions. Nevertheless, measurement errors can occur, when jumping from promoter to neutral to detractor. The method proposed therefore should only be applied, if original NPS survey data is available, as in Figure 5-7, and interpolating the score is possible.

PART TWO:

LEAN SIX SIGMA FOR SOFTWARE

CHAPTER 6: FUNCTIONAL SIZING

The following chapter introduces functional sizing and explains some methods for functional size measurement. Functional sizing is the base for all Six Sigma metrics that pertain to software, and to services. Functional Points Counting exists for many decennials; however, with ISO/IEC standardization and the replacement of software coding by software services, the paradigm changed. The focus shifts towards sizing models for the software under consideration; models, that partly describe software services and partly software code that needs to be written or rewritten.

Several sizing examples are given, and it is explained how to count the functional size of UML ™ sequence diagrams based on the international standard ISO/IEC 19761.

6-1 FUNCTIONAL SIZING OVERVIEW

The following chapter provides a brief introduction for some selected measurement methods of software functional size. Some text material and the sample originate with permission from ISBSG, the International Software Benchmarking Standards Group (Hill, 2010).

6-1.1 THE ISO/ IEC STANDARDS

The series of standards ISO/IEC 14143 consists of the following parts, under the general title *Information technology — Software measurement — Functional size measurement:*

- Part 1: Definition of concepts (ISO/IEC 14143-1:2007, 2007)
- Part 2: Conformity evaluation of software size measurement methods
- Part 3: Verification of functional size measurement methods [Technical Report]
- Part 4: Reference model [Technical Report]
- Part 5: Determination of functional domains for use with functional size measurement [Technical Report]
- Part 6: Guide for use of ISO/IEC 14143 series and related International Standards

ISO/IEC 14143-1:2007 defines the concepts of *Functional Size Measurement* (FSM). These concepts overcome the limitations of earlier methods by shifting the focus away from measuring how the software is implemented to measuring size in terms of the functionality required by the user.

ISO/IEC 14143-2:2011 establishes a framework for the conformity evaluation of a Candidate FSM Method against the provisions of ISO/IEC 14143-1. It describes a process for conformity evaluation of whether a Candidate FSM Method meets the (type) requirements of ISO/IEC 14143-1 such that it is an actual FSM method; i.e. they are of the same type. It also describes the requirements for performing a conformity evaluation to ensure repeatability of the conformity evaluation process, as well as consistency of decisions on conformity and the result.

ISO/IEC TR 14143-3:2003, ISO/IEC TR 14143-4:2002 and ISO/IEC TR 14143-5:2004 are technical reports now on the way to obsolescence in view of the fundamental change that occurred to ICT technology, while ISO/IEC 14143-6:2012, the guide for using these standards, has recently been updated.

The following methods are compliant to the ISO/IEC 14143 series of international standards:

- ISO/IEC 20926:2009 Software and systems engineering — Software measurement — IFPUG functional size measurement method 2009 (ISO/IEC 20926:2009, 2009)
- ISO/IEC 19761:2011 Software engineering — COSMIC: a functional size measurement method (ISO/IEC 19761:2011, 2011)
- ISO/IEC 29881:2008 Information technology — Software and systems engineering — FiSMA 1.1 functional size measurement method (ISO/IEC 29881:2010, 2010)
- ISO/IEC 24570:2005 Software engineering — NESMA functional size measurement method version 2.1 — Definitions and counting guidelines for the application of Function Point Analysis (ISO/IEC 24570:2005, 2005)
- ISO/IEC 20968:2002 Software engineering — Mk II Function Point Analysis — Counting Practices Manual (ISO/IEC 20968:2002, 2002)

This book will not cover the last two ISO standards.

6-1.2 FIRST GENERATION SOFTWARE MEASUREMENTS

Albrecht (Albrecht, 1979) has invented *Functional Size Measurement* as a measurement method for software value back in 1986. His heritage is still vividly governed by the International Function Points User Group (IFPUG) and many local user software metrics user communities such as DASMA in Germany, FiSMA in Finland, NESMA in the Netherlands, UKSMA in the UK, GUFPI/ISMA in Italy, ICTscope.ch in Switzerland; to name only a few. Function Points are famous because they serve as a reference base for early project estimation, and increasingly for contractual agreements in software contracts. However, project cost estimation is not the only interest in Six Sigma. Six

Sigma targets at reductions of variations and at defect containment – including variations, and defects in estimations that also benefits from transfer functions used for measuring the quality of project estimations, see *Chapter 12: Effort Estimation for ICT Projects*. Contrary to this, first generation FSM counting methods focused on cost prediction.

6-1.3 WHAT IS FUNCTIONAL SIZING?

Functional Size Units (FSU) measure the size of a software product or service, by evaluating the size of the functional user requirements that are, or will be, supported or delivered by the software or service. In the following, we often refer just to "software" but always understand that today's software services are based on various kind of techniques, be it coding, connecting and interfacing services, based on premises or in the cloud.

In simple terms, FSU measures what the software must do from an external user perspective, irrespective of how the software is constructed, or what service is used. In the same way as the size shown on a building's floor plan is expressed in units of square meters, an FSU reflects the size of the software's functional user requirements, irrespectively whether the building is constructed on premises or as a prefabricated house. Most software – no matter how large or how small – addresses "user" requirements. Consequently, you can measure it by FSU. Some practitioners use the analogy of a "black box" to describe how to measure software by functional size, independent of the physical inner workings of the software.

In the 20th century, developers wrote most software services from scratch, creating code in some programming language. In the 21st century, the paradigm has changed. More and more software services rely on service-oriented architectures, connecting services, creating value from combining different software services; e.g., listing restaurants and shops that are next to the geographical position of the requestor. Such services have a measurable functional size. The measurement relies on counting functional size units just alike the transactional software systems built in the last century. Thus, functional sizing applies to both, software based on some coding, and software services, where no code is available and technical constraints remain hidden. The standards based on ISO/IEC 14143 work for both; code is irrelevant. "Software" in the context of functional sizing always refers to both, bespoke software and software services, irrespectively.

6-1.4 The Key to Functional Sizing is to "Think Logical"

Unlike lines of code, functional size units are independent of the physical implementation, languages and services used to develop the software. The relationship between function size units and software development compares with square feet and construction. Table 6-1 illustrates some of these comparisons.

Table 6-1: Functional Size Units Paradigm (courtesy ISBSG)

Metric	Construction Units of Measure	When is it important to measure?	IT Units of Measure	When is it important to measure?
Estimated project size	Square meters	Floor plans stage	Functional Size Units	Requirements or contract stage
Unit cost & overall cost	€ per square meter & total €	Construction & contract negotiation	€ per function point & total €	Go or no-go development decision
Estimated work effort	Person months	Throughout construction or whenever change occurs	Hours or person months	Throughout development or whenever change occurs
Size of change orders	Square meters & € (impact)	Whenever change identified	CFP, € or hours (impact)	Whenever change identified

Functional sizing counts from a logical user perspective, based on functional user requirements. This is a paradigm shift for developers who are skilled in programming and physical configuration management; less so for architects that understand how to combine different services. It does not matter, whether it takes a thousand lines of C# code and eight subroutine calls, or a hundred lines of Java script, or a one-line call to some cloud service to perform a business function; the functional size count is the same, because the user functionality is the same.

6-1.5 Functional Size and Work Effort

Just as the quality of the raw materials, the piping configuration, the configuration of the floors and the overall layout, affects the work effort required to plumb a house, so too will the programming environment, the services used, and other attributes affect the time it takes to develop the software or service. However, regardless of house design and construction, the floor plan size of the house stays the same. With software, the software size is independent of the language, skills, physical configuration, and other factors used in its development. Functionality, and consequently functional sizing, ignores many related quality aspects such as safety, security, and legal compliance.

To take the analogy further, in the case of house construction, the building costs of a house of hundred square meters depend upon whether it is timber, brick veneer or double brick. A single floor home typically will be cheaper per square meter to build than a double floor, or split level, but this too varies with geography, building codes, and construction methods. For software development, the building cost of a system with 1000 FSU depend upon the environment, available services, interfaces, middleware, programming language, the platform and many other variables. For system integration or today's cloud functionalities based on services, there is much less dependency from the programming languages than from the organizational or legal complexity encountered when integrating these services.

6-1.6 THE LOGICAL BOUNDARY

One of the first steps in functional sizing is to identify the *Logical Boundary* around a software application. This boundary separates the system or software from the user domain – users can be people, things, objects, other software applications, departments, other organizations, etc. As such, the software may span several physical platforms and include both batch and on–line processes. The boundary around systems and software does not identify the implementation of the physical system. It rather shows how an experienced user would view the system or software. This means that a single application boundary can encompass several hardware platforms; e.g., cloud, mainframe, PC hardware and smartphones used to provide access to an application, would all be included within the application boundary.

6-1.7 TYPES OF FUNCTIONAL SIZING

There are three major types of measuring functional size:

1) Portfolio Sizing

   This count is the size of an installed base application. The base size in FSU is a point-in-time snapshot of the current size of an application. This number is useful whenever comparisons are required between different applications; e.g., defects regarding base FSU.

2) Project Sizing (Development, Enhancement, or Maintenance Project Sizing)

   This count reflects the size of the functional area developed by a development project or "touched" by an enhancement project. An enhancement project count is the result of summing the new functions added in the project, and the functions removed from the application by a project, plus the functions changed by the project.

3) Test Sizing, Defect Counting and Defect Prediction

This is the major functional sizing application area in the Lean Six Sigma discipline. Functional size is indispensable for applying Six Sigma statistical tools to software services or operations. The caveat is that such sizing must be a measure in a linear vector space. Ratio scale metrics are such measures; low/medium/high scales such as in ISO/IEC 20926 IFPUG are not ratio scales.

6-1.8 WHAT IS INVOLVED IN FUNCTIONAL SIZE COUNTING?

The basic steps involved in functional sizing include:

1) Determine type of count, whether it is a new development project count, an enhancement project count, or an application or base count aiming at software maintenance.
2) Identify the counting scope and the application boundary. i.e., what functions must this software perform? This corresponds to creating a context diagram for the application or project.
3) Determine the count according the method selected.

Every method includes a measurement process description that follows the steps requested by ISO/IEC 14143-2.

6-2 ISO/IEC 20926:2009 IFPUG FUNCTIONAL SIZING

We use the denomination *IFPUG Function Point*, abbreviated IFP, for the FSU in ISO/IEC 20926. This is the unit of measure for functional size within the international standard ISO/IEC 20926:2009; referred as IFPUG 4.3.1. In the past, there used to be a confusing distinction between *Unadjusted Function Points* (UFP), and a derivative kind of adjusted function points calculated by multiplying with *Value Adjustment Factors* (VAFs); however, finally this is gone now. We will not comment that any further in this book. Complete details are available in the International Functional Size Unit Users Group (IFPUG) Counting Practices Manual 4.3.1, published by IFPUG January 2010 (IFPUG Counting Practice Committee, 2010).

The following five functional components of the software evaluate for the count according the ISO/IEC 20926 IFPUG rules based on the user requirements:

1. Internal Logical File (ILF)

- IFPUG 4.3.1: user recognizable group of logically related data or control information maintained within the boundary of the application being measured.

- A user identifiable group of logically related data that resides entirely within the applications boundary and is maintained through External Inputs.

2. External Interface File (EIF)

- IFPUG 4.3.1: user recognizable group of logically related data or control information referenced by the application being measured; however, maintained within the boundary of another application.

- A user identifiable group of logically related data that serves as reference purposes only. The data resides entirely outside the application and another application maintains it. The External Interface File is an Internal Logical File for another application.

Figure 6-2: The IFPUG Model

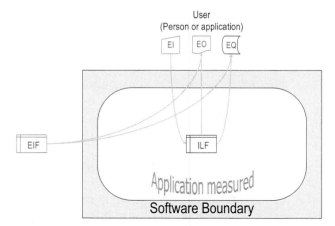

3. External Input (EI)

- IFPUG 4.3.1: elementary process that processes data, or control information sent from outside the boundary.

- An elementary process in which data crosses the boundary from outside to inside. This data may come from a data input screen, electronically or another application. The data can be either control information or business information. If the data is business information, it maintains one or more internal logical files. If the data is control, information it does not have to update an internal logical file.

4. External Output (EO)

- IFPUG 4.3.1: elementary process that sends data or control information outside the boundary and includes additional processing logic beyond that of an External Inquiry.
- An elementary process in which derived data passes across the boundary from inside to outside. The data creates reports or output files sent to other applications. These reports and files originate from one or more internal logical files and external interface file.

5. External Inquiry (EQ)

- IFPUG 4.3.1: elementary process that sends data or control information outside the boundary.
- An elementary process with both input and output components that result in data retrieval from one or more internal logical files and external interface files. This information crosses the application boundary. The input process does not update any Internal Logical Files and the output side does not contain derived data. None of the following can be involved in the process: Calculations, derivation of data, update of one or more ILF, altering of system behavior (e.g., browsing the data).

Note that the arrows in Figure 6-2 refer to *File Type Referenced* (FTR); see section *6-2.3 Transaction Complexity*.

6-2.1 FUNCTIONAL POINT COUNTING COMPONENTS

The five types of functional components are not the same as physical components. For example, *Internal Logical Files* might have nothing to do with physical files or data sets. ILF refers to the logical, persistent entities maintained through a standardized function of the software. In other words, ILF are the stand–alone, logical entities that typically would appear on all *Entity Relationship Diagrams* (ERD). For example, a human resources application will maintain all associate or employee data. This entity counts as an ILF. Such an application typically would rely on user accounts to prevent misuse; however, another application (e.g., Microsoft's *Active Directory*) maintains these accounts. Thus, this entity counts as an EIF.

Another illustration of counting functional components is when referring to EI (*External Inputs*), which are the logical, elementary business processes that maintain the data on all ILF, or which control processing. The logical business process of adding all associate would be one user function, and therefore in functional sizing we would count one EI. The size in function points for this one EI would be the same, regardless of how we physically implemented it, because in every implementation it performs one logical user function. Par example, the count for *Add Associate* is the same regardless of the number of physical screens, keystrokes, batch programs or pop–up data windows needed to complete the process.

6-2.2 FUNCTIONAL SIZE COUNT IN ISO/IEC 20926

The IFP count needs two distinct steps:

- Count the data function types: ILF, which are logical data groups maintained within the application boundary, and EIF, used for reference by the application.
- Count the transactional function types Els, which are data entry processes and controlled inputs; EO, (e.g., reports with calculations) and EQ, (e.g., retrieval of stored data by inquiries from one or more ILF/EIF).

ISO/IEC 20926 provides several simple matrices to determine whether a function is *Low*, *Average* or *High*, based on *Data Element Types*, (DET; user recognizable, non–repeated data fields), *Record Element Types* (RET; subsets of user recognizable data), and *File Types Referenced*, (FTR; number of logical data groupings, (i.e., ILF and EIF), required to complete a process).

The ISO/IEC 20926 Table 6-3 summarizes the number of function points assigned to each function type. Following the IFPUG guidelines, count and rate all the identified functions and add the function points together. The resultant number is the functional size in units of *Function Points* (FPs).

Table 6-3: ISO/IEC 20926 Function Points (IFP) Count per Function Type

Function Type	Low	Average	High
External Inputs (EI)	×3	×4	×6
External Outputs (EO)	×4	×5	×7
External Queries (EQ)	×3	×4	×6
Internal Logical Files (ILF)	×7	×10	×15
External Interface Files (EIF)	×5	×7	×10

6-2.3 TRANSACTION COMPLEXITY

To determine the function points assigned to each of the five transaction types, we must know its complexity as shown in Table 6-4.

- **FTR: File Type Referenced**
  A FTR is a file type referenced by a transaction. An FTR refers either to an internal logical file (ILF) or to an external interface file (EIF).

- **DET: Data Element Types**

  A DET is a unique user recognizable, non-recursive (non-repetitive) field. A DET contains information that is dynamic and not static. A DET can invoke transactions or can be additional information regarding transactions. If a DET is recursive then only the first occurrence of the DET counts, not every occurrence.

  In the case of a table, DET is the number of columns holding different data fields. In case of objects, DET is the number of attributes including methods.

- **RET: Record Element Type**

  A RET is user recognizable sub group of data elements within an ILF or an EIF. It is best to look at logical groupings of data to help identify them.

  In a table that contains one sub-table, we count two RET. The fixed part counts as one RET, each variant count as an additional RET.

  In an object, the RET is the number of different user recognizable methods, where each overloading counts as another RET.

Table 6-4 demonstrates how the ratings are determined using the transaction complexity and summarizes the IFPUG 4.3.1 rules for counting unadjusted functional size units.

Table 6-4: Transaction Complexity

FTR's	Data Element (DET)		
	1 – 4	5–15	> 15
0 - 1	Low	Low	Ave
2	Low	Ave	High
> 2	Ave	High	High

EI Complexity Table

FTR's	Data Element (DET)		
	1 – 5	6–19	> 19
0 - 1	Low	Low	Ave
2 - 3	Low	Ave	High
> 3	Ave	High	High

EO / EQ Complexity Table

RET's	Data Element (DET)		
	1 – 19	20–50	> 50
1	Low	Low	Ave
2 - 5	Low	Ave	High
> 5	Ave	High	High

FTR Complexity Table

Rating	Values		
	EO	EQ	EI
Low	4	3	3
Average	5	4	4
High	7	6	6

EO / EQ / EI Rating
for transaction function types

Rating	Values	
	ILF	EIF
Low	7	5
Average	10	7
High	15	10

ILF and EIF Rating
for data function types

Dasgupta, Cigdem and Symons noted that the number of functional components, plus FTR, DET and RET would yield an interesting metric for software size (Dasgupta, et al., 2015).

6-2.4 LIMITATIONS

The ISO/IEC 20926 counting method has the big disadvantage that it is not linear. Thus, it is no ratio scale, as stipulated by Saaty (Saaty, 2003) and the ISO standard on QFD (ISO 16355-1:2015, 2015) for using statistical tools and methods. If the rating in Table 6-3 jumps from *Low* to *Average* to *High*, the IFP count jumps as well. Thus, measuring application **A** and measuring application **B** does not necessarily yield the same sum of IFPUG Function Points (IFP), as when measuring a single application that provides exactly the functionality of both **A** and **B**. Although it is arguable that such linearity is almost impractical for software or it no longer counts for large functional sizes, it is a major flaw if you try to apply statistical methods to software sized with IFPs. There is no unit value; a single function point does not exist, and is not defined. All usual statistical methods, formulas and key indicators are strictly speaking not applicable.

For large and complicated software systems, the ISO/IEC 20926 count remains capped. This means that with increasing project size, functional size grows sub-proportional. Gencel and Bideau explain the reason for this as follows (Gencel & Bideau, 2012): "In the IFPUG method, if either DET or FTR/RET significantly exceeds the boundary values, the (transaction) complexity (see Table 6-4) of the observed function is still rated as "high," although the underlying complexity is in theory increasing on to infinity". On the other hand, an ISO/IEC 20926 count yields unreasonable high counts for very small apps or embedded software (Fehlmann & Kranich, 2014-1). The reason or this observation might be the same: any EO yields at least three IFP according Table 6-4. This is another reason why the ISO/IEC 20926 counting method has limited applicability in a Lean Six Sigma context, where statistical methods require a ratio scale count. Nevertheless, doing an ISO/IEC 20926 count improves requirements elicitation by detecting missing requirements; see (Fehlmann, 2011).

Another important limitation is that the ISO/IEC 20926 method has only limited applicability for today's service-oriented architectures, because it is difficult to measure layered architectures and because it hardly models two-way communications. While ISO/IEC 20926 identifies communication input from an external application as an EIF, the response to the external application would be an EO or EQ, visually unrelated to the EIF. The upcoming examples will demonstrate it.

Nevertheless, the weakness lies in the counting rules for complexity only; the IFPUG model still has its merits. Since a few years, IFPUG promotes a new standard for non-functional sizing called the *Software Non-functional Assessment Process* (SNAP), now out in release 2.0 (IFPUG Non-Functional Sizing Standards Committee, April 2013). It operates on the usual ISO/IEC 20926 IFPUG model; it only assigns other sizing metrics, targeted at *Non-functional User Requirements* (NFR). This raises the question, whether the actual IFPUG method should become an assessment method for sizing traditional management information systems, besides other assessment methods that for instance aim at counting defects in software, based on the IFPUG model.

It is noteworthy that SNAP counts model elements that are not counted for functional size. Thus, the notion "IFPUG model" is not precise enough since it is not clear whether to include code data as ILF, for instance. It is preferable to use the notion *Transaction Map* for the graphical representations of the IFPUG Model introduced in Figure 6-2. This allows to graphically represent model elements even if they are not considered for the functional size.

6-2.5 A FUNCTIONAL SIZING CASE STUDY

Without getting into the specifies of how to rate a specific component as low, average or high, the following example illustrates the basic steps to arrive at the function point count for an application or development project. This example originates from the ISBSG's publication *Practical Project Estimation,* first published in March 2001, now available in its third edition (Hill, 2010).

The functional user requirements are:

- Create an application to store and maintain employees consisting of the following data fields: name, employee ID, rank, street address, city, state, post code, date of birth, phone number, office assigned, date the employee data was last modified.

- The application must provide a means to add new employees, update employee information, terminate employees, and merge two employee records.

- The application must provide a weekly report that lists the employees whose information has changed during the past week and a total number of changed employees.

- The application must provide a means to browse employee data, including the list of corporate apps that the user can access through the organization's Active Directory.

No requirements are for data validation. Thus, it is safe to assume all edits use hard coded, not modifiable, data. For a more modern, user-centric form of these requirements see *Table 7-4: Functional User Requirements recorded as User Stories in the Grant Rule Format.*

The functional components based on the above, include:

1) One *Internal Logical File* (ILF), for the employee data because it is a persistent logical entity maintained by the application. Based on an evaluation of the data elements and logical record types, (contained in the IFPUG Counting Practices Manual Release 4.3.1): This ILF categorizes as *Low* and is worth seven IFP.

2) Four *External Input* (EI) processes (each whose primary intent is to maintain data in the ILF): One EI each for *Add Employee*, *Update Employee*, *Terminate Employee* and *Merge Employee Records*. Assuming each one is of *Low* complexity, (each requires only one logical entity and requires less than 16 data elements), and each EI would be worth three IFP for the totally 12 IFP.

3) One *External Output* (EO) process: The weekly report falls into the category of external output because its primary intent is to present data to the user and there is a calculation involved. Typically, the report would consist of less than 20 data elements and require both the employee logical file and the *Active Directory*. Based on the counting rules, this external output classifies as *Average* and worth five IFP.

4) One *External Query* (EQ) process: The process to browse the employee data would be classified as an external query because its primary intent is to present data to the user, data is retrieved both from an ILF and an EIF, and there are none of the following involved: calculations, derived data, ILF update or alteration of system behavior. Based on the number of data elements, (less than 20), and the number of logical files accessed, (the employee ILF and the *Active Directory* EIF), this EQ would be classified as *Average* and worth four IFP.

5) One *External Interface File* (ILF): The active directory provides the unique username and the account list where he or she has access. Upon employee termination, there is a control flag set when accessing its existing accounts, but no accounts are closed or any fields in the *Active Directory* updated. Thus, there is only one RET, rated Low, and it counts for five IFP.

6) Total of components from the above points 1) through 5) equals 33 IFP.

Figure 6-5 shows the IFPUG count for employee data study. The connecting lines represent the FTR connecting transactional function types with data functions types. The count is recorded in the objects representing the transactional function type; the data function type record RET and DET. The IFPUG graphical model is used for explaining functional sizing and presenting results in a way business users understand.

Figure 6-5: Example Functional Size Units Counting Components

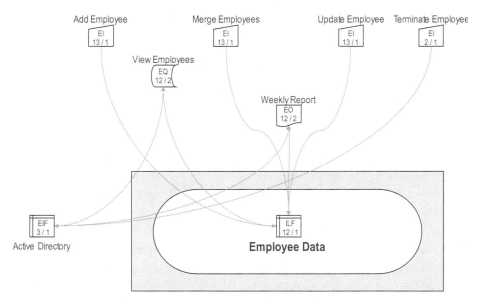

If you assume a PDR of 9.1 PDR (hours/IFP) according ISBSG data base release 10 for *New Development* of database applications (Hill, 2010), then the 33 IFP would equal 300 hours development work, or roundabout 38 days. This total does not include managerial overhead or project management, but serves as a guideline for estimating development effort. However, in today's world of complex ICT, there are more cost drivers than functional size only, and some have become more dominant. We will discuss this in *Chapter 12: Effort Estimation for ICT Projects*.

6-3 ISO/IEC 19761 COSMIC FUNCTIONAL SIZING

The ISO/IEC 19761 COSMIC measurement method originates from the perceived weakness of measuring real time software by the transaction–based ISO/IEC 20926 framework. Its first version was published in 1998, thus more than twenty years later than the IFPUG method, and is now available in the fourth updated version (COSMIC Measurement Practices Committee, 2015).

6-3.1 The COSMIC Functional Size Measurement Method

The principles behind COSMIC are:

- The *Functional User Requirements* (FUR) generate *Functional Processes*. A functional process is "a set of data movements, representing an elementary part of the Functional User Requirements for the software being measured, that is unique within these FUR and that can be defined independently of any other functional process in these FUR…" (COSMIC Measurement Practices Committee, 2015).

- Software manipulates pieces of information, designated as data groups, which consist of data attributes. Consult Figure 6-6 for the data group flow.

- Functional processes involve sub-processes, concerned with movement – Entries (E), eXits (X), Reads (R), and Writes (W) – and transformations of data groups.

- The functional size of a functional process is directly proportional to its number of data movements.

- The functional size of an application is the sum of the sizes of its functional processes.

Therefore, the COSMIC Function Point unit CFP defines a ratio scale and makes functional sizing available for analysis with statistical tools.

Sizing an application with COSMIC involves identifying the functional processes; identifying data groups; and identifying and counting the movements of data groups. There is no upper limit on the data movement count. Each functional process involves at least two data movements: An Entry needed to trigger the functional process, and something happening such as an eXit or a Write. The unit of measurement in COSMIC is equivalent to one single data movement type at the sub-process level.

Figure 6-6: COSMIC Counting Model

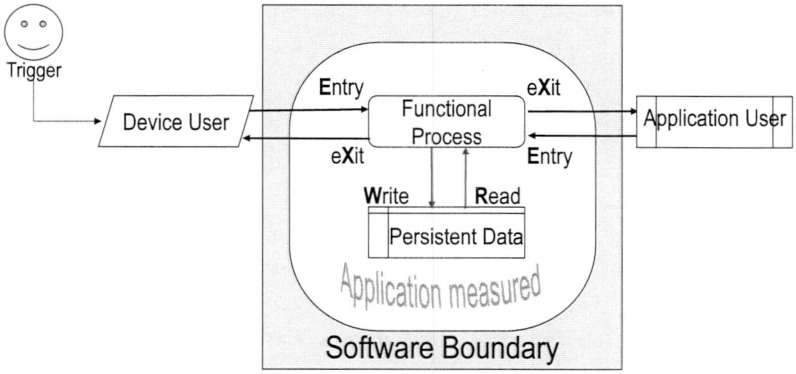

6-3.2 THE COSMIC MEASUREMENT PROCESS

Figure 6-7 shows the relationship of the three phases of the COSMIC method. The indicated chapter numbers correspond to the COSMIC Manual V4.0.1 (COSMIC Measurement Practices Committee, 2015).

Figure 6-7: COSMIC Measurement Process

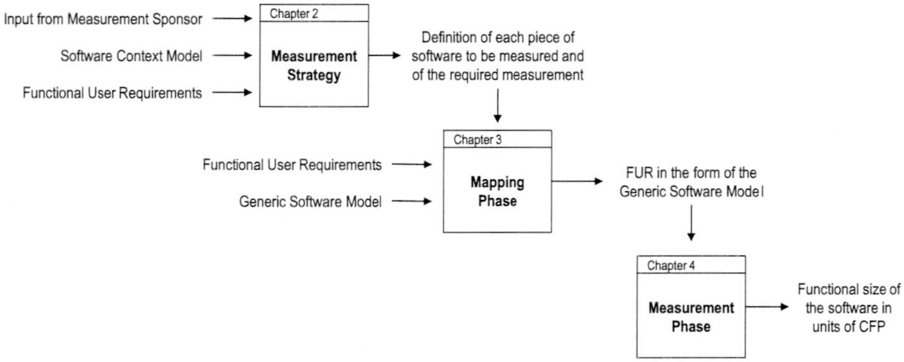

The three COSMIC measurement process phases are:

- The *Measurement Strategy* phase, defining purpose and scope of the measurement. The Software Context Model is then applied so that the software to be measured and the required measurement are unambiguously defined (chapter 2).
- The *Mapping Phase* in which the generic software model is applied to the FUR of the software to be measured to produce the COSMIC model of the software that can be measured (chapter 3)
- The *Measurement Phase*, in which actual sizes are measured (chapter 4).

Readers who wish to learn how to apply the COSMIC method should refer to the manual (COSMIC Measurement Practices Committee, 2015). The following sections explain the principles in the mapping and measurement phase as needed for this book.

6-3.3 THE LEVEL OF GRANULARITY OF THE MEASUREMENT

The *Level of Granularity* is the expansion of the description of a single piece of software (e.g., a statement of its requirements, or a description of the structure of the piece of software) such that at each increased level of expansion, the description of the functionality of the piece of software is at an increased and uniform level of detail.

The *Functional Process Level of Granularity* is the description of a piece of software at which the functional users

- are individual humans or engineered devices or pieces of software (and not any groups of these), AND
- detect single occurrences of events that the piece of software must respond to (and not any level at which groups of events are defined).

With this definition, it is possible to use the COSMIC measurement method for measuring services at the level of granularity that they exhibit, without going in the details, how to implement these services. Precise functional size measurement of a piece of software requires known FUR at a level of granularity that allow identifying its functional processes and data movement sub-processes.

6-3.4 IDENTIFICATION OF APPLICATIONS

A system consists often of more than just one application. It depends from the purpose, from organizational responsibilities, or any other need, to distinguish different deliverables for performance measurement, effort estimation or for software contract purposes (COSMIC Measurement Practices Committee, 2015).

Thus, the split into applications is somewhat arbitrary and pragmatically. It should not affect measurement results except if different levels of granularity make measurements incomparable. The ISO/IEC 20926 IFPUG method uses a similar concept for application splitting.

6-3.5 IDENTIFICATION OF TRIGGERING EVENTS

An event, recognized in the FUR of the software being measured, that causes one or more functional users of the software to generate one or more data groups, each of which will subsequently be moved by a triggering Entry. A triggering event cannot be sub-divided and has either happened or not happened. Note that clock and timing events can be triggering events.

6-3.6 IDENTIFICATION OF FUNCTIONAL PROCESSES

The following three criteria qualify for a *Functional Process*:

- A set of data movements representing an elementary part of the FUR for the software being measured, that is unique within that FUR and that can be defined independently of any other functional process in that FUR.
- A functional process may have only one triggering Entry. Each functional process starts processing on receipt of a data group moved by the triggering Entry data movement of the functional process.

- The set of all data movements of a functional process is the set needed to meet its FUR for all the possible responses to its triggering Entry.

The FUR for a functional process may require one or more other Entries in addition to the triggering Entry.

6-3.7 IDENTIFICATION OF DATA GROUPS

A *Data Group* is a distinct, non-empty, non-ordered and non-redundant set of data attributes where each included data attribute describes a complementary aspect of the same object of interest. A data attribute is the smallest parcel of information, within an identified data group, carrying a meaning from the perspective of the software's functional user. The identification of data attributes is useful but optional in the COSMIC functional sizing measurement method.

A data group is characterised by its persistence, defined as follows: *Persistence* of a data group is a quality describing how long the data group is (to be) retained in the context of the FUR.

Three types of persistence are distinguished:

- *Transient*: The data group does not survive the functional process using it.
- *Short*: The data group survives the functional process using it, but does not survive when the software using it ceases to operate.
- *Indefinite*: The data group survives even when the software using it ceases to operate.

In practice, COSMIC only distinguishes 'Transient' from 'Persistent' (i.e., short or indefinite) data groups. Data persistence is a characteristic used to help distinguish between the four types of sub-processes, as further explained in section 6-3.8 of this chapter.

Once identified, each candidate data group must comply with the following principles:

a) Each identified data group shall be unique and distinguishable through its unique collection of data attributes.
b) Each data group shall be directly related to one object of interest in the software's Functional User Requirements
c) A data group materializes within the computer system supporting the software.

6-3.8 IDENTIFICATION OF DATA MOVEMENTS

A *Data Movement* is a base functional component, which moves a single data group type. It moves one or more data attributes belonging to a single data group. It occurs during the execution of a functional process. There are four sub-types of data movement: *Entry*, *eXit*, *Read* and *Write*, each of which includes specific associated data manipulation. A data movement type is equivalent to an ISO Base Functional Component Type (BFC Type); see (Buglione & Cencel, 2008).

Short definitions of the four sub-types of data movement read according the Measurement Manual Version 4.0 (COSMIC Measurement Practices Committee, 2015) as follows:

1. **Entry (E)**

   An **Entry (E)** shall move a single data group describing a single object of interest from a functional user across the boundary and into the functional process of which the **Entry** forms part. If the input to a functional process comprises more than one data group, each describing a different object of interest, identify one **Entry** for each unique data group in the input.

   An **Entry** data movement includes all data manipulation to enable a data group to be entered by a functional user (e.g., formatting and presentation manipulations) and to be validated.

2. **eXit (X)**

   An **eXit (X)** shall move a single data group describing a single object of interest from the functional process of which the **Exit** forms part across the boundary to a functional user. If the output of a functional process comprises more than one data group, identify one **Exit** for each unique data group in the output.

   An **eXit** data movement includes all data manipulation to create the data attributes of a data group to be output and/or to enable the data group to be output (e.g., formatting and presentation manipulations) and to be routed to the intended functional user.

3. **Read (R)**

   A **Read (R)** shall move a single data group describing a single object of interest from persistent storage to a functional process of which the **Read** forms part. If the functional process must retrieve more than one data group from persistent storage, identify one **Read** for each unique data group that is retrieved.

   A **Read** data movement includes all computation and/or logical processing needed to retrieve a data group from persistent storage.

4. **Write (W)**

   A **Write** (W) shall move a single data group describing a single object of interest from the functional process of which the **Write** forms part to persistent storage. If the functional process must move more than one data group to persistent storage, identify one **Write** for each unique data group that is moved to persistent storage.

   A **Write** data movement includes all computation and/or logical processing to create or to update a data group to be written, or to delete a data group.

All data movements conform to the following four characteristic criteria:

- Do all data attributes and associated data manipulations describe only one object of interest?
- If separate data movements are identified, is there a FUR requiring it?
- Is the data movement moving a data group with its data manipulation unique?
- Is this data movement identified without respect to value-associated behavior?

The notion of "object of interest" is somewhat misleading, as it is not to be confounded with an object in the object-oriented design terminology. The COSMIC manual defines it as "any 'thing' that is identified from the point of view of the Functional User Requirements. It may be any physical thing, as well as any conceptual object or part of a conceptual object in the world of the functional user about which the software is required to process and/or store data" (COSMIC Measurement Practices Committee, 2015). The term is coined to avoid reference to some specific software engineering method.

We will later use that term to identify any object that a functional process exchanges data with, be it a device, another application, or the data store within the system boundary.

6-3.9 APPLYING THE MEASUREMENT FUNCTION

The COSMIC measurement unit, 1 CFP (*Cosmic Function Point*), is the size of one data movement. The COSMIC measurement function lists the identified data movements in a table and counts all of them.

The following principles hold for the aggregation of measurement results:

a) For each functional process, the functional sizes of individual data movements are the aggregation into a single functional size value by summing up the corresponding sizes.

$$Size(Functional\ Process) =$$
$$\sum Size(Entry_i) \ + \sum Size(eXit_i) \ + \sum Size(Read_i) \ + \sum Size(Write_i)$$

b) For any functional process, the functional size of changes to the FUR is the aggregation of the sizes of the corresponding modified data movements for the following formula.

$$Size\big(Change(Functional\ Process)\big) =$$
$$Size(Added\ Data\ Movements) +$$
$$Size(Modified\ Data\ Movements) +$$
$$Size(Deleted\ Data\ Movements)$$

c) The size of any change to a piece of software within a defined scope shall be obtained by aggregating the sizes of all changes to all functional processes for the piece, subject to rules d) and e) below.

d) Sizes of pieces of software or of changes to pieces of software within layers are additive only if measured at the same functional process level of granularity of their FUR.

e) Sizes of pieces of software and/or changes in the sizes of piece of software within any one layer or from different layers are additive only if it makes sense to do so, for the measurement.

Consequently, COSMIC yields linear ratio scale measurements and thus can be used for Lean Six Sigma and statistical purposes.

6-4 ISO/IEC 29881 FiSMA FUNCTIONAL SIZING

The method ISO/IEC 29881:2010 FiSMA 1.1 is intended for use by those persons associated with the acquisition, development, use, support, maintenance, and audit of software. FiSMA 1.1 assesses functional user requirements. It measures the functional size of a piece of software from the perspective of the users. It looks at the software architecture where it identifies 28 different types of *Base Functional Components* (BFC).

6-4.1 CHARACTERISTICS

These BFC services transport data, like the data movements in COSMIC that are the constituents of functional processes. However, different to COSMIC, they receive a weight based on their type: for instance, with interactive input (here: i1) the number of output data presented (n), of writing references (w) and of reading references (r) yields *FiSMA Function Points* (FFP) according the equation (6-1):

$$FFP = m * \left(0.2\ \ + \frac{n}{5} + \frac{w}{1.5} + \frac{r}{2} \right) \tag{6-1}$$

where $m = 1,2,3$ stands for one, two or three possible interaction types of CrUD (Create–Update–Delete), and FFP for FiSMA Function Points.

Figure 6-8: ISO/IEC 29881 FiSMA Function Point Model

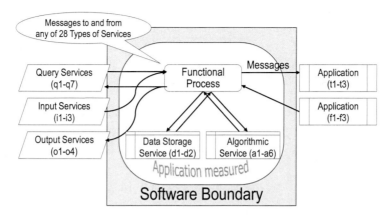

The 28 message types and its corresponding formulas are:

- Interactive end-user navigation and query services (q1-q7)
  - $FFP = 0.2 + n/7 + r/2$, applicable to the following BFC
    - Icon (q1)
    - Login and logout dialogs (q2)
    - Menus (q3)
    - Selection lists (q4)
    - Data inquiries (q5)
    - Report or routine generation dialogs (q6)
    - Browsing lists (q7)
- Interactive end-user input services (i1-i3)
  - $FFP = m * (0.2 + n/5 + w/1.5 + r/2)$, applicable to the following BFC
    - 1-functional input dialog, $m = 1$, supporting only one of the three CrUD operations (i1)
    - 2-functional input dialog, $m = 2$, supporting two out of three CrUD operations (i2)
    - 3-functional input dialog, $m = 3$, supporting all three CrUD operations (i3)
- Non-interactive end-user output services (o1-o4)
  - $FFP = 0.2 + n/5 + r/2$, applicable to the following BFC
    - Output forms (o1)
    - Reports (o2)
    - E-Mails and text messages (o3)
    - Monitor screen output (o4)

- Interface services to other application (t1-t3)
  - $FFP = 0.5 + n/7 + r/2$, applicable to the following BFC
    - Messages to other applications (t1)
    - Batch records to other applications (t2)
    - Signal to devices or other applications (t3)
- Interface services from other applications (f1-f3)
  - $FFP = 0.2 + n/5 + w/1.5 + r/2$, applicable to the following BFC
    - Messages from other applications (f1)
    - Batch records from other applications (f2)
    - Signal from devices or other applications (f3)
- Data storage services (d1-d2)
  - $FFP = 1.5 + n/4$, applicable to the following BFC
    - Entities or classes (d1)
    - Other record types (d2)
- Algorithmic and manipulation services (a1-a6)
  - $FFP = 0.1 + n/5 + r/3$, applicable to the following BFC
    - Security routines (a1)
    - Calculation routines (a2)
    - Simulation routines (a3)
    - Formatting routines (a4)
    - Database cleaning routines (a5)
    - Other manipulation routines (a6)

The sum of all part FFPs constitute the total. The count can rely on an architecture diagram only and is therefore easily automatable; a simple spreadsheet might provide the formulas. FiSMA also yields linear ratio scale measurements and makes the method suitable for statistical purposes.

6-4.2 ADVANTAGES OF ISO/IEC 29881 FISMA

Counting according ISO/IEC 29881 FiSMA is world leading about speed. The software architecture provides all information needed for counting. Automation is possible for every tool that models the software architecture; under the condition that the 28 message types are identified correctly. This is then less a question of the tool as of the qualification and rigorousness of its users.

6-4.3 CHALLENGES

The FiSMA method is widely used in Finland only. Trainings in English are available, in German or Dutch as well; however, the method remains somewhat out of the main stream.

One reason for its limited success might be that software architecture changes so fast. Looking at the definition of ISO/IEC 29881 FISMA, it reflects the kind of software projects implemented in the Zeros of this century. Today's projects involving all kind of mobile device Apps and the Internet of Things look already quite different and need extra interpretation.

The factors in the formulas probably need adjustment from time to time. The introduction clause of ISO/IEC 29881:2010 says, "The correctness of counting parameters and thus the usefulness of a FSM method can be evaluated based on the correlation between functional size and effort under similar environmental and technical circumstances and quality requirements. This kind of evaluation may indicate a need to justify the counting parameters used to derive functional size…" Thus, it is a general, parameterized functional size measurement method for all types of software. It also claims to be valid for all types of systems. Validity is a transfer function from the $\langle n, r, w \rangle$ – parameter count space into effort.

Obviously, calibrating such a method by standardized effort under all existing and coming technologies is a challenge. Six Sigma requires a calibration method that also makes defect counting possible and is reproducible over time.

6-5 HOW TO CHOOSE THE BEST SIZING METHOD

Functional size measurement based on function points counting is different from what the VIM (ISO/IEC Guide 99:2007, 2007) calls a measurement as neither software nor the underlying user requirements are the direct objects of the count. The objects are rather models of the software. Such models exhibit certain aspects of the software, and others remain hidden.

As with a model of a house (Figure 6-9), it is only possible counting what the model exhibits; e.g., doors, windows, roof windows, while other important aspects like heat-

Figure 6-9: House Model

ing, water pipes, isolation, etc. remain hidden. Most architects therefore use more than just one model to explain their building plans. Software engineering is different, as building plans do not help much to elicit requirements. Models, in contrary, do. Model-based design and development has become very popular, especially in embedded software. However, not all models provide the metrics needed for Lean Six Sigma in Software. Only the IFPUG model is popular for counting, usually to predict development effort.

6-5.1 MODEL ACCURACY

A model allows exact, digital counts of the model elements. The question is whether the model is accurately representing the software under scrutiny. Thus, measurement accuracy based on counting model elements refers to the model validity in view of the measurement goals and not to the counting rules. Abran uses the term *Validation* of the model against *Verification* of the measurement context to discuss measurement accuracy (Abran, 2010, p. 34). Counts are something else than measurements.

Every function point counting method states, "defining the goal" of the count as its primary process step of the method. Building the model depends from the stated goal. In addition, the model might not be accurate with respect to this goal. The challenging problem is that the model accuracy is not expressible in the same dimensions as the measurement units themselves. It makes little sense to say that some COSMIC model has "maybe a few data movements more or less". For IFPUG, it sounds more reasonable to take all complexity scores as low, then all as high, to get something resembling a measurement range. Nevertheless, this too is not sound from a metrological point of view; essentially, this means counting an incomplete model, lacking information. This has nothing to do with a lack of precision.

Thus, measuring the model means counting. There is no precision range because the count is digital. Measuring characteristics of software or services should come with a precision indication. The precision depends from how well the model can reflect the true characteristics of the software or services. For instance, counting one data movement to get an authentication from some service provider might be appropriate for cost estimation, if the functional user's viewpoint is to understand what happens. If the programmer needs to setup such an *Application Programming Interface* (API), he might see a bunch of data movements going forth and back, setting up a communication session and establishing a communication protocol. Thus, the measurement precision of the count depends from the measurement goal and the viewpoints of the functional users.

6-5.2 WHICH METHOD IS BEST?

Although functional sizing still is of major impact for software project estimation, cost estimation is not the only reason for functional sizing. Available benchmarking data focuses on effort prediction; still most of them collected with the IFPUG method, although COSMIC is rapidly gaining ground. Without other cost drivers, the correlation between functional size and effort is far from convincing; see Fehlmann & Kranich (Fehlmann & Kranich, 2012-1).

On the other hand, IFPUG is difficult to communicate with modern software engineers. They experience the IFPUG terminology as old-fashioned and cannot map it to how they develop software today. COSMIC performs better in automotive or telecom environments (Bagriyanik, et al., 2014), also for embedded systems in the *Internet of Things* (IoT), while business and enterprise management applications still benefit from the transaction modeling done with IFPUG.

Another difficulty is that the IFPUG transaction map seems awkward when modeling service-oriented architectures. If an EIF used in a system is maintained by another application, it is understood that any acknowledgment for the data received is part of the EI, EO or EQ transaction that use the EIF to some extent. However, if the system wants to tell the other application maintaining the EIF relevant data, for crossing the border, it must use an EO or EQ. The model does not exhibit two-way communication between a system and the services it uses; thus, information exchange remains somehow on a meta-description level.

There is a trend towards automatic functional size measurement based on models such as UML 2.0, or code generation tools such as Simulink; see e.g., (Oriou, 2014). While for COSMIC, automatic counting tools seem widely available, corresponding tools for IFPUG rely on simplified counting rules.

6-6 METROLOGY AND MEASUREMENT ACCURACY

Every measurement has a precision range that tells how reliable the measurement is. In contrary, a count does not have a range of dispersion. Thus, is functional sizing a measurement? Or rather a count, as expressed with the traditional term *Function Point Counting* that some years ago, was used instead of functional sizing? How does functional sizing compare to the standards of metrology?

Transfer functions map a count into some measurement, simply because a count – be it IFPUG, COSMIC, or else – is by itself not a measurement in the sense of the metrology standards. These standards are the International Vocabulary of Metrology (VIM), see (ISO/IEC Guide 99:2007, 2007), and the Guide to the Expression of Uncertainty in Measurement (GUM), (ISO/IEC CD Guide 98-3, 2015), set by the *Bureau International des Poids et Mesures* (BIPM). Every measurement must have its range of variation, indicating the measurement's precision. However, a functional size has no stray area. Although it is precise in the sense (ISO/IEC Guide 99:2007, 2007, p. 25), that quantity values obtained by replicating measurements agree, if the counter does not violate counting rules, the variation remains unknown. Thus, if the sources of counting errors remain unexplored, it is unclear what a count measures.

Albrecht (Albrecht, 1979) positioned the function point counting method as a transfer function mapping EI, EO, EQ, ILF and EIF into something equivalent for business value. However, business value is difficult to measure (Bakalova, 2014). Function point counting became popular because it proved to be equivalent to development effort under conditions that the ISBSG data collection practice describes (Hill, 2010).

On the other hand, COSMIC maps data movement counts into value created by integrating various knowledge gained from data sources and applications. This makes COSMIC attractive for measuring mobile apps and software services (Abran, 2010). The creators of COSMIC looked at embedded systems that nowadays emerged into the Internet of Things. Data movements providing interconnectivity add value for today's web of things, including the investigation and prevention of defects, safety and security flaws created today with software.

6-6.1 THE ROLE OF TRANSFER FUNCTIONS IN MEASUREMENTS

In view of section 6-5.1 the fundamental step of measurement is to define the counting goal. To get the measurement precision right, the measurement goal must exhibit a clear profile, which means the goal is measurable by the relative weights of the topics addressed. Such a profile is a *Goal Profile*; see for instance (Fehlmann & Kranich, 2011-2). It is advisable to rely on more than one count alone. In the traditional application scenarios for functional sizing, the measurement goal is to correctly predict cost of software development projects, see for instance Fehlmann and Kranich's IWSM paper of 2012 (Fehlmann & Kranich, 2012-1). Other goals could be predicting defect density as in the same authors' IWSM paper of 2014 (Fehlmann & Kranich, 2014-1).

In both cases, getting a goal profile is straightforward. The controls are the various counts applied for the project. Measuring the transfer function might be less straightforward, as the impact of each count on the measurement goals might either need consensus among experts (the QFD-way) or a regression analysis among similar projects previously done (the Six Sigma way). Data for such measurements might not be readily available.

6-6.2 A SAMPLE FUNCTIONAL SIZE MEASUREMENT

For this example, consider four different approaches for counting functional size in a project:

- An IFPUG count, to get the number of transactions and interfaces right;
- A COSMIC count, for understanding the data communication architecture;
- Some suitable non-functional sizing count; e.g., SNAP or the Buglione-Trudel Matrix, see *Chapter 12: Effort Estimation for ICT Projects*;
- An estimate for sizing legal constraints. Legal Size is assumed to measure counting regulations and legal constraints, to which the software must be compliant.

Figure 6-10 reflects a sample project having the following three measurement goals.

Figure 6-10: Measurement Goals for Sample Project

Customer's Needs Topics	Attributes			Weight	Profile	
y1 Predict Functionality	Boundary right	Right model	Complete Coverage	**35%**	0.57	
y2 Estimate Quality Effort	Required quality	Legal constraints	Stakeholder value	**48%**	0.77	
y3 Predict Overhead Cost	Project Management	Stakeholder Management	Project Marketing	**17%**	0.28	

The transfer function might look as shown in Figure 6-11 below.

The estimator might have used several functional sizing methods and had confidence into them regarding the measurements goals. That level of confidence depends from the project and may vary. The QFD matrix cells contain the level of confidence expressed by the estimator as correlation values. The measurement accuracy is the convergence gap, here 0.05, or 5% in relation to the unit length of the profiles.

Figure 6-11: Transfer Function for Measurement Accuracy

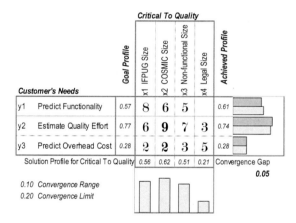

The solution profile combines several functional sizing methods into one measurement; thus, it becomes possible to define a functional sizing unit per transfer function. The counts contribute to measurement precision in specific ways.

Figure 6-12 shows the estimation controls resulting from the transfer function. The impact measurements in this case are expert judgments in the QFD way. The solution profile for counts indicates how much weight each count has concerning the measurement goals. In this case, it refers to the consolidated cost estimation for the whole project. Thus, transfer functions make functional size counts compliant to metrology standards. The convergence gap reflects measurement precision regarding the chosen measurement goals.

The solution profile (The solution profile combines several functional sizing methods into one measurement; thus, it becomes possible to define a functional sizing unit per transfer function. The counts contribute to measurement precision in specific ways.

Figure 6-12) shows the contribution that each of the counts adds to the measurement goals. Thus, measurement precision is not with regard to the count but with regard to the measurement goals; in accordance with the standards for measurement imposed by the *Bureau International des Poids et Mesures*, the VIM (ISO/IEC Guide 99:2007, 2007) and the GUM (ISO/IEC CD Guide 98-3, 2015).

The solution profile combines several functional sizing methods into one measurement; thus, it becomes possible to define a functional sizing unit per transfer function. The counts contribute to measurement precision in specific ways.

Figure 6-12: Contribution Profile for the Various Counts Applied

Critical To Quality Topics	Attributes		Weight	Profile	
x1 IFPUG Size	ISO/IEC 20926	Transactions count	30%	0.56	
x2 COSMIC Size	ISO/IEC 19761	Count functional processes	33%	0.62	
x3 Non-functional Size	SNAP	Buglione-Trudel-Matrix	27%	0.51	
x4 Legal Size	Applicable Laws	Site Constraints	11%	0.21	

6-7 CONCLUSION

Functional sizing according one of the ISO standards is a count of model elements, be it COSMIC functional processes, IFPUG transactions or FiSMA base functional components. The transfer function provides the expected accuracy, based on the confidence assessment by the estimator. Thus, functional sizing yields a measure with confidence interval.

While there is an abundance of software models, the functional sizing models have the unique advantage that there is a defined level of granularity. UML use cases for instance are not fit for saying anything about functional size, because not compliant to ISO/IEC 14143-1:2007. In practice, you can experience all kind of use case specifications. There is no standard. Size measurements with IFPUG, COSMIC, and other compliant models in turn come with some well-defined level of granularity. This makes functional size models fit for use in agile software development.

As we will see, application splitting poses some problems for counts according ISO/IEC 20926 IFPUG, because these counts do not easily add up. The IFPUG scale is not a ratio scale. This affects its suitability for counting small mobile apps and the Internet of Things.

With that caveat, each of the functional sizing methods have its advantages; it might be that someday, or after reading this book, the IFPUG committee also comes up with an alternative ratio scale for measuring the size of the otherwise very valuable IFPUG transaction map. The current low/medium/high once was helpful when automatic sizing was not yet the rule; with automated sizing, complexity of an elementary transaction or function is easily measurable with an interpolated curve.

On the other hand, in view of the role of sizing for other measurement purposes than effort estimation, why not make the complexity measurement just another assessment, probably in view of effort estimation? The *Software Non-functional Assessment Process* (SNAP) already works that way (IFPUG Non-Functional Sizing Standards Committee, April 2013), and the need for something like a *Software Security Assessment Method* (SSecAM) or a *Systems Safety Assessment Method* (SSafAM) is already emerging. Both the IFPUG and the COSMIC model, as well as FiSMA, and others possibly as well, could serve for measuring various aspects of software. The modeling tools for IFPUG and COSMIC transaction maps and data movement maps that were used for this book (and available to the readers for download) are already prepared to accommodate SNAP, security and safety assessments. For the latter two, internationally accepted counting or assessment rules have yet to be defined and agreed.

If this book motivates people finally to stop using non-ratio scale metrics, and make their counts suitable for statistical evaluation, and functional size measurement conformant to the metrology standards, then this discipline would make the most prominent step forward for this decennium at least. The metrology standards are the GUM and the VIM (ISO/IEC CD Guide 98-3, 2015) and (ISO/IEC Guide 99:2007, 2007). Unfortunately, SNAP also uses the low/medium/high assessment type, making SNAP points as uncompliant to the metrology standards as the traditional IFPUG function points. Only ISO/IEC 19761 COSMIC is fully compliant. Thus, there is room for innovation in the software measurement community.

CHAPTER 7: THE MODERN ART OF DEVELOPING SOFTWARE

The paradigm inherited from traditional software projects – count size initially, possibly mid-term, and at the end of development – is not suitable for agile development since each time a software artifact has been completed, functional size measurements must be updated and actualized, see Cohn (Cohn, 2005). Multi-layered software works in systems involving people, devices, networks, and computers and meets high levels of security standards.

Developing software is iterative or even agile – in both cases it is hard to predict how a system looks when it is completed because the system's (commercial) success depends heavily from quickly changing market demands. Developing according a plan is becoming the exception – usually, a vision must do. The requirements grow as the system is growing, and during development, they become understood better and better by its stakeholders.

7-1 USING UML 2.0 SEQUENCE DIAGRAMS IN AGILE

The concept of data movement is very close to the concept of messages between entities – classes, objects, data stores – in UML™ (Ambler, 2004). Data groups, in turn, resemble objects. Thus, why not counting software based on *Sequence Diagrams?* Sequence diagrams are an UML™ notation for communicating dynamic behavior of software among stakeholders. According Donald Bell, "the sequence diagram is used primarily to show the interactions between objects in the sequential order that those interactions occur. ...An organization's business staff can find sequence diagrams useful to communicate how the business currently works by showing how various business objects interact (Bell, D., 2004)."

Drafting sequence diagrams is mainly a matter of available media. In most cases, whiteboards and walls simply are not big enough, and good design tools are expensive and not easily available. Scott W. Ambler recommends using sequence diagrams foremost of all other UML artifacts in agile software development teams (Ambler, 2004). If available, the team appreciates them for facilitating mutual understanding with the sponsor and the users, and spotting design weaknesses early.

Sequence diagrams allow the development team to communicate effectively with business people, adopting the end user's viewpoint to look how data moves through the system. Implementation details can clarify what might affect quality features of the finished software product such as performance. Many teams use them to identify risks for performance bottlenecks, especially in real-time software applications.

In agile development, the *User Story* is the format of choice to communicate FUR. We use the term user story in such context, although user stories do not necessarily refer to functional requirements; non-functional requirements might give rise to user stories as well. Agile teams use user stories to reflect the customer's needs in a convenient way. Thus, when creating the sequence diagram for a user story, tasks needed for implementing the user story become immediately apparent, see section *7-1.3 : Sequence Diagramming in User Stories*. Tasks derived from user stories constitute *Story Cards*. Story cards organize work in agile development among developers. The functional story cards correspond to the messages – or data movements – that send entities, or data groups, among objects. Thus, the effort to create a sequence diagram is awarded not only by a better common understanding of the story, but also by an indisputable decomposition of a user story into story cards, at least for the functional part. Other story cards describe tasks that are not functional, such as defining a database schema, or making a user interface attractive. The latter might end up with additional functionality as well; however, the original task is not specifying any specific functionality.

With story cards, you can use Kanban to manage software development. Kanban is not agile but the pull principle is more effective than the usual push because it avoids idling and motivates developers to finish a story card to get attractive new cards.

For the non-functional part of user story implementation, the sequence diagrams also give hints; e.g., for identifying critical entities for performance, such as data sinks being overwhelmed with data for storing after some high-volume data movements (Fehlmann & Kranich, 2011-1).

7-1.1 Sizing Sequence Diagrams

UML™ 2.0 sequence diagrams identify class objects that occur as data groups in functional processes, and messages that implement data movements. Levesque et. al. (Levesque, et al., 2008) propose that each *Use Case* is a functional process. The actors of the use case are the users. The entities of the sequence diagram are the data groups. The data movements correspond to the messages among the entities of the sequence diagram. The authors claim that results of this approach compared to expert counts exhibit a difference of 8% only. See Marín (Marín, et al., 2008) for comparisons with various other sizing methods for conceptual models.

The problem with *UML Use Cases* is that their granularity is not well defined. Most often, use cases host several functional processes serving several different functional users. Some teams use the term *Use Case* concurrently with *Epics*. Epics are the initial form of user stories that finally will evolve into a collection of user stories. This better corresponds with the author's experiences. Throughout this book, we will use the term *Use Case* concurrently with *Epics*, not following Levesque et. al.

Identifying data movements between data groups according the ISO/IEC 19761 international standard is a rather small additional effort when drafting sequence diagrams. However, not all classes identified by an object-oriented design appear in functional processes, because some classes do not directly implement FUR but rather respond to technical requirements, representing viewpoints usually not shared with users. The distinction between classes appearing in functional processes, and "technical" entities, is not a problem in practice; it refers to the level of granularity explained in section 6-3.3 for IFPUG models. Further guidance is available in the COSMIC reference material (COSMIC Measurement Practices Committee, 2015).

For sizing UML sequence diagrams, it is useful to use the paradigm of a map, like the transaction maps introduced in 6-2.4 for the IFPUG graphical models. Maps allow a quick overview of software or systems without unnecessary details. *Data Movement Maps* resemble UML sequence diagrams. Missing are

- Option combination fragments (combined fragments in UML 2.0)
- Optional return messages (if data is moved, they are not optional in COSMIC)
- Sending messages to itself – used in UML sequence diagrams to draw attention to the fact that the object's life line is a functional process indeed.

An UML sequence diagram is no COSMIC count in itself – functional processes need being identified among the UML objects, and persistent stores and devices as well. Data movement maps do this, and allow rapid COSMIC counts by mapping the objects and classes involved in the UML use case scenario and the sequence of messages exchanged onto COSMIC model elements. Thus, data movement maps characterize the functionality of the scenario.

7-1.2 EMBRACING CHANGE WITH USER STORIES

Ambler (Ambler, 2004) points out "agile teams will identify new stories during construction iterations, split existing stories when you realize that they're too large to be implemented in single iteration, re-prioritize existing stories, or remove stories that are no longer considered to be in scope. The point is that stories evolve over time just like other types of requirements models evolve." Sales and support people may identify further enhancement requests during the production phase and then forward them to a development team as they are working on an upcoming release.

7-1.3 SEQUENCE DIAGRAMMING IN USER STORIES

Jenner (Jenner, 2011) proposed functional sizing of UML diagrams in view of automatic function point counting based on ISO/IEC 19761 COSMIC. In Agile, the recommended practice is drawing UML sequence diagrams (Marín, et al., 2008) on the back – or attached – to story cards, as the late Grant Rule proposed 2010 in a note, see (Fagg & Rule, 2010). Diagrams create a common understanding and allow identifying risks with the planned software at an early stage. Although they did not propose data movement maps, the graphical representation proposed in Figure 7-1 is close enough.

Figure 7-1: User Story with FSM according ISO/IEC 19761, cit. (Fagg & Rule, 2010)

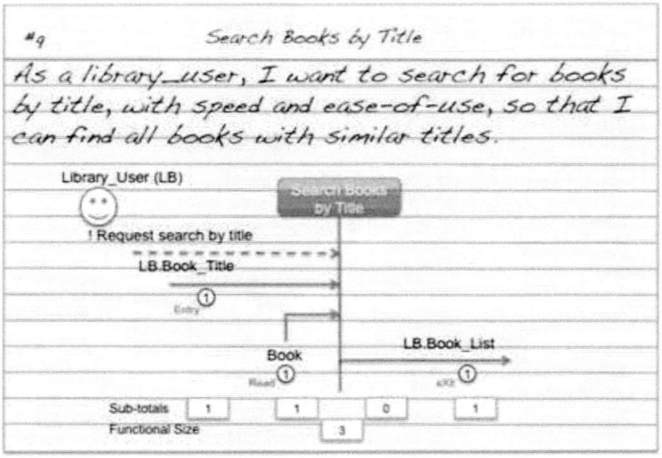

Rule and Fagg presented this sample story card in their famous paper mentioned above; shown in Figure 7-1. The change needed to integrate functional size measurement into user stories is to adapt the usual template (Schwaber & Beedle, 2002) somewhat, by relating its structure to the international standard ISO/IEC 19761 for functional sizing.

In addition, they proposed to optionally specify required qualities, because non-functional quality requirements relate most often to some specific additional functionality; conformant to IFPUG's Software Non-functional Assessment Process SNAP (IFPUG Non-Functional Sizing Standards Committee, April 2013).

Instead of the textbook template (Cohn, 2005):

As a … [stakeholder role] …
I want to … [perform an action / record some information] …
[With some frequency and/or quality characteristic] …
So that … [description of value or benefit is achieved].

The template for user stories reads according Grant Rule (Fagg & Rule, 2010):

> As a … [functional user] …
> I want to … [respond to an event / retrieve some data] …
> [With some frequency and/or quality characteristic] …
> So that … [description of value or benefit is achieved].

With just these small adaptations, agile teams adopt the international standard ISO/IEC 19761 using a sizing measurement method that yields comparable results across time, across teams, across projects, and across organizations. Developers, product owners, and their customers thus have a low-cost, effective tool to use when estimating, tracking progress, and assessing value-for-money. We refer to user stories stated in the way above as "in the Grant Rule format". The frequency/quality characteristics part is optional. We always use this format for user stories.

Note that the COSMIC guideline (Buglione, et al., 2011) only proposes three parts: "As a <user type>, (part I) – I want to <feature or functionality>, (part II) – so that <value or expected benefit>" (part III), not mentioning the optional quality aspects.

Figure 7-2: Search Book by Title as Data Movement Map

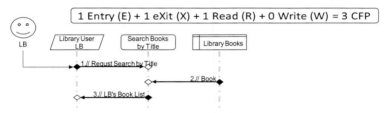

Figure 7-2 shows the User Story as a data movement map with the three data movements mentioned, with the functional process "Search Books by Title", as above, and with a device object "Library User LB" and a persistent store "Library Books". Contrary to Figure 7-1, the diagram identifies also the other two objects and shows how they interact with the rest of the system – for instance with establishing the book inventory, not shown here – but does not restrict the functional object to one single functional process.

7-1.4 FUNCTIONAL SIZING AND SOFTWARE PROJECT MANAGEMENT

According the guideline mentioned, a well-written user story describes one functional process in COSMIC. For the programmers, implementing this user story from Figure 7-1 might induce seven tasks. Firstly, there are three elementary data movements: 1) Enter book title, 2) Read from data store containing existing books, and 3) Exit with book list; secondly, four quality-related tasks: 4) Include subtitle search, 5) Speedy search, 6) Forgiving grammar checks, and 7) Pattern matching search. We might write up to seven story cards for this user story.

As demonstrated by this small example, functional sizing models also yield work breakdown structures, consisting of functional and non-functional tasks. The model used for functional sizing allows identifying work items and make them available for the agile team as story cards. The more business-oriented approach of IFPUG also yields a work breakdown structure; however, the rather technically oriented approach of COSMIC better describes what developers must do. Thus, both approaches contribute to software project management.

Another aspect seems also typical: the four tasks referring to non-functional requirements outnumber the three functional tasks. Moreover, what is even worse, the efforts for the non-functional tasks might be considerably higher than for the functional tasks, where a suitable development framework might speed up development and lower effort needed.

7-2 COUNTING DATA MOVEMENT MAPS WITH ISO/IEC 19761

This section is about automatic counting of sequence diagrams. As some of the counting rules are highly context-sensitive, any automatic count are approximations, -because context is usually non-automatable, see, e.g., Fehlmann and Kranich (Fehlmann & Kranich, 2011-1), and Jenner (Jenner, 2011).

7-2.1 OBJECTS OF INTEREST

Counting with ISO/IEC 19761 becomes simple and straightforward, once you use data movement maps. A few precautions are necessary:

- UML objects with lifelines represent objects executing functional processes, or hosting persistent storage, and various devices;
- Functional processes are executed by UML *Functional Objects* only;
- Devices include (human or machine) device users and peer or layered applications;
- Data movements always touch UML functional objects; either in- or outbound;
- One functional object in UML may host more than one COSMIC functional process.

Consequently, you cannot connect devices directly with storage, as you might do in UML. Any read or write always includes the functional process that is in control. This limitation is not a real one; it just means splitting such data movements into two, adding a little bit more rigor to the data movement maps. No device ever can read or write directly into a store without going through a functional process, be the data store as simple as it might be.

7-2.2 FUNCTIONAL PROCESSES

This last precaution above, namely that one UML object might represent several functional processes in COSMIC, holds if the UML object executes more than one user story in a sequence. In the data movement map, the UML functional object appears only once with one lifeline only, even if executing several distinct functional processes according ISO/IEC 19761. It is tempting to identify functional UML objects with use cases or epics – however, this too is also not an established standard in software engineering.

A typical case is an UML object representing distributed intelligent things; e.g., a smartphone central processor unit executing more than one functional process, or a sensor both capturing and consolidating data. One single UML functional process can host all smartphone processes. However, since each functional process executed by some functional object needs an extra trigger to start, multiple functional processes on the same lifeline are easy to distinguish. Thus, we do not enforce separate functional objects per functional process in data movement maps.

According COSMIC 4.0, such triggers might originate from a functional user; however, they also might originate from a system clock or from reading a counter in persistent storage indicating for instance that after a certain number of times an event must occur. If credentials are to be re-assessed from a human user every 50 times only, assuming the human user did not change in between, then the trigger is a Read from the counting storage triggering a conditional functional process to run.

7-2.3 TRIGGERS

The ISO/IEC 19761 standard requires a trigger for every functional process. UML does not have triggers. COSMIC avoids counting internal processing that do not respond to a FUR, even if data movement maps might display such internal processing. Thus, if a data movement map displays internal processing, it has no trigger, and therefore this should not count for the functional size.

In addition, data movement maps allow for loops: data movements whose origin and destination are the functional process itself. Such loops show internal processing, such as sorting before showing to a device user. An UML object with loops likely represents functional processes. The COSMIC counting rules do not count internal processing while in the data movement map they have a meaning and are fine.

7-2.4 DATA GROUPS

ISO/IEC 19761 requires that a data movement moves data groups, and if there is more than one data movement moving the same group to the same functional process, it counts only once. Since ISO/IEC 19761 rules do not distinguish between different persistent data store objects, as a data movement map does, a UML diagram shows data movements often twice or more when fetching similar data moved as the same data group. Such multiple moves, perfectly legal in data movement maps, must not be counted as two data movements in ISO/IEC 19761; the rules require that the data group moved is unique. Therefore, if you move the same data group to two different persistent data stores in a data movement map, still only one data movement adds to the ISO/IEC 19761 count.

The problem rarely pops up in practice. In data movement maps, a persistent store object that uses redundant storage is but one store. Creating two identical data stores in a single functional process would model implementation details, not functionality, and violate the 'modularity' principle when drawing data movement maps. Thus, it is safe to assume that different data stores in data movement maps accommodate different data groups, making the collection of all persistent storage objects in data movement maps representing persistent store in ISO/IEC 19761, related to the **R**eads and **W**rites in a functional process.

7-2.5 PERSISTENT STORE

ISO/IEC 19761 does not name the store associated to some functional process; it names data groups instead. Data groups are implementation independent. However, an UML diagram identifies storage as objects and uses them to move data between functional processes. Different storage objects normally accommodate different data groups but not all storage objects are persistent.

This was a difficulty with COSMIC versions prior to 4.0 but now is no longer.

7-3 THE EMPLOYEE DATABASE SIZED WITH ISO/IEC 19761

The following count uses the same sample case as in section 6-2.5 : A Functional Sizing Case Study. ISO/IEC 19761 and ISO/IEC 20926 counts do not differ much if they refer to a transactional application.

7-3.1 THE PURPOSE

The purpose for the count is to explain the ISO/IEC 19761 counting principles, for comparison with ISO/IEC 20926.

7-3.2 THE SCOPE

The application layer view shown in Figure 7-3 looks like the ISO/IEC 20926 count shown in Figure 6-5; however, the transactions convert to functional processes.

There is but one functional user: the HR clerk that has access to all employee data within the organization. This layer does not update the active directory; only upon termination, a control flag indicates that the IT department must close accounts upon termination date.

Figure 7-3: Scope of the Employee Database Count: Application Layer

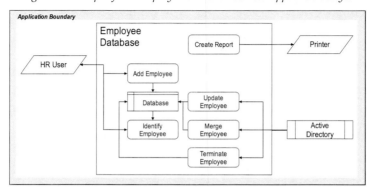

Figure 7-3 shows the application overview. The big olive box represents the application. Devices are placed to the left, usually, and other application to the right. The application box includes functional processes and persistent storages belonging to them. They are not always shown explicitly; here for instructional purposes only.

7-3.3 FUNCTIONAL USER REQUIREMENTS

The functional user requirements (FUR) from section 6-2.5 read when stated in the *Grant Rule* format as shown in Table 7-4 below. Presenting requirements this way, in the form as user stories, supports requirement elicitation. This means that making the requestor explicit – here the Human Resource Department clerk only – and separating functional and quality requirements clearly helps identifying the full functionality needed. several quality attributes call for extra functionality reflected in both the IFPUG and the COSMIC model.

More on requirements elicitation follows in *Chapter 9: Requirements Elicitation*.

Table 7-4: Functional User Requirements recorded as User Stories in the Grant Rule Format

	Label	As a ... [Functional User]	I want to ... [get something done]	Such that ... [quality characteristic]	So that ... [value or benefit].
1)	Add Employee	HR clerk	to store and maintain employees	the following data fields can be retrieved: name, number, rank, street address, city, state, post code, date of birth, phone number, office assigned, date the employee data was last modified	I always have control what employees we currently care for
2)	Update Employee	HR clerk	update employee information in case anything changes	I have the actual employee data and keep track of marital status and salary updates	I know their rights
3)	Terminate Employee	HR clerk	terminate employees	access to IT is blocked immediately without losing previous data	I can retrieve information later if needed
4)	Merge Employees	HR clerk	can detect and correct doubletons	both data sets get merged	I do not keep redundant information
5)	Browse Employees	HR clerk	can answer questions quickly	my reputation as being responsiveness increases	I keep data under control
6)	Weekly Report	HR clerk	get a weekly work report on my output device	such that a summary of all weekly transactions is available	I can manually check against weekly salary payments whether all changes had been reflected

7-3.4 FUNCTIONAL PROCESSES

Among the ten objects of interest, we find six functional processes:

- Add Employee – Adds a new employee to the data base
- Update Employee – Changes some data about an employee
- Terminate Employee – Marks an employee record as referring to an ex-employee
- Merge Employees – Corrects a doubleton identified by merging data
- Browse Employees – Creates a list of all employees selecting them by wildcard
- Create Weekly Report – Lists all employees changed during the week that passed

See the Figure 7-3 for an overview. Each functional process has its own functional object.

7-3.5 DATA GROUPS

There are five data groups:

- Employee Data – Name, ID, rank, address, city, state, post code, date of birth, phone number, office assigned, date the employee data was last modified; termination date
- Account Data – Username and list of applications that user has access to
- Employee ID – Name, or ID, identifying an employee entry in the data base
- Date & Time – Start and end date for weekly reports
- Weekly Report Data – Contains employee and account data

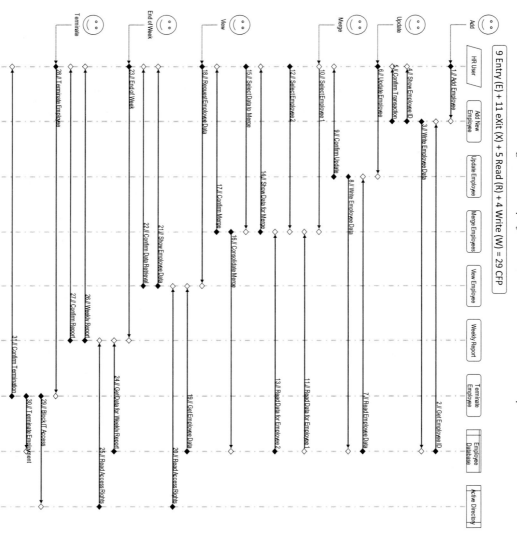

Figure 7-5: Employee Database as a Data Movement Map

7-3.6 Data Movements

According COSMIC, 29 data movements count as CFPs; additionally, one data movement accesses data for merging employees that does not add to the COSMIC count.

The data movement map makes clear that for transactional application, an ISO/IEC 19761 count is significantly more complicated than the ISO/IEC 20926 model. It also requires more detail: for instance, in ISO/IEC 20926 it is irrelevant how an employee terminates. With ISO/IEC 19761, it becomes an issue, because termination, as shown in the ISO/IEC 20926 model, could be included in the updating an employee functional process if it just means setting a termination flag (which is equivalent to deleting a record in the employee database). The system must communicate termination to the Active Directory. This makes clear what the advantage of an ISO/IEC 19761 count is. It uncovers technical requirements that otherwise might go forgotten; see Fehlmann (Fehlmann, 2011), and Trudel (Trudel, 2012).

The total count is 29 CFP, compared with 33 IFP according IFPUG 4.3.1. This corresponds to the observation that ISO/IEC 19761 counts tend to be like ISO/IEC 20926 counts for transactional applications. However, this is not at all true for real-time projects, mobile services, or embedded systems.

7-4 The Sphygmomanometer Case

The following application examples are theoretical in nature and aim at repulsing two common misconceptions:

- There is no generally applicable conversion factor between ISO/IEC 19761 COSMIC and ISO/IEC 20926 IFPUG counts;
- Functional size is a cost driver among others – others might dominate modern service-oriented applications;

and a conjecture:

- ISO/IEC 19761 COSMIC is better suited to model, and thus size, functionality of modern software services, especially on mobile platform and appliances belonging to the Internet of Things.

There has been some very interesting work on conversion factors between the COSMIC and the IFPUG method. However, none looks convincing, as both methods model software differently and consequently measure different things. The most interesting contribution doubtless has been Çiğdem Gencel's and Carl Bideau's preliminary findings presented at the IWSM 2012 in Assisi (Gencel & Bideau, 2012).

7-4.1 The Sphygmomanometer – a Medical Instrument

The Sphygmomanometer is a commonly used medical instrument, stand-alone, for measuring blood pressure. It helps understanding the challenges when counting mobile apps.

A Sphygmomanometer consists of a sleeve that can be pumped full of air, a measurement instrument with pulse sensor and valve for evacuating the sleeve. The blood circulation is blocked up to the extent that pulse no longer is sensed (systolic compression), then the pressure is release until it can no longer be heard (diastolic compression). These values converted into mmHg yield the blood pressure. The instrument is targeted for home use and the sample sphygmomanometer shown in Figure 7-6 is not connected to the Internet; it even lacks any data connection. It is autonomous indeed.

Figure 7-6: Sample Sphygmomanometer

A close inspection of the instrument shows three different functional users: The patient measuring its blood pressure; the doctor interested in result series, and the instrument itself that needs to know the time zone and expected time format for stamping the results with date and time.

The instrument has an internal clock as all computers do; however, the time zone and the time format is configurable by the user, since Internet time is inaccessible. Thus, there are three applications in a Sphygmomanometer:

- A measurement unit, with air pump, pressure sensor, pulse listener, motor pump and air valve;
- A data collection unit that collects measurement series and calculates average pulse, SYS and DIA pressure;
- An internal clock that remembers date and time.

The three applications work independently from each other and sizing should be separate according both ISO/IEC 20962 IFPUG and ISO/IEC 19761 COSMIC.

7-4.2 Functional Size according ISO/IEC 20926 IFPUG

The first application acquires data from the "patient", which is the main purpose of the Sphygmomanometer.

It allows starting the measurement process, sensing the pulse, deleting data – otherwise, if the "patient" does not delete the measurement data, it accepts and keeps the data for storage and for accumulation with other measurement data. The time zone is maintained externally. Battery status is an external output (EO) without referencing anything relying only on the status of the sphygmomanometer.

An interesting point in Figure 7-7 is whether the acquired measurement data is persistent and maintained by the measurement equipment.

Figure 7-7: Acquire Data for Sphygmomanometer – Patient's View

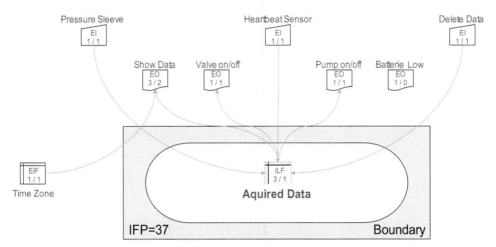

There is an indication that it is an ILF indeed: pressing the on/off button cancels the acquired measurement data; otherwise, the data recording application (Figure 7-8) includes it for processing the arithmetic mean of all prior measurements. Moreover, the patient expects the measurement results (i.e., the diastolic and the systolic value) being remembered by the instrument for further processing. This is an explicit functional user requirement. Thus, it does not matter that usually there is no hard disk available in such a device – a battery powered solid-state memory will do because implementation details do not matter for the count. Such a configuration is widely used in instruments, and sizing must consider it.

Less astonishing is the recording data part for the "doctor" – who can be the same physical person as the "patient" – using the acquired data for collecting and calculating the mean; see Figure 7-8. There is a calculated output (EO) – the mean among all measurements – and a browsing transaction (EQ) that allows viewing prior measurements. Two transactions maintain the internally stored measurement data (ILF) – an EI for saving results and an EI for deleting certain measurements; e.g., outliers.

Figure 7-8: Data Recording in Sphygmomanometer – Doctor's View

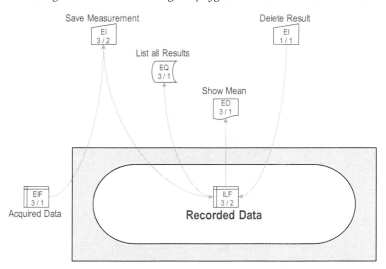

Setting the time zone (Figure 7-9) is simple – still, it adds 14 IFP to the total count. This is obviously too much in relation to the previous two application counts. An estimation based on average ISBSG data (Hill, 2010) would yield something like 19 days' effort for implementing the time zone setting; this is exaggerated even if re-utilizing some existing software is impossible.

Figure 7-9: Setting Time and Time Zone in the Sphygmomanometer

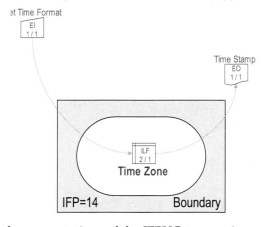

While the graphical representations of the IFPUG transaction map yield value, there is a problem with the size. The total size of all three applications is 37 IFP + 25 IFP + 14 IFP = 76 IFP; if counting all three applications as a single apple-cation – as would be compulsory with IFPUG 4.2 and earlier – the result would be 66 IFP. This is a significant difference. The reason is that certain elementary transactions such as the time zone appear once as an ILF and then again as an EIF in another application. The IFPUG functional size is not additive when combining applications. Moreover, due to the complexity factors, measurements are not on a ratio scale; they hop from one to another level. This is a difference felt quite a bit for small, mobile applications.

7-4.3 FUNCTIONAL SIZE ACCORDING ISO/IEC 19761 COSMIC

The sphygmomanometer consists of the same three applications – *Acquire Data, Record Data* and *Set Time Zone* – for ISO/IEC 19761 COSMIC as well. Contrary to the ISO 20926 IFPUG approach, the split in applications does not affect the total functional size count. The application that acquires data contains three functional processes, one for measurement initiation and battery status control, one for the measurement process including pump and valves, and the third one for accepting or rejecting the measurement result for inclusion in the average calculation. In addition, the doctor's view application also has two independent functional processes, both triggered by separate triggers. One is for calculating and presenting the average measurement, the other is for removing obsolete data from the calculation of the average.

In the data movement map (Figure 7-11) the *Set Time Zone* application is on top with one trigger, followed by the patient's view with three triggers and the doctor's view with two triggers, corresponding to the functional processes.

Figure 7-10; Sphygmomanometer Overview with Three Applications

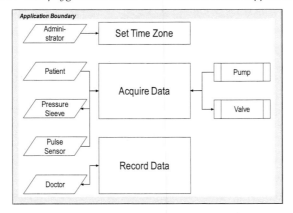

There are six functional processes in total for all three sphygmomanometer applications. For data movement maps, it does not matter whether take all three applications together or split them, since functional processes cannot communicate directly without an intermediary data store, or device.

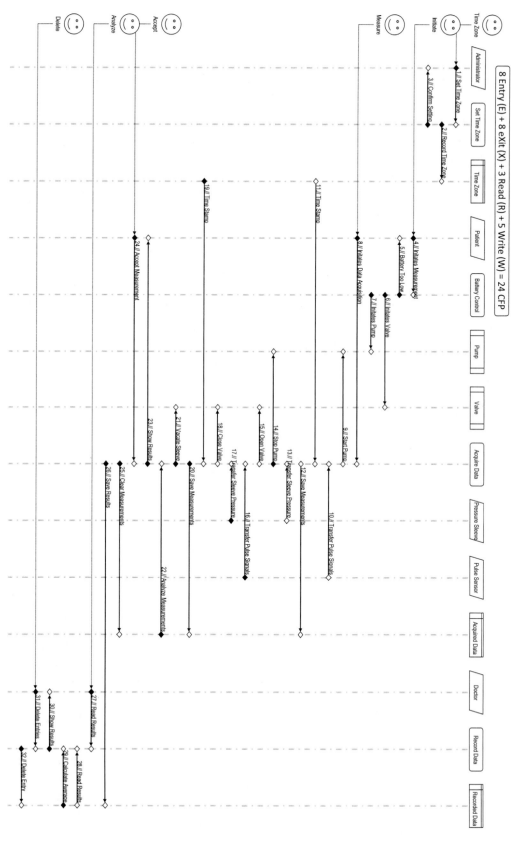

Figure 7-11: Three Sphygmomanometer Applications in One Data Movement Map

8 Entry (E) + 8 eXit (X) + 3 Read (R) + 5 Write (W) = 24 CFP

7-4.4 COMPARING THE COSMIC AND THE IFPUG APPROACH

The sphygmomanometer case shows that both approaches have their value. The IFPUG transaction map shows the structure of the applications residing in the sphygmomanometer instrument clearly; also, distinguishes reusable elements such as setting the time zone – equally for all isolated instruments – or the data average calculation and reporting feature from the data acquisition application, which is specific to the sphygmomanometer.

The IFPUG count in turn is exaggerated and near to useless; the data movement map model better reflects how the six functional processes of the sphygmomanometer collaborate. The COSMIC count works well for comparisons with other, more complex sphygmomanometer instruments that possibly interface with the Internet or other electronic devices.

7-5 THE WEB TICKET SHOP CASE

Another example that is more complicated but very common and known to almost everyone using public transportation is a web shop for purchasing travel tickets. In Europe, all railways operate such ticket shops on the Internet with the aim of saving sales channel cost by closing the traditional ticket tellers at stations. However, each railway has invented its own way of selling and validating travel tickets over the Internet; thus, customers hardly adapt to this new sales channel. While ticket automates serve the same purpose, they are much more expensive to build and operate; nevertheless, customers still prefer them because they change less often than a web portal.

7-5.1 THE TICKET SHOP AS AN IFPUG TRANSACTION MAP

The system consists of two application, one for selecting the ticket and another for executing payments. The payment application is independent from the ticket shop and is useful for any other shop or payment purpose as well.

The ticket sale application in Figure 7-12 relies on a few external interfaces, including timetable, station list, and the ticket shop – an application service that is located somewhere in the IT infrastructure of our travel company, servicing the ticket shop portal on the web and many other ticket sales channels.

The time-out elementary process does not relate to any of the internal or external files – it is a system clock triggering a time out in case the user abandons the ticket purchase without logging out or explicitly canceling the process.

The portal lacks many features that users expect, such as remembering the travel habits of the purchaser. Since there is a login EQ, tracking the users' habit would allow proposing her or his frequent choices – choices not reflected in the model, for the sake of simplicity. Not considered – out of counting scope – are the authentication and transaction parts of the payment service that likely go over international credit card exchange services, the marketing department, accounting, and more. The *Selected Ticket* as a means of transporting routing information from one application into another. The *Payment Order* is a separate EO, targeted towards the payment application.

Figure 7-12: Transaction Map for the Ticket Sale Application

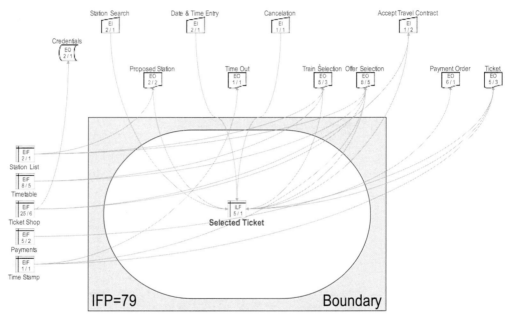

The IFPUG count for the ticket sale application yields 79 IFP, which seems high for this simple and basic ticket shop; the payment application shows 32 IFP, not reflecting the high complexity when setting up such an application. Note that the *Selected Ticket* ILF of the ticket sale application now appears as an EIF in the ticket payment application. Consequently, when drawing the boundary different and looking at the ticket app as a single application, the size is not exactly the sum of the sale and payment apps, namely 77 IFP only.

The payment part looks shows basic functionality only, see Figure 7-13:

Figure 7-13: Transaction Map for the Payment Application

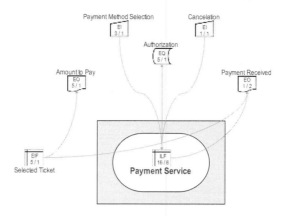

7-5.2 THE TICKET SHOP AS A DATA MOVEMENT MAP

The ticket shop in the data movement map shows more technical details how data moves through the two shop applications; however, the full representation of all data movements is probably not needed to understand how this works.

The overview in Figure 7-14 shows how the two applications interact.

Figure 7-14 Ticket Sale & Payment Applications Overview

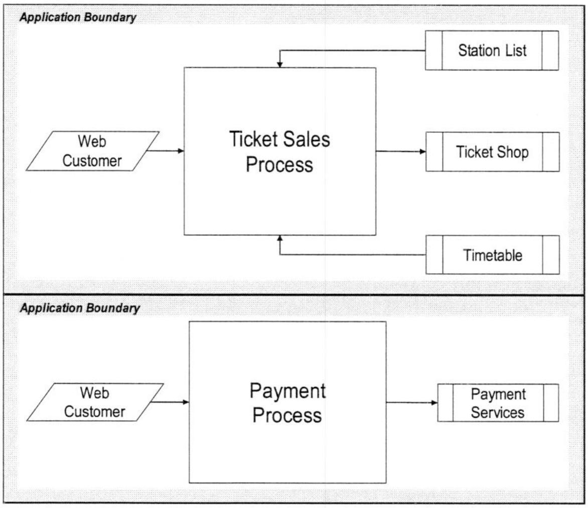

The data movement maps show significant less function points than the IFPUG counts, better reflecting the basic functionality modeled.

Figure 7-15: Data Movement Map for the Payment Process for Selling Tickets

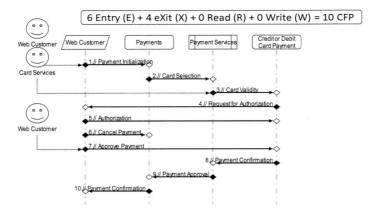

Figure 7-15 shows the payment process; compare with Figure 7-13.

Figure 7-16: Data Movement Map for the Ticket Sale Application (Extract)

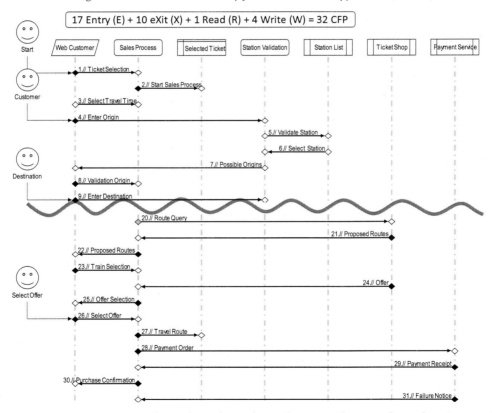

Figure 7-16 is an extract from the ticket sale application, showing how data is routed for providing the basic functionality as needed.

The total functional size is $10 + 32 = 43$ CFP; significantly less than with IFPUG. However, when looking at the data movements, it becomes clear that establishing those data movements can become quite costly; while getting data from the ticket shop and the station list might prove easy within the IT of the travel service provider, connecting to the payment service might induce more than just a few technical challenges. Security and privacy might affect total cost and effort quite a bit, and this is perfectly visible in the data movement map.

Nevertheless, the IFPUG transaction map is friendlier to humans with limited technical skills; while the IFPUG count is near to meaningless for functional sizing.

7-5.3 THE FAST TICKET APPS

A little application thought as a *Ticket App* best explain the concepts. The app shall allow regular customer in a hurry to buy tickets quickly for trains or buses taking them home without fiddling with ticket machines or a teller – especially useful for late home-comers when the latter are mostly closed at late evenings and the former typically out of order when needed. Customers might voice their needs as "Give me a ticket subito! Without much ado and questions if I simply want to go home".

The app is not providing full ticket shop services; only a predefined, limited set of destination stations can be used, and the boarding station is selected automatically based on the GPS services of their smartphone. The app runs on smartphones and tables.

7-5.3.1 USER STORIES

Using suitable methods for the goal profile (see e.g., *Chapter 5: Voice of the Customer by Net Promoter®*), user stories read as follows:

Table 7-17: Functional User Requirements recorded as User Stories in the Grant Rule Format

	Label	As a ... [Functional User]	I want to ... [get something done]	Such that ... [quality characteristic]	So that ... [value or benefit].
1)	Prepare Destinations	Traveler	can store my preferred destinations	they are valid for the Ticket Shop	acquiring tickets becomes easy
2)	Find Route	Traveler	locate nearest station with GPS	that's being served right now	I immediately can see whether it's right

	Label	As a ... [Functional User]	I want to ... [get something done]	Such that ... [quality characteristic]	So that ... [value or benefit].
3)	Authenti-cation	Transportation Service Provider	give user access to their preferred payment options	all payments are traceable in Ticket Shop	they can manage spending
4)	Issue Ticket	Traveler	get a valid ticket	with settings from Ticket Shop	I no longer must pay fees when caught with-out tickets

The solution is a fast ticket app based on previous agreement to automatic billing. It uses predefined payment services – this could be the Telecom or some pre-authorized credit card institute – and predefined destination stations. The app then finds the nearest boarding station – still served by transport services at the time of demand – using the smartphone's GPS service. Then it proposes the ticket solution to its user. The user only needs one click to agree to the proposed ticket and to using the predefined payment service. Against a full-service ticket shop on the web, this is significantly less functionality, at least seen by the user; however, the complexity of the service exceeds that of the ticket shop, because it interconnects with many more additional services and devices. Moreover, it relies on preparations made beforehand.

The fast ticket app uses the SIM card certificate discussed further in section 7-6 for *Federated Identity Services*. Authentication is by using the SIM card in the smartphone.

7-5.3.2 THE TRANSACTION MAP

The transaction map in Figure 7-18 looks somewhat strange. The *Travel Contract* entry does not update any internal file – however, it is a legal requirement that contracts are recorded. One can argue that this EI belongs to another application, namely some previously accepted terms & conditions that govern the use of the fast ticket app.

Moreover, *Enable Payment* is a single EQ, despite that it makes two very different things: Enabling authentication services via the SIM card certificate, and authorize credit card, or mobile bill, payment. However, the IFPUG 4.3.1 manual says, "Do not split an elementary process with multiple forms of processing logic into multiple elementary processes" (IFPUG Counting Practice Committee, 2010).

Figure 7-18: Transaction Map for the Ticket App

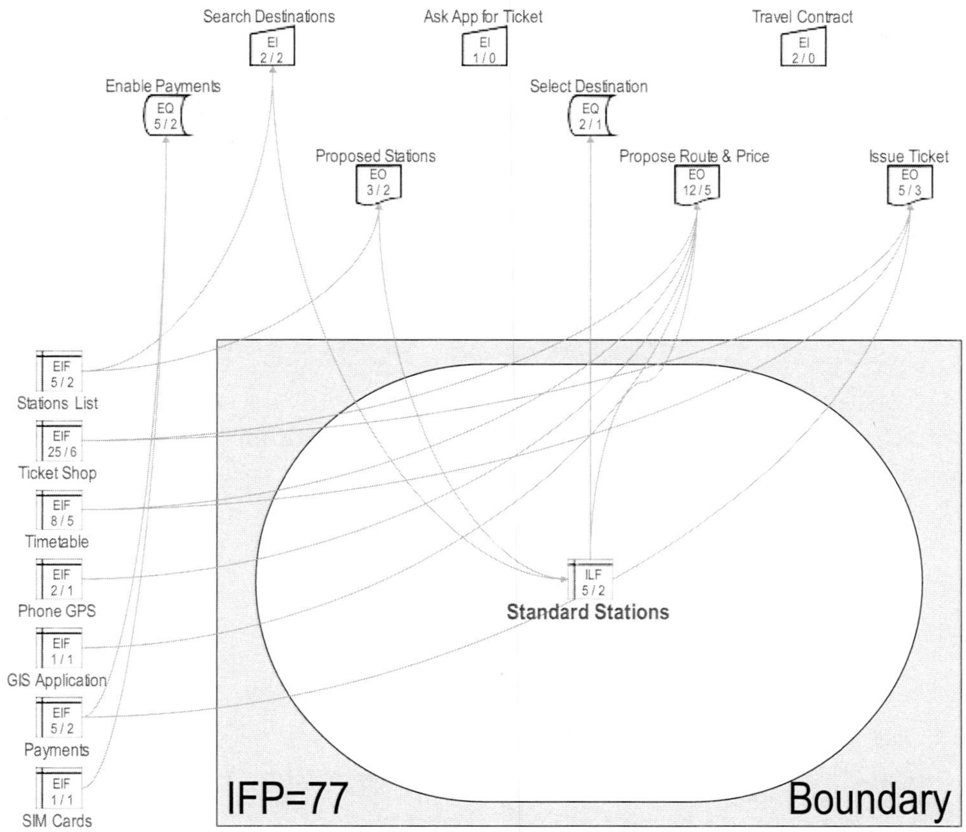

However, the worst thing is that the IFPUG size seems smaller for the fast ticket app than for the ticket sale application on the web portal (section 7-4.4). The fast ticket app is 77 IFP and therefore smaller than the ticket sale application which counts 79 IFP. Although we expect the fast ticket app delivering less functionality to the user than the full web portal; from a technical view, hiding the functionality seen by the app user means significantly more software functionality providing the users' contextual knowledge.

7-5.3.3 THE DATA MOVEMENT MAP

The COSMIC model in turn looks simpler and less complicated (Figure 7-19). The data movement map depicts this very simple application software. With functional size measurements, the usual methods of Six Sigma for construction and operation of software solutions are applicable. Measuring code is useful for assessing quality regarding maintainability or portability; however, it is not telling much about variations in software services. Services are assessable based on their functionality only.

Figure 7-19: Fast Ticket App Overview

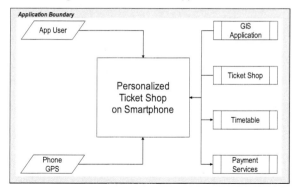

ISO/IEC 19761 COSMIC counts 38 CFP and the data movement map identifies six functional processes. The extract in Figure 7-20 shows how the user prepares its standard destinations and authorizes an authentication service to approve automatic payments from her or his account. This makes one of the crucial app features immediately visible to the development team and to all stakeholders.

Figure 7-20: Data Movement Map for the Fast Ticket App (extract)

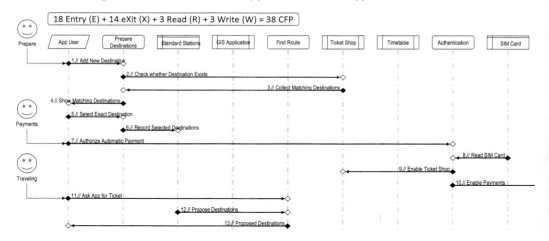

Moreover, not only this data movement has special impact on the app project, there are a few more – not shown here – that are of special importance regarding reliability, security, privacy or theft protection. For instance, payment received must trigger issuing the ticket, and the travel company must check payment authorization against the telecom service provider's list of stolen or blocked SIM cards.

Such aspects are perfectly visible and measurable in data movement maps, again underlining that serious software engineering starts with both a transaction and a data movement map, yielding both an IFPUG and a COSMIC count.

7-6 MOBILE ID: USE SMARTPHONE FOR AUTHENTICATION

This application aims at establishing the telecom company as a *Trusted Third Party* (TTP) for providing identity services. Users no longer must remember, or manage, thousands of passwords; instead, they can use their smartphone for authentication. It works even for conventional mobile phones, as all communication uses encrypted GPS messages (Short Message Service – SMS). Providers of web portals that need to authenticate users simply can ask the Mobile ID service of the telecom. That saves them maintaining a password repository and managing credentials of its subscribers. An application residing on the *Subscriber Identity Module* (SIM card) asks the user for credentials, the only ones to remember, and even can keep authentication valid across web portals or other secure applications. For e-Businesses and m-Business, or financial services, this is an attractive offering.

7-6.1 USER STORIES AS FUNCTIONAL USER REQUIREMENTS

Four functional users provide the project team with four user stories:

Table 7-21: Functional User Requirements recorded as User Stories in the Grant Rule Format

	Label	As a ... [Functional User]	I want to ... [get something done]	Such that ... [quality characteristic]	So that ... [value or benefit].
1)	Recognize me	Internet User	get known by the app	nobody else can access my data	they know my settings, assets and preferences
2)	Authenticate	Web Portal	authenticate a user name	nobody else can misuse the user's credentials	private data remains secure
3)	Create Certificate	Internet User	get authenticated	without harassment	I do not have to enter the password each time
4)	Confirm Validity	Third-Party Trustee	make sure the holder of the SIM card is still in possession of its smartphone	she or he needs proper smartphone security protection and stays at its place	I'm sure the righteous SIM card holder did not get robbed

These user stories do not cover initiation, updates or termination of the authentication service; however, since SIM cards move across phones, no dependencies exist from hardware platforms.

7-6.2 MOBILE ID WITH GPS SENSING

The current Mobile ID solution that is in operation since 2013 with Swisscom in Switzerland, requests password credentials for every authentication request, avoiding creating a SIM certificate that risks straying uncontrollably through the Internet; however, it still has the big advantage that users no longer need remembering hundreds of passwords except the one for the Mobile ID.

The solution presented here goes one logical step further: the smartphone tries to determine whether it needs to ask for the Mobile ID password or whether it is unnecessary. For instance, if the user sits at his desk and visits a few web portals without moving away, it is sufficient if she or he enters the password once at the beginning; it remains valid for a certain amount of time and if the user does not move away from the cell. A cell covers a few square kilometers; if the cell position combines with the GPS position that covers a few square meters, it is safe to assume that the smartphone remains in the possession of its owner who originally entered its credentials at this place. Even when traveling in trains, GPS might be capable to identify the railway track and thus allow access to web portals undisturbed by repeated password requests. Any rapid movement of the mobile phone shall also reset credentials; a train emergency break thus resets credentials as does if a stranger swiping away the smartphone.

7-6.3 MOBILE ID PEER SERVICE ARCHITECTURE

The three application boundaries constitute three peer services that are located on some web portal sing the Mobile ID for authentication, on the Telecom's service cloud, and on the user's smartphone.

Figure 7-22: Mobile ID Get Credentials, Service, and SIM Applications Overview

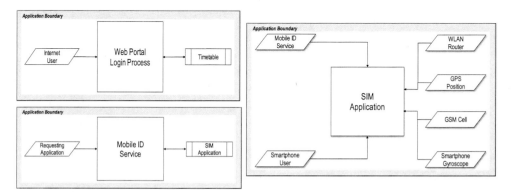

The four user stories transform into three applications; see Figure 7-22 :

- The *Mobile ID Get Credentials* application, a plug-in for web portals or other services requiring user authentication that try the mobile ID service first and ask for the local credentials only in case of failure, usually for a preset password;
- The *Mobile ID Service*, the Telecom's trusted third-party *Identity Service* that receives requests for the credentials application and checks with the SIM application whether the user can be authenticated;
- The *Mobile ID SIM Application*, a software residing on the SIM card in a smartphone that checks with the smartphone and its services whether the smartphone is still in the possession of its lawful owner. If there is doubt; e.g., rapid movements, shakes or change of both GSM cell position and GPS position, the smartphone user must enter its credentials to re-validate the Mobile ID.

Note that while in motion (in a train, plane or bus) the GPS position changes while the GSM cell remains stable, if the vehicle is equipped with a GSM repeater. WIFI senders provide the same geographic information as GSM cell antennas. In such cases, change of GPS position should not block user authentication. As a side effect, the SIM application could also serve as theft protection.

Registration with the Mobile ID service is not part of the functionality shown in this section.

7-6.4 THE MOBILE ID GET CREDENTIALS APPLICATION

The application will not ask for the credentials related to the SIM card identity. Instead, it uses the traditional credentials kept locally with the registered users. This application does not maintain the user registry.

Figure 7-23: Mobile ID Get Credentials Application

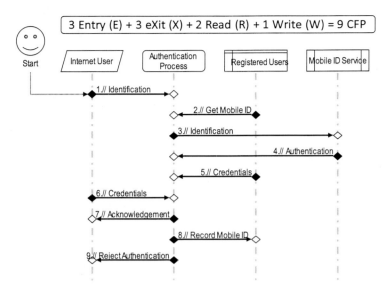

In case of reject, the user learns whether the Mobile ID failed, be it because of entering a wrong password, or while the SIM card was blocked.

7-6.5 THE MOBILE ID SERVICE APPLICATION

Figure 7-24: Mobile ID Service Application

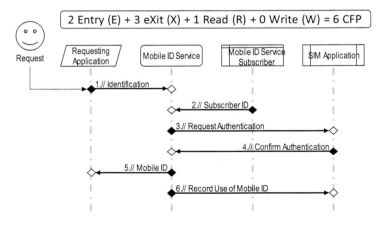

This application service is typically located within the premises of the Telecom acting as a TTP. The Telecom is not in possession of any password; the smartphone alone oversees authentication.

Figure 7-25: Mobile ID SIM Application as a Data Movement Map

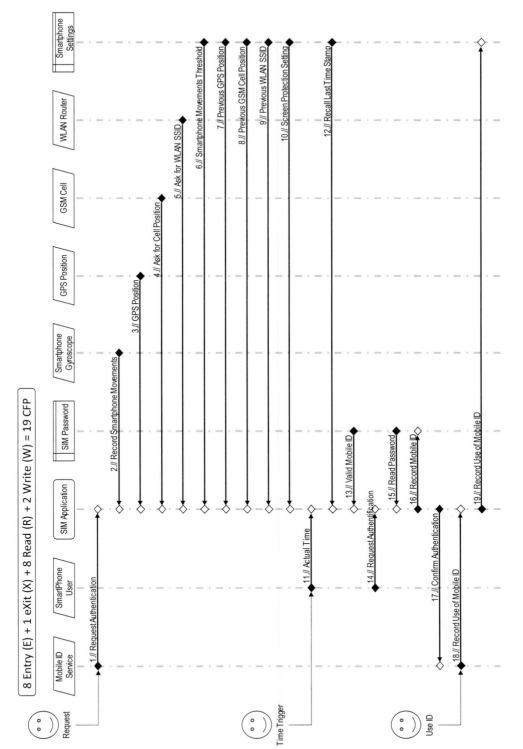

7-6.6 THE MOBILE ID SIM APPLICATION

The SIM application resides on the SIM card and thus travels with the user into the user's new smartphones. It must interface to some of the smartphones services. This is a technical challenge, given the many different platforms in use for mobile phones. The critical features of the Mobile ID SIM Application are accessing the gyroscope of the smartphone, GPS services if available, and asking GSM and WIFI services for actual data. Then, it stores data that later will be needed for comparisons regarding other authentication requests.

Thus, the SIM application can determine whether the smartphone has moved in between two requests. Furthermore, a time trigger might force the password request. In Figure 7-25, three functional processes make up the SIM application, although all three reside on one single lifeline for the functional object "SIM Application". The time stamp triggers the guarded conditional frame requesting password authorization from the users for authentication, in case too much time elapsed or any of the position parameters changed dramatically.

There is only one device for both the smartphone with its time trigger and the smartphone user, requesting authentication. Most often, the user does initiate authentication by her or his own will. The application starts authentication because the user requests another functionality. In case the position parameters remain within limits, new authentication might not be needed. Not part of the application shown are any additional functionalities for setting thresholds, maximum time delays between authorization requests, or a list of apps that always need a password check.

It is noteworthy that the functional size measured for the Mobile ID system tells not very much for the cost expected to build such as system. Major cost drivers are legal constraints, security, privacy, technical complexity and number of stakeholders involved. Moreover, when looking at smartphones, the number of different operating systems, display size and technologies supported for making the Mobile ID successful affect the development cost budget much more than functionality alone.

7-7 CONCLUSION

Functional size measurement is not a discipline of its own; it directly connects to feature design and requirements elicitation. The size is only a side effect from the model chosen. It always makes sense to use more than one model; this allows to distinguish different aspects in the model, and to detect other kind of missing requirements.

It is obvious that project estimation is not the only goal for functional sizing. The cases where functional size is the major driver for project costs are nowadays rather limited; they still exist when looking at technical, or military, applications that have a clearly stated functionality goal and cannot just reuse or redirect already existing functionality to new devices and use cases.

Linking functional size measurement to data movement maps and transaction maps opens a wide range of new applications to functional sizing. Not only do data movement maps explain new software better to users, sponsors, and developers, they also serve as a base for identifying risks such as squeezed performance, developing test cases, and measuring test coverage, test effectiveness and defect density because data movements are easily recognizable as constituent elements in test cases, and defects found.

The most interesting, and urgent, new area for software measurement are security and safety, as already mentioned in the conclusions to *Chapter 6: Functional Sizing*. It will be further discussed in *Chapter 13: Dynamic Sampling of Topic Areas*.

CHAPTER 8: LEAN & AGILE SOFTWARE DEVELOPMENT

SW Development has been the pilot undertaking to collect experience with adoption of transfer functions in economics. Transfer functions help with Agile SW Development, Avoiding Muda (無駄), and Succeeding with Use Cases. The Buglione-Trudel Matrix adapts the power of QFD for lean & agile development team.

Estimating agile projects is another benefit originating from QFD, using transfer functions to manage the software development process.

8-1 INTRODUCTION

Six Sigma orientates products and services towards measurable customer values. It provides tools and methods to eliminate and prevent defects whereby anything that diminishes customer values is termed a defect.

Contrary to common belief, a Six Sigma approach for Software has not much to do with bugs found by engineers: non-adherence to specification for instance does not automatically imply that customers see this as a defect. Specifications may be wrong, or may not be of any value to a customer, and hence non-adherence to specification will not be a defect. This makes defect counting difficult, especially in software, as only the customer value counts. Weaknesses that engineers classify as minor may end up as major defects in customer's perception, and vice versa. Customers might ignore major technical weaknesses of software if it does not hurt business. However, in traditional software development environments, the customer viewpoint is not readily available to the development team.

Thus, it is necessary to make the views of various stakeholders visible to engineers, product owners, customers, users, and other stakeholders. Agile development methods try to prioritize sprints according these views. This chapter introduces the *Buglione-Trudel Matrix* (BT-matrix) that combines two transfer functions mapping user stories to functional and non-functional requirements, showing the convergence gap between expected and achieved profiles. This application of Six Sigma transfer functions maps process controls onto the process response. For the functional part, the transfer function consists of data movements with the ISO/IEC 19761 COSMIC measurement method. For the non-functional part, it is more like story points in agile: the team agrees on its impact on business drivers.

Finally, we develop a method for cost prediction in Lean Six Sigma software development projects.

8-2 LEAN & AGILE SOFTWARE DEVELOPMENT

We assume the reader familiar with the agile software development paradigm; see for instance (Schwaber & Beedle, 2002). In this chapter, we look at four pain points in agile:

- Sizing user stories
- Measuring Progress and define conditions when work is finished
- Translating user stories into story cards
- Prioritizing user stories

For these problem areas, Six Sigma provides proven techniques fitting well into an agile development approach. Current techniques used in Agile Teams rely on Delphi techniques (Cohn, 2005) for sizing *Story Points*. However, story points are non-reproducible agreements in development teams and do not refer to sizing, rather to effort.

While the most popular project management method for Agile seems to be Scrum, Six Sigma for Software fits best to *Lean* and *Test-Driven Development*, as stated in (Poppendieck & Poppendieck, 2007) or (Beck, 2000). Some Scrum representatives seems not to see any value in Lean Six Sigma (Roriz Filho, 2010).

8-2.1 DETECTING 無駄 (MUDA) BY TRANSFER FUNCTIONS

Key to detecting muda is the sound understanding what brings value to the customer. However, business value is an often-heard but not well-understood notion. Bakalova (Bakalova, 2014) has shown that in agile development, when apparently, business value should govern priorities and selection of work items, it remains vague and subject to huge variation, even over the duration of an agile project. Nevertheless, we know that some values are very persistent, even for centuries. Denney (Denney, 2005) calls them *Business Drivers*.

8-2.2 BUSINESS DRIVERS AND FUNCTIONAL USER REQUIREMENTS

Denney (Denney, 2005) used *Quality Function Deployment* (QFD) to design use cases, so that they match business drivers. The notion of business drivers refers to "something that provides a clear sense of why a project is being undertaken and the ultimate value it will provide; it's a force to which businesses must respond and drives a business's direction", says Denney. Examples for business drivers include "Reliability of a system" or "minimize risk by exploring technology not well understood". Ideally, business drivers are win-win in nature, providing value for both supplier and customer. Business drivers matter most when designing software, and Denney (Denney, 2005) gives many examples how to select use cases that best sustain them, using QFD.

Business drivers are mostly non-functional in nature. *Functional User Requirements* (FUR) describe what a system is supposed to do, in terms of what input they expect and what output they produce after computation. FUR transform business drivers into requirements reflecting current technological possibilities. FUR can be as simple as stating "I want to have my data available for a certain time" or "I want to send a message to my friend in Macau, China". Business drivers state how things must look, and how things must be done.

Agile methodologies apply *User Stories* to describe what functionality to implement, as opposed to UML that wraps such specifications into *Use Cases*. Use cases and user stories are not the same, while we use the terms FUR and user story interchangeably, depending upon context agile or not. Some agile organizations transform use cases into *Epics* for initial requirements, and elaborate epics later into a sequence of user stories, and user stories in turn might be broken down even further into work items that fit into a sprint. Such items are represented as *Story Cards*. Thus, a story card can represent one user story, or two or more work items that together make up a user story. It is the smallest kind of work item in an agile project.

We are aware that there are many different notations, especially with story cards. Story cards describe the work needed to implement a user story, eventually split along technologies, part functionality, or functional and non-functional parts, such as user interface design, and input/output functionality, whatever suits best.

The agile team agree on the minimum set of user stories needed to implement business drivers. We will see that a relatively simple Six Sigma transfer function exist that measures such coverage, identifying gaps between functionality and business drivers.

8-2.3 PROFILING BUSINESS DRIVERS

To distinguish muda story cards from lean story cards, we need to know how much business drivers and user stories matter. A measurement method for customer values could qualify software based on predicting market acceptance. Metrics based on such measurements would help developers to add attractive characteristics to their software. Such metrics could become an important aspect in software development contracts, and would support senior management, marketing professionals and product managers in assessing existing software and detecting weaknesses. See *Chapter 4: Quality Function Deployment* for methods how to analyze Voice of the Customer.

8-2.4 TRANSFER FUNCTIONS FOR ROOT-CAUSE ANALYSIS

In this case, we use transfer functions to identify the relevant user stories. In Figure 8-1, the transfer function maps user stories onto *Business Drivers* (BD), assuring functional effectiveness for the user stories selected. The convergence gap is a measure for it; i.e., if zero, it means that the user stories cover functionality exactly according the customer's priorities expressed by the business driver profile. The larger the gap, the more unnecessary functionality might get implemented.

It is obvious for lean & agile practitioners; how valuable such a metric is. It would help them not only prioritizing their sprints right, it also could help customers and product owners to detect missing functional effectiveness early in a project.

Figure 8-1: Transfer Function Mapping User Stories to Business Drivers

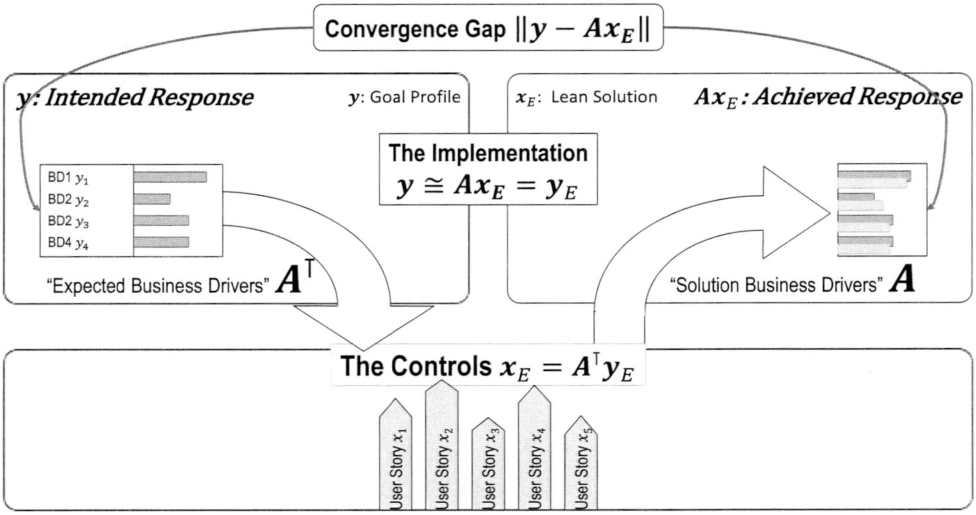

There are two different ways how to measure the transfer function: one method is to measure the functionality if contributes to each of the business drivers; the other is to measure the impact of story cards on business drivers. The first method, called *Cellar Convergence,* focuses on functionality, the latter method is the *Sundeck Convergence* looking at the impact the work items represented in story cards have on meeting business drivers and thus business expectations.

Before proceeding, it must be made clear how teams measure functionality for Lean Six Sigma and Agile. In principle, any internationally recognized functional sizing method will do. However, some particularities apply. The well-known ISO/IEC 20926 IFPUG method for functional sizing (IFPUG Counting Practice Committee, 2010) suffers from the problem that it is not additive; i.e., the size of two components measured together is not equal to the sum of the two components measured separately. In many cases, the method ISO/IEC 19761 COSMIC is preferred because of its additive property. As example, Figure 8-4 shows the functionality of a small app that allows a user accessing the SIM card of his smartphone for use as a personal certificate, usable for identification or authentication.

8-2.5 FUNCTIONAL EFFECTIVENESS

Given a business profile, and the ease of functional sizing given by data movement maps, it is tempting to look for *Functional Effectiveness*. Do the data movements cover all needs of the customer, as expressed by the FUR, or user stories? Functional effectiveness is easily measurable; it simply means assessing which data movements contribute to what goal target, and then compute the convergence gap. A software is functionally complete, if the convergence gap closes.

The question whether a software is functionally complete has practical value. While missing functionality hints at missed business drivers, sometimes functionality is required that does not contribute to some business driver; maybe other reasons call for it. Then the convergence gap closes only if those other requirements are part of the goal profile.

Thus, the question is interesting in both cases: why some software is functionally complete or not. In practice, checking for functional effectiveness is a means to detect both missing functionality and excess functionality; consequently, it is a metric of high interest for Lean Six Sigma practitioners.

To assess functional effectiveness, it suffices to count how many data movements support business value. However, such an assessment is not straightforward; for instance, what exactly is the business value for a data movement connecting the Telecom user management information with its SIM cards to the travel companies *Customer Relationship Management* (CRM) application?

8-2.6 USER STORIES AND FUNCTIONAL USER REQUIREMENTS

A word of caution: *User Stories* (USt) and *Functional User Requirements* (FUR) as defined in ISO/IEC 19761 COSMIC are not necessarily the same. While in a technically oriented example such as the Mobile ID presented in section 7-3.6 in *Chapter 7: The Modern Art of Developing Software* user stories come identical to FUR, in general this not the case. In many cases, development teams must transform user stories into functional requirements before they are ready for implementation. Agile teams do that regularly and keep talking about "user stories", even if their improved level of granularity allows proceeding to the next sprint for implementation. UML best practices design use cases, and although use cases normally contain sequence and actions diagrams, they in general define more than just one FUR.

We prefer in this book to use the term user story over use case for functional and non-functional user requirements. The measurement methods and complexity handling are the same for agile and more traditional UML development approaches. Often, UML and agile work well together.

As already mentioned, the generic problem with use cases is that granularity is not standardized. Hus, the critical step is, user stories must exhibit a standard level of granularity that makes them functional processes in the meaning of ISO/IEC 19761 COSMIC. They need a functional user and perform a task that fulfills her or his FUR. This establishes a quality criteria for user stories making them fit for implementation. In practice, agile teams often split user stories until they describe only one functional process.

In such cases, it is more appropriate to distinguish business drivers (BD) from functional user requirements (FUR) and mapping the FUR to the BD. Then, the FUR describe functionality supporting the BD, and if the FUR transfer to the BD with a small convergence gap, the complexity of counting data movements that effectively support business diminishes considerably.

For practical purposes, identifying FUR with user stories works well, if the user story is at a level of granularity required by ISO/IEC 19761 COSMIC like the FUR. As mentioned, agile teams insist on such a granularity before they approve a user story for implementation. Then, best practices in agile software development converge with COSMIC standard definitions what underlines the statement that ISO/IEC 19761 COSMIC is particularly helpful for agile software development – not only because it works well with data movement maps. In fact, it is possibly the method of choice for measuring velocity in agile development – ISO/IEC 19761 COSMIC allows identifying the functionality added when implementing non-functional user requirements easily. Acknowledging that the initial functionality defined at project inception will not be the final one, they simply count added data movements, preferably automatically when checking in implemented code. Thus, the initial functionality is subject to change during the sprints and related requirements elicitation activities of the software development team. This will be further elaborated in *Chapter 9: Requirements Elicitation*.

8-3 A Sample Travel Helpdesk Project

Assume the transportation company from section 5-4 in *Chapter 5: Voice of the Customer by Net Promoter®* wants to enhance its helpdesk operations and make it fit for today's social media environment. It uses an agile approach for implementing improvements.

8-3.1 Requirements

Table 8-2: User Stories for the Travel Helpdesk in the Grant Rule Format

	Label	As a ... [Functional User]	I want to ... [get something done]	Such that ... [quality characteristic]	So that ... [value or benefit].
1)	Q001: Customer Story	registered customer of the travel company	be recognized based on the SIM card in my smartphone and my trip history	I can pay my travel with the payment method recorded in my user profile	I will not lose connections due to payment procedures and get a suitable alternative offered
2)	Q002: Helpdesk Story	Helpdesk staff	identify a client without having to ask for the name or get credentials, regardless whether calling by phone, sending e-mail, or when chatting	I can charge service fees or ticket sold according the clients' preferred payment method	in case of urgency I can help more customers faster and more effectively
3)	Q003: Enrollment Story	non-registered customer	can get enrolled quickly and easily with my credit card and my smartphone	I can grant access securely and exclusively	I can use the services of my travel service provider immediately
4)	Q004: Social Story	socially committed person	plan, book and amend my travels on short notice	I can amend and cancel any travel plans on short notice any time of day or night	I always can be where I have just the most fun or best work to do
5)	Q005: Get SIM Certificate	user of laptop computer or smartphone	store my SIM certificate that I need for authentication	certificate records are in the usual certificate store provided by my operating system	I can safely communicate with my travel provider's helpdesk

Based on the business driver assessment shown in Figure 8-5. The basic idea is to provide an efficient and effective helpdesk service in case of delays, or missed connections. This is the vision behind the five user stories in Table 8-2.

Since time matters, and typically they must serve many customers at a time, they invest into technology including mobile ticketing, assuming a large part of their customers use such devices. Additionally, they want to persuade customers moving to these technologies and stop buying tickets at station tellers. All user stories employ the *Grant Rule* format (Fagg & Rule, 2010).

Figure 8-3: Application Overview with five Functional Processes for the Helpdesk Application

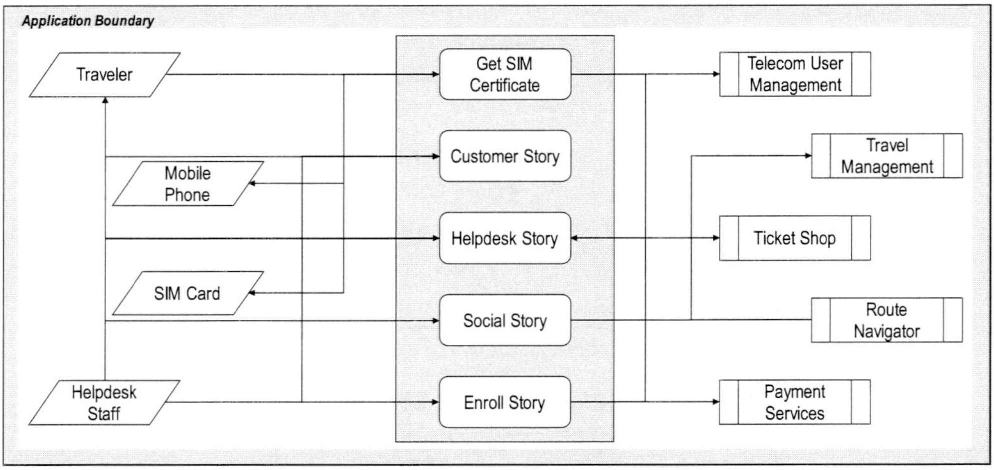

The functional process "*Get SIM Certificate*" allows the user to obtain a certificate for signing any commitment, such as a replacement ticket or a re-booking.

The telecom provider owning the SIM card issues this certificate, as shown in Figure 8-4. The telecom company acts in this case as a trusted third party, providing a federated identity certificated for users, similar to the Mobile ID example in section 7-6.2 of *Chapter 7: The Modern Art of Developing Software.*

Following the *Guideline for the use of COSMIC to manage agile projects* (Buglione, et al., 2011), each of the functional processes should correspond to one user story. However, that does not work out exactly so when considering implementation. In fact, although the user story *Q005: Get SIM Certificate* is fine as a FUR, the *Q004: Social Story* does not specify functionality very clearly. It can be understood as an NFR, requiring fast reaction of the helpdesk in case the customer wants to change travel plans fast. However, this requirement is not fulfilled with simply adding more powerful computers, it might require additional functionality that will only discovered after a few sprints by the development team, including the sponsor. The *Q005: Get SIM Certificate* user story alone will not do.

In this example, an initial count, based on the project vision, revealed a total of 110 CFP. The final count is expected to be significantly higher.

Figure 8-4: Extract from Data Movement Map for Functional Process "Get SIM Certificate"

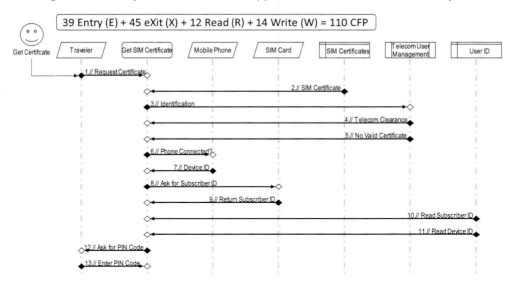

8-3.2 BUSINESS DRIVER PROFILE

The transportation company got the following profile shown in Figure 8-5. This business driver profile originates from the sample NPS analysis in section 5-4 in *Chapter 5: Voice of the Customer by Net Promoter®*.

Figure 8-5: Business Driver Profile for the Helpdesk Project

Topics		Attributes			Weight	Profile	
a) Helpdesk	BD1 Responsiveness	Fast Response	Few Selections	Known Issue	16%	0.34	
	BD2 Be Compelling	Able to resolve issue	Committed to help	Knowledgeable	20%	0.44	
	BD3 Friendliness	Keeps calm	Unagitated	Cool	19%	0.41	
b) Software	BD4 Personalization	Knows who's calling	Knows travel plans	Knows delays	30%	0.64	
	BD5 Competence	Can figure out solution	According my preferences		15%	0.33	

Software development is a knowledge acquisition journey, where stakeholders learn to know how to express the problem in such a precise way that even dumb machines, networks and clouds do it right. The agile team consists of sponsors, business analysts, product owners, developers, and testers, all speaking different languages and seeing priorities differently. However, all humans are willing to communicate but need visual communication that stirs imagination and creativity: Thus, how can we visualize lean eigenvector solutions to the team? In addition, in what respect is lean & agile software development more than value-driven prioritization of agile?

8-3.3 FUNCTIONAL EFFECTIVENESS

Functional effectiveness counts cell by cell in the transfer function mapping user stories to business drivers. Figure 8-6 shows the data movements that ensure *BD3: Friendliness* in the user story *Q002: Helpdesk Story*.

Figure 8-6: Assessing Functional Effectiveness for BD3: Friendliness in Q002: Helpdesk Story

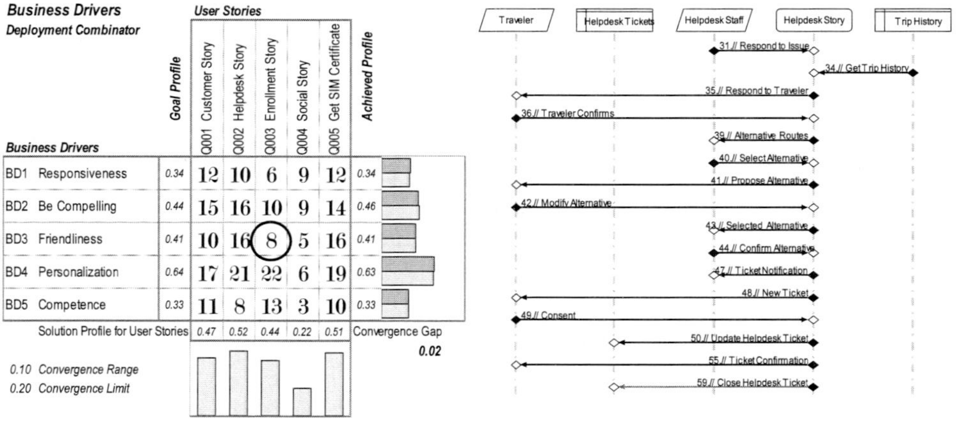

Such an assessment conducted cell by cell reflects the user view; it counts those data movements that are user recognizable in view of business driver BD3: Friendliness. The assessment is an expert judgement. This is acceptable, since functional effectiveness is not always possible, and possibly does not employ all data movements. This is because of technical constraints that enforce additional functionality not directly contributing to some of the business drivers. However, the question, which one does not contribute, is of highest interest, as it might open an opportunity to save on 無駄 (muda).

8-3.4 VISUALIZATION OF MUDA FOR AGILE TEAMS

This section compares with Fehlmann and Kranich's paper (Fehlmann & Kranich, 2014-2) focusing on early project estimations for agile projects; however, from a different perspective.

User stories translate into many story cards. There are two different kind:

1. Story Cards that predominantly implement functionality
2. Story Cards that predominantly implement non-functional qualities

Figure 8-7 is a predominantly functional story card that implements part of the functionality shown in Figure 8-4 needed to access a SIM card in a mobile phone.

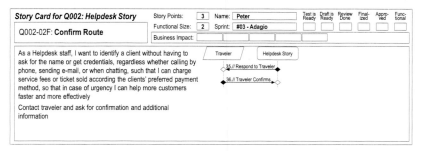

Story cards record a task description as short as possible, story points assigned by the agile team, functional size counted with ISO/IEC 19761 COSMIC, and specific business impact on the business drivers. The team defines specific business impact as the contribution of some story card to one of the business drivers. Impact might be zero, if the task is purely functional – a must-be task required by some user story or FUR, see Figure 8-7 – or if there is no contribution that is specific to one of the business drivers identified.

Figure 8-8: Sample Impact Story Card – not yet Assigned

A story card is a *Functional Story Card* if it has no specific impact on one or several of the business drivers. Story cards that have such an impact are *Impact Story Cards*. For an example, see Figure 8-8. Impact story cards may well contain functionality; however, most often such functionality is unknown at the beginning of the project. Typically, the development team devises together with the sponsor and product owner what additional functionality to add to impact cards. Thus, total functional size will grow during the development cycle. Other story cards might address dedicated effort not involving any functionality but required by business.

For all impact story cards, business impact is weighted, for instance with 1 to 6 points, to distinguish between low and high impact. The team can agree on a suitable scale.

No one expects that functionality alone will give enough power to the business drivers of our transportation company, so they know in advance that additional impact story cards will be needed to make the helpdesk project fly. The sample card in Figure 8-8 adds no additional functionality yet; however, when it comes to its sprint and implementation, the team might as well invent additional functionality that was impossible to anticipate at project inception stage. Therefore, non-functional requirements play an eminent role for project effort estimation. With impact story cards, actual cost can be controlled and measured.

8-3.5 DETERMINING BUSINESS IMPACT FOR STORY CARDS

The process of assigning business impact on story cards links to assigning story points. In most cases, story points and impact go in parallel; the more story points the higher is business impact. However, we expect the relationship not linear, since doubling the effort rather seldom also doubles business impact. The relationship we adopt in practice is with the Fibonacci sequence that is quite popular for assigning story points. If there is no functionality, the impact simply is the Fibonacci sequence number. If there is functionality, the story points will consider the related effort, and the relationship will not be so simple; see section 8-3.10 in *Chapter 8: Lean & Agile Software Development*. Typical impact values are one (yellow cards in Figure 8-9, three (blue cards) and six (red cards).

8-3.6 THE BUGLIONE-TRUDEL MATRIX

The *Buglione-Trudel Matrix* (BT-matrix) provides a different view on story cards. It visualizes functional and non-functional aspects of software development, as explained in (Fehlmann, 2011-2) and (Buglione & Trudel, 2010). This combination allows tracking effort spent on non-functional or quality tasks, and compare it to work performed on functionality. Many of these tasks become apparent during requirement elicitation. Traditional effort estimations rely on functionality measurements and handle non-functional requirements in a completely different way, detached from the specific project. Therefore, they rapidly become obsolete. The lower part of the BT-matrix, called the *Cellar*, contains the functional story cards. The non-functional, quality-related story cards reside in the upper half, the *Sundeck*. The BT-matrix contains two transfer functions; one mapping user stories into business drivers by means of the functional effectiveness transfer function – referenced in the matrix cellar – and another one mapping user stories into business drivers by means of the business impact recorded on story cards. The sundeck transfer function is explicit in the BT-matrix. Figure 8-11 (initial) and Figure 8-12 (final) show the convergence gaps calculated by the transfer function.

In Figure 8-9, it is obvious that the total impact does not satisfy all business drivers and thus does not provide the correct business value. The team can identify weak spots in the BT-matrix and get an indication where improvements add impact. The goal profile for the business drivers shows over- and underachievement by comparing the two profiles.

Figure 8-9: Initial Buglione-Trudel Matrix

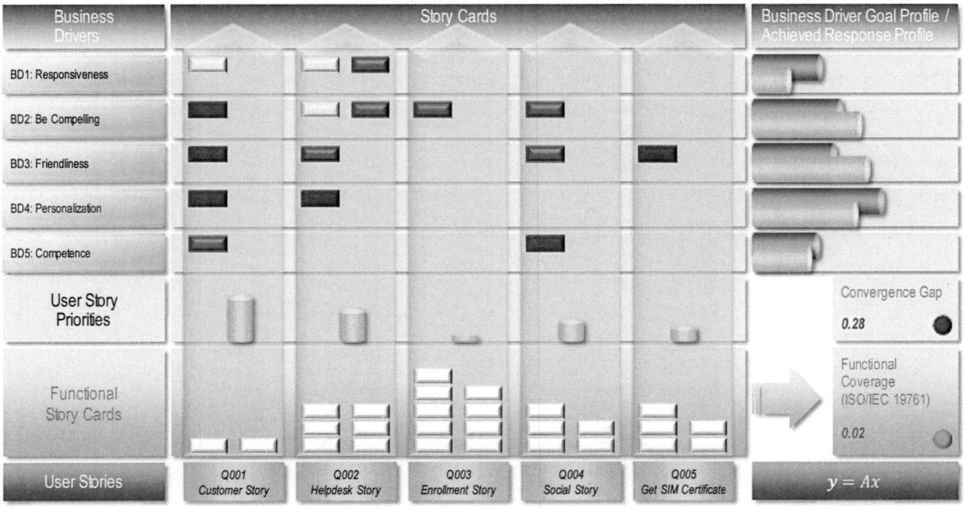

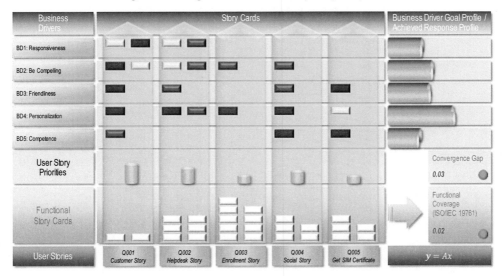

After a few sprints, the Buglione-Trudel Matrix has been adjusted by the team to meet the goal profile of the business drivers (Figure 8-10):

Figure 8-10: Buglione-Trudel Matrix after a few Sprints

The two profiles are regular (see section *2-4.5 Regularization*); that means, if the yellow bar overtakes the green bar, it stands for a relative overachievement. Values are not absolute. The business driver *BD3: Friendliness* is not necessarily an overachievement; it is relatively strong only because some of the others are too weak in the focus.

For instance, in our sample project, functional requirements drive user interface design not alone; quality aspects including corporate design and ease of use are as important as any additional functionality. In the sample case shown in Figure 8-8, the added functionality was storing preferences such that the customer can select entries from previous choices, avoiding unnecessary typing and clicking.

The total impact of a correlation matrix cell is the sum of all impact points for story cards recorded in the matrix cells contributing to the respective user story. The achieved profile results from applying the sundeck matrix to the eigencontrols, the solution profile $x = A^{\mathsf{T}} y_E$.

Figure 8-11: Sample Sundeck Tracking Control Transfer Function – initial

Business Drivers / Deployment Combinator	Goal Profile	Q001 Customer Story	Q002 Helpdesk Story	Q003 Enrollment Story	Q004 Social Story	Q005 Get SIM Certificate	Achieved Profile
BD1 Responsiveness	0.34	1	4				0.19
BD2 Be Compelling	0.44	6	4	3	3		0.54
BD3 Friendliness	0.41	6	3		3	6	0.57
BD4 Personalization	0.64	6	6				0.51
BD5 Competence	0.33	3			6		0.29
Solution Profile for User Stories:		0.73	0.52	0.11	0.34	0.23	Convergence Gap
Total Business Impact: 90							0.28

0.10 Convergence Range
0.20 Convergence Limit

Functional size is measurable from the data movement maps and recorded on the story cards, see Figure 8-7; the team agrees on business impact, just like on story points for effort estimation. When user stories split into story cards, we postulate with Ambler (Ambler, 2004) to do this along functionality; i.e., data movements must not split among story cards. For assigning tasks to developers, this seems natural.

This rule forbids separating an algorithm, part of a functional process, from all its input and output data movements. At least one Entry, eXit, Read or Write must belong to the functional process implemented by some story card. This rule is not only helpful for development, also for testing, because it allows writing unit tests per story card – another very useful lean practice in software development.

Figure 8-12 shows the calculation of the convergence gap for the final BT-matrix. Like Figure 8-11, it is a standard QFD matrix with the peculiarity that the matrix cells count impact of story cards.

Figure 8-12: Sample Sundeck Tracking Control Transfer Function – final

Business Drivers Deployment Combinator	Goal Profile	Q001 Customer Story	Q002 Helpdesk Story	Q003 Enrollment Story	Q004 Social Story	Q005 Get SIM Certificate	Achieved Profile
BD1 Responsiveness	0.34	7	4				0.33
BD2 Be Compelling	0.44	7	4	3	3		0.44
BD3 Friendliness	0.41	6	3		3	6	0.42
BD4 Personalization	0.64	6	9	6	6	1	0.66
BD5 Competence	0.33	3			6	6	0.30
Solution Profile for User Stories:		0.65	0.52	0.27	0.42	0.25	Convergence Gap
Total Business Impact: 89							0.03

0.10 Convergence Range
0.20 Convergence Limit

The Buglione-Trudel Matrices (both Figure 8-9 and Figure 8-10) do not show the details of the cellar matrix indicating functional effectiveness – for simplicity, but also because developers do not need looking at it, normally. The cellar matrix is significantly simpler, as there is no business impact assessed by experts. It is sufficient to assign data movements in a data movement map to one of the user stories. Following ISO/IEC 19761 COSMIC, this is a side effect of the count.

Figure 8-13: Deming Chain for Business Driver's Coverage and Functional Effectiveness

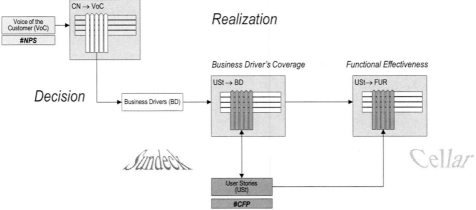

Figure 8-13 shows the Deming chain for business drivers and functional effectiveness. The two transfer functions can be the same, but usually are not. If they are the same, they convey a system whose business drivers are purely functional. Such systems exist but are not the normal case.

The solution profile for business driver's coverage and for functional effectiveness usually are not the same. This is because the two transfer functions represent different measurements. The first solution profile is the goal profile for prioritizing story cards; the second is the goal profile for testing.

8-3.7 SIX STEPS TO COMPLETION

The Six Steps to Completion method in Six Sigma for Software is the *Definition of Done* for lean & agile software development. It has two additional steps, extending traditional agile: The *Test-Driven Development* (TDD) approach is compulsory as starting step, and there is an explicit approval step. Either the team or the product owner, or both, approve. The Six Steps to Completion measure progress on story cards by assigning a completion rate to each card.

Figure 8-14: Six Steps to Completion

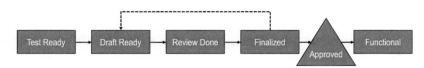

- *Test is Ready* stage (10%): do we have a unit test to start with?
- *Draft is Ready* stage (30%): does a full draft exist? i.e., is the deliverable complete?
- *Review Done* stage (15%): has the deliverable been peer reviewed?
- *Finalized* stage (20%): usually, some improvement opportunities need a fix.
- *Approved* stage (15%): team and sponsor agree that the result is complete.
- *Ready for Use* stage (10%): other stakeholders or users rely on the work delivered.

The Six Steps to Completion measurement method allows for an unbiased assessment of the progress done in a sprint, and early identification of obstacles. It is a proven way to structure daily scrums (Fehlmann, 2005-1), and it creates the awareness that completing a draft code is not more 30% of the total work required. This creates correct perceptions.

The story cards contain all information needed for self-organized agile teams, including references to the matrix cell, to functional size, to effort estimation and actual effort spent. Change effort needed because of new insights, new needs, or refactoring, is also recorded; it adds to actual effort. Progress indicators according Six Steps to Completion are marked green when completed, yellow when currently at work, and red if blocked by obstacles; see Figure 8-7. Since all information is measurable, together they constitute a set of simple, visually communicated metrics that all stakeholders easily understand.

8-3.8 THE KANBAN CHART

Six Steps to Completion steps yield burndown charts but also *Kanban*; see Figure 8-15. It organizes, assigns and tracks sprints with the development team. The *Waiting* column consist of story cards that already have been sized and estimated and are ready to be implemented, but are not yet included in the current sprint. One sprint consists of the story cards in the five columns *Test Ready* until *Approved*.

Figure 8-15: Kanban Table after a few Sprints

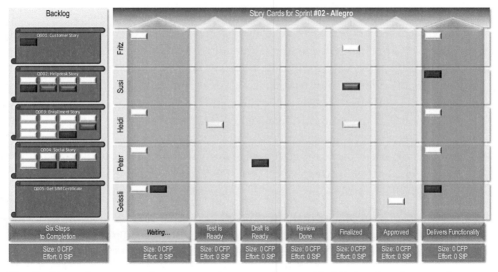

The *Backlog* column contains user stories, sometimes called *Epics*, that the team has not yet split into story cards and are not yet ready for implementation. The team knows the functionality of these user stories if they performed a functional size count to identify functionality; e.g., following (Buglione & Cencel, 2008) and (Buglione & Trudel, 2010).

The story cards generate the input to the sprint matrix chart; unlike Scrum (Schwaber & Beedle, 2002), we need the story cards for visualizing additional information, and thus a purely paper-based Kanban will probably not do well for this purpose, except maybe for very small projects. Ideally, both the BT-matrix and the Kanban table are views on the story card collection, and all these views are shareable with the team.

The story cards also hold applicable test cases; this works as well with globally distributed development teams. Visualization backed by metrics is probably the best one can get.

In the experience of the author, it is vital for team communication to print the story cards out and let the team manually, hands-on, modify them. The modifications are easily recorded back on the electronic cards; this is a task for the project office.

8-3.9 HARVESTING DEVELOPER'S INTELLIGENCE

The development team sees its story card arranged with business drivers on the BT-matrix. Initially, response will not match expected response, because one or several of the business drivers are not well covered. The convergence gap opens. However, in the practice of the author (Fehlmann, 2011-1), the profile comparison on the right-hand side of the BT-matrix tells easily where the deficit – or the overachievement – occurs.

The matrix cells evaluate as follows: the developers distribute impact points to all story cards recorded in the matrix cells, indicating how much they contribute to each business driver. This happens initially, at planning poker time, after the daily scrum meetings, and the team might revise it whenever needed. The developers use color marker buttons that they can distribute; red cards represent value 6, blue cards represent value 3, and light green cards represent value 1. Intermediary values are permitted, although if used they should be backed by other considerations such as aligning impact to story points, as already mentioned. Story cards in the "functional" cellar receive no impact points, as functional user requirements are not specific to any of the business drivers. The team places story cards with specific impact always in the sundeck, although their functionality is part of functional effectiveness.

This visualization is immediately understandable for developers and customers alike. Areas with strong correlation in the matrix contain much red, indicating important focus topics for reaching the stated goal. Areas that contribute less remain with lighter color or even white. Developers will come up with ideas how to balance the matrix and reduce the convergence gap.

The development team adds story cards according importance for the customer, such that the BT-matrix becomes balanced. The convergence gap metric is a quality indicator for the quality of the software product. A small convergence gap indicates that all business drivers are addressed, and none over-achieved.

8-3.10 CONTROLLING AGILE DEVELOPMENT

Meeting the customer's business drivers means to implement a solution that fits the goal profile for the business drivers. The sundeck transfer function maps the business impact of user stories onto business drivers. Business impact are the reasons why non-functional story cards have been included to the user story during one of the sprints. The team chooses and agrees these detailed reasons with the customer, as part of requirements elicitation. They were unknown before in full detail.

The convergence gap measures how well the chosen story cards match the business driver's goal profile. If the gap opens, the team can identify which aspects need more attention, and place additional story cards. In contrary, if some aspects are over-fulfilled, planned story cards can be removed or new, brilliant ideas rejected just on the fact that they apparently do not add new value for the customer. The sundeck of the BT-matrix serves for balancing the efforts with the needs of the customer, in a well-understood, visual manner. This makes agile software development lean. It blocks waste effectively from becoming part of a sprint.

8-3.11 CONTROLLING FUNCTIONAL COVERAGE

Meeting functional user requirements is different. Data movements implement FUR, not business impact. Data movements from a data movement map describe the functionality needed. The transfer function is simply measurable by counting the number of ISO/IEC 19761 COSMIC data movements that contribute to some FUR in a user story. The total number of data movement counts in the cells of the cellar matrix is larger than the total functional size, as there are many data movements serving more than one FUR. Such a transfer function implements the set of user stories by data movements as necessary. The cellar shows to what extend goals meet FUR priority. Waste planned functionality opens the convergence gap widely. If so, there is the possibility to save effort by removing part of the functionality as originally planned. Usually the cellar is more predictable and stable than the sundeck and less prone to adding or changing requirements.

The data movement map in Figure 8-4 shows a part of the cellar, referring to the functionality needed for the helpdesk and the certificate story. Part of the data movements appear on Figure 8-7 as well.

8-4 EARLY PROJECT ESTIMATION THE SIX SIGMA WAY

Today's lean and agile software developers use *Story Points* to predict the effort required for implementation, and plan sprints. Agile development methods outperform older approaches because they embrace requirements elicitation. Developing software is a knowledge acquisition process. However, it is hard to predict how long such a knowledge acquisition process will last, and what its cost will be.

Nevertheless, the need for software increases and *Information & Communication Technology* (ICT) has become essential for all but very few industries. Thus, getting reliable predictions what software will cost at the end is mission-critical for many organizations. Agile methodologies have but limited ways answering this challenging question.

8-4.1 APPLYING TRADITIONAL ESTIMATION APPROACHES TO AGILE

Traditionally, estimators use either *Macro* or *Micro Estimation* approaches. Macro estimations rely on historical benchmarking data such as the ISBSG database (Hill, 2010). Parametric tools such as Galorath or QSM allow fitting historical data to today's projects. Micro estimation try to identify all tasks needed to get the work done in a *Work Breakdown Structure* (WBS), size them, estimate the effort, and then add risk avoidance, mitigation and retention tasks as needed. In either case, functional size measurement provides the base for any decent estimation method; see the AACE International Recommended Practice No. 74R-13 (American Association of Cost Estimators, 2014).

The family of international risk management standards ISO/IEC 31000 defines risk as the "effect of uncertainty on objectives" (ISO/IEC 31010:2009, 2009), thus causing the word "risk" to refer to positive possibilities as well as negative ones. Project managers add the cost of the selected risk reduction strategies and risk contingency to the project budget to cope with risk exposure, for instance following the recommended practice (American Association of Cost Estimators, 2008-2). Usually such an approach works well if the domain is sufficiently well known and the project scope known in advance.

However, in most software projects, requirements elicitation is part of the project, and requirements change while performing the project. The amount of change is unknown in advance and difficult to predict using risk management techniques. Attempts to identify cost drivers for ICT projects are promising (Beni, et al., 2011); however, predicting and measuring cost drivers is hard and requires sophisticated techniques; see (Fehlmann & Kranich, 2012-1), and *Chapter 12: Effort Estimation for ICT Projects.*

8-4.2 EFFORT PREDICTION IN AGILE METHODOLOGIES

If there is no plan, there cannot be an estimate for the planned project. However, from Cohn (Cohn, 2005) the agile community learned how to deal with the uncertainty of projects without help of a plan. The basic learnings are that a) how to use story points to predict effort for a *Story Card*, or work item, in a sprint, and b) that ideal size and effort are somewhat orthogonal, and the two issues not being confused. The driver for setting priorities and story card selection should be the value created for the customer.

In Scrum, it is the task of the sponsor to decide what value creation means. Throughout this book, story card refers always to a work item that fits into one single sprint. User stories might split into more than one story card, if they do not fit into one sprint.

Agile masterminds like Cohn (Cohn, 2005) spread the idea that focusing on customer needs and business value avoids producing waste; however, reality is that project sponsors often find it difficult identifying value. New research (Bakalova, 2014) has shown that identifying value creation in agile projects is not as easy or straightforward as it might appear. While values remain stable, the value creation process changes over the lifetime of a project. That makes prediction of total effort even harder. For instance, the value of being able to move physically remains equally high over time. However, it depends on the available means and technologies, how such value is created: Horse carriages in the 19th century, cars and planes in the 20th and high-speed trains for the 21st century. Software projects face the same kind of technical evolution; however, in years not centuries.

Combining effort prediction methods with measuring business value or customer needs thus becomes interesting. Six Sigma has a long experience in uncovering and measuring customer needs, even hidden needs and requirements not (yet) consciously expressed and outspoken by customers. The *Quality Function Deployment* (QFD) discipline is the tool of choice in the Six Sigma toolbox for defining market strategy, for product management and improvement, requirements elicitation, and other aspects of customer orientation (Akao, 1990).

8-4.3 BUSINESS DRIVERS AND CUSTOMER NEEDS

Result of the VoC analysis is a *Priority Profile* describing the relative priorities among the topics of interest; see Figure 8-5. For software development, topics of interest can be functionality of software, or other rather non-functional characteristics that contribute to the success of the software product. The term customer needs often refers to functionality, whereas business drivers has been introduced by Denney (Denney, 2005) to denote primarily non-functional aspects needed to make a product successful. However, the distinction is fluent: non-functional software requirements regularly become functional requirements, when implemented, and account for a significant part of the so-called scope creep that affects almost all software projects. The advantage of the agile approach is the ability to cope with scope creep in a sensible way.

8-4.4 CALIBRATING STORY POINTS

A few initial story card tasks of the functional type can calibrate story points as a team-individual metric, and business impact per story card is an expert criterion measured on some ratio scale agreed by the team; e.g., 0 to 6.

For functional story cards in the cellar, the correlation between effort – estimated by story points – and functional size is linear. See Hill (Hill, 2010) and many others for studies that confirm proportionality between effort and functional size.

For non-functional story cards on the sundeck, this is unlikely. In contrary, business impact relates to story points. Business impact on the ratio scale 1, …, 6 correspond to the Fibonacci sequence 1, 2, 3, 5, 8, 13, 21 … Thus, a story card with 13 story points has business impact of six. This reflects the observation that doubling the effort does not necessarily double the effect. In turn, business impact can split among various business drivers; this is the only aspect that requires special attention. What further underpins the assumption that business impact depends from story points alone is the observation that user stories with many story points, no functionality and low business impact probably never qualify for implementation. This is another Lean aspect of the approach. Thanks to the robustness of transfer functions, this metrics work for any sequence used for story points in story cards, and any length of story point sequence. We only must count them, and assign to the right business driver.

Thus, functional story cards calibrate non-functional cards. This obviously work only if we measure functionality with a software size measure that is linear, since functional size splits to several story cards. ISO/IEC 19761 COSMIC meets that criterion. Both story cards and functional size refer to data movements.

8-4.5 EARLY ESTIMATION BY A QFD WORKSHOP

Tracking agile projects includes tracking efforts because of the dependency between effort and business impact on the sundeck, and effort and functional size in the cellar. Thus, before starting a project, experts from both business and development can predict the sundeck of the BT-matrix using classical QFD workshop techniques, while the cellar is predicable by transforming user stories into data movement maps, and using a benchmarking database. In classical QFD workshops, the experts agree on numbers in all matrix cells, representing the business impact needed per user story for the sundeck without bothering for the unknown details that later will be written into story cards. This is micro estimation, based on knowledge about the work breakdown structure in a project, and fully replaces the traditional Gantt chart approach; see *Chapter 12: Effort Estimation for ICT Projects.*

The preconditions for early estimations based on QFD are:

- Knowing the user stories (epics);
- Knowing the business driver's goal profile;
- Knowing FUR and their goal profile;
- Knowing the team's velocity; i.e., how many story points fit into one sprint;
- Have story points calibrated to functional size.

The last point refers to *Story Point Delivery Rate* (sPDR): the number of story points needed to implement one COSMIC function point; i.e., one data movement. If data movements touch different environments; e.g., when exchanging data with the SIM card provider, calibration might not equal among all data movements in the project. The ISBSG database provides guidance in identifying the various types of industry and application environment dependencies (Hill, 2010).

The quality of estimation for both parts of the BT-matrix is immediately perceivable by looking at the conversion gap. Like the developer team when planning the sprints, it is visually perceivable which parts need special attention. If the convergence gap closes, the matrix shows a transfer function that solves the problem. It might not be the best one, or the only one, or the cheapest, but it is a solution.

The cellar matrix prediction is less difficult as there are no "soft" impact evaluations involved, just functionality. For the cellar, the functional effectiveness matrix (see left side of Figure 8-6) tells whether planned functionality covers customer's needs – the business driver profile in our case. A small convergence gap indicates in this case functional completeness; thus, the functional size model likely describes what the customer really wants, reducing the risk of huge functional changes during development. For Lean Six Sigma in software development, this tool is indispensable to avoid excess and waste functionality be included in a software project.

Excellent tools exist that allow deriving cost out of functional size, if a benchmark is available with known final costs exists. The *International Software Benchmarking Standards Group* (ISBSG) collects and assesses industry data (Hill, 2010). This database, now in release 13, provides the best and affordable benchmark for doing macro estimations. Since story cards exhibit their functional size (preferable according ISO/IEC 19761 COSMIC) and effort linearly depend from size, if not disturbed by additional non-functional requirements, each story card has an estimation target connected to it, by the ISBSG database.

Since story cards also carry story points, it is possible to calibrate story points against functional size. This yields cost (€) per story point (*StP*). Thus, all we need for predicting the cost of a BT-matrix is how many story points we will have on the sundeck.

$$Cost \ per \ Story \ Point = \frac{Total \ Cost \ in \ €}{Total \ Size \ in \ StP} \tag{8-1}$$

Equation (8-1) predicts cost for the cellar of the BT-matrix.

8-4.6 THE SUNDECK MATRIX PREDICTION

Since at the beginning of a project the sundeck details remain unknown, since individual story cards do not yet exist, we can at least predict a relative profile of sundeck story cards.

Figure 8-16: Sample Sundeck Prediction QFD

Business Drivers		Goal Profile	Q001 Customer Story	Q002 Helpdesk Story	Q003 Enrollment Story	Q004 Social Story	Q005 Get SIM Certificate	Achieved Profile
BD1	Responsiveness	0.34	9	1				0.35
BD2	Be Compelling	0.44	9	3	3			0.44
BD3	Friendliness	0.41	3	3		3	9	0.42
BD4	Personalization	0.64	3	9	3	9	1	0.65
BD5	Competence	0.33	3	1	3	3	3	0.30
Solution Profile for User Stories			0.63	0.51	0.23	0.45	0.30	Convergence Gap
Total Business Impact: 81								0.03

0.10 Convergence Range
0.20 Convergence Limit

Figure 8-16 shows a sample sundeck QFD, showing a matrix that corresponds to the sundeck part of the Buglione-Trudel matrix in Figure 8-10. The total business impact value per story card yield the cell value in the transfer function, thus the cell numbers represent total business impact. Experts can predict business impact using standard QFD workshop techniques.

The experts distribute the correlation values where they expect some additional work needed to cope with the expectations of the customer. They anticipate the story cards that the team will invent during the sprints.

Sure, such a guess can become terribly wrong, and there is no way to allocate absolute costs levels directly; however, the convergence gap avoids completely unrealistic guesses. In our case, even the solution profile for the user stories came almost right in view of Figure 8-10. However, in general, other solutions exist and the development team must choose among the alternatives.

The experts doing the QFD workshop used the traditional 1-3-9 scale for their prediction; obviously, you can fine-tune such a prediction if you allow more granularity but doing so might not necessarily increase prediction accuracy.

As usual, the convergence gap of 0.03 shows the vector length differences between the goal profile vector and the effective profile vector, represented in the graph on the right. Thus, it tells the experts whether they considered all influencing factors.

For a micro estimate, it is possible to perform *Design of Experiments* (DoE), by selecting some crucial cells, typically with a high correlation value, and ask the development team how many story points they think this cell will need. The result is another ratio, story points per correlation value:

$$StP \ per \ QFD \ Cell = \frac{Cell \ Size \ in \ StP}{Correlation \ Value} \tag{8-2}$$

Multiplying the average story points according equation (8-2) with the total of all correlation values predicts the total cost of implementing the sundeck of the BT-matrix. This is a micro estimation approach (see section 12-1.4 in *Chapter 12: Effort Estimation for ICT Projects*); however, the prediction strongly depends from identifying relevant cells to calibrate equation (8-2).

8-4.7 COMPLETING THE COST PREDICTION BY MICRO ESTIMATION

With equations (8-1) and (8-2), the cost of development – or better: the number of sprints needed to complete software development – can be estimated. For this, we calculate the ratio between story points assigned to predominantly NFR story cards from the sundeck and predominantly FUR story cards form the cellar.

Multiplied with the functional size CFP_{FUR}, equation (8-3) yields an equivalent functional size CFP_{NFR} predicting additional effort needed to implement the NFR, in terms of functional size:

$$CFP_{NFR} = CFP_{FUR} * \frac{Total \ StP \ for \ NFR \ cards}{Total \ StP \ for \ FUR \ cards} \tag{8-3}$$

Note that there are story cards that both have impact and functionality. Over time, most story cards on the sundeck will do that. Such story cards count for both totals, *Total StP for NFR cards* and *Total StP for FUR cards*. The dividend in equation (8-3) will grow as the project progresses and compensate for the growth of CFP_{FUR}.

Comparing two projects requires adjusting by the *Total Business Impact* – the sum of all cell values for the respective QFD matrix – in order compensate for the difference in size of non-functional requirements. To make two QFD matrices comparable in terms of business impact, all cell values must be normalized to a scale, usually 0 to 9. For instance, it is possible to compare the prediction QFD in Figure 8-16 with the actual tracking QFD in Figure 8-12. During implementation, the total business impact increased from 81 (Figure 8-16) at prediction to 89 (Figure 8-12), while tracking progress in the project.

$$CFP_{NFR_{P2}} = CFP_{FUR_{P2}} * \frac{Impact_{P2}}{Impact_{P1}} * \frac{Total\ StP\ for\ NFR\ cards}{Total\ StP\ for\ FUR\ cards} \qquad (8\text{-}4)$$

Thus, a prediction is possible based on the functional size of the new project P2, $CFP_{FUR_{P2}}$ and its total business impact, regarding an existing project P1.

It is convenient to define an *NFR Extension Factor* as follows:

$$NFR\ Extension\ Factor = \frac{1}{Impact} * \frac{Total\ StP\ for\ NFR\ cards}{Total\ StP\ for\ FUR\ cards} \qquad (8\text{-}5)$$

This extension factor can be used conveniently to predict the NFR equivalent for functional size as shown in equation (8-6):

$$CFP_{NFR} = CFP_{FUR} * Impact * NFR\ Extension\ Factor \qquad (8\text{-}6)$$

The total functional size of a project, based on the initial count made for the project vision, is thus:

$$CFP = CFP_{NFR} + CFP_{FUR} \qquad (8\text{-}7)$$

Based on the *Project Delivery Rate* (PDR), in hours per function point, usually taken from the ISBSG database (Hill, 2010), the total effort is then easily predictable.

$$Total\ Effort\ in\ Hours = CFP * PDR = (CFP_{NFR} + CFP_{FUR}) * PDR \qquad (8\text{-}8)$$

The sample estimation shown below predicts 8 sprints, based on an average duration of sprints of 9.1 Days length and a *Team Power* of 4.5 for the Helpdesk project shown before.

The only reference project used in Figure 8-17 are the first two sprints in the Helpdesk project shown; thus, the PDR is not applicable and has been replaced by an assumed 12 h/CFP. The PDR proposed by the ISBSG database (Hill, 2010) refers to apps development with performant frameworks and is certainly not applicable here.

8-4.8 QUALITY MANAGEMENT BY COST PREDICTION

An interesting consequence of the Eigenvector approach is that QFD Prediction Workshops cannot have an arbitrary outcome but they need to agree on one eigenvector corresponding to one of the m eigenvalues (see *Chapter 2: Transfer Functions*), where m is the dimension of the goal profile. There are not infinitely many solutions how to meet customer's needs and the corresponding business drivers. Consequently, the finite number of possible solutions is worth being investigated.

Figure 8-17: Agile Effort Prediction for Helpdesk Project

Effort Prediction	Reference	Manual	Selected	ISBSG
Project Delivery Rate (PDR):	74.46 h/CFP	12.00 h/CFP	12.00 h/CFP	4.25 h/CFP
NFR Extension Factor (CFP/Impact):	0.76%		0.76%	
Team Power:	5	4.5	4.5	
Average Sprint Duration:	9.1 Days		9.1 Days	
Hours per Day:	8.0 h	7.2 h	7.2 h	
Reference Functionality:	110 CFP		110 CFP	
Predicted Functionality:		110 CFP	110 CFP	
Predicted Impact:	81		81	

Predicted	FUR Size	PDR	Hours	NFR Size
Functional	110.00 CFP	12.00 h/CFP	1320 h	67.32 CFP
Non-functional	67.32 CFP	12.00 h/CFP	808 h	
Total	**177.32 CFP**		**2128 h**	
		294 h/Sprint -->	8	Sprints

Selecting cheaper or more expensive solutions is important to business strategy and the QFD team can do this transparently and on purpose. The total business impact serves as a metric. For instance, when looking at the helpdesk example (Figure 8-16), two other possible solution strategies can be selected, corresponding to two other eigenvalues. The most expensive solution usually is found by filling in as many cells of the matrix as possible, the least expensive by filling as few as possible. The solutions might be feasible and desirable or not; this is left to the business strategy owner to decide.

For example, a high-end solution to the helpdesk project might come out of a QFD prediction matrix as shown in Figure 8-18:

Figure 8-18: High-end Sundeck Prediction QFD for Helpdesk

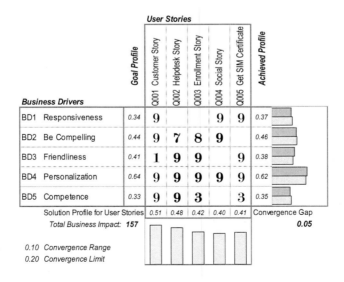

The total business impact of 157 corresponds to 10 instead of 8 sprints and to almost equal distribution of quality-related extra effort for all five user stories.

Figure 8-19: Low-end Sundeck Prediction QFD for Helpdesk

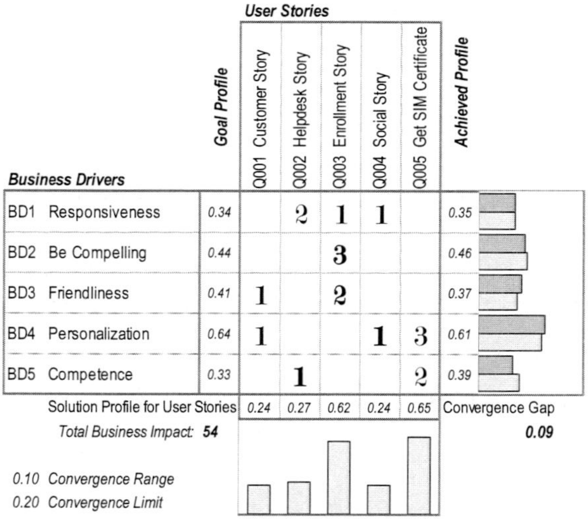

Business Drivers		Goal Profile	Q001 Customer Story	Q002 Helpdesk Story	Q003 Enrollment Story	Q004 Social Story	Q005 Get SIM Certificate	Achieved Profile	
BD1	Responsiveness	0.34		2	1	1		0.35	
BD2	Be Compelling	0.44			3			0.46	
BD3	Friendliness	0.41	1		2			0.37	
BD4	Personalization	0.64	1			1	3	0.61	
BD5	Competence	0.33		1			2	0.39	
Solution Profile for User Stories			0.24	0.27	0.62	0.24	0.65	Convergence Gap	
Total Business Impact: **54**								**0.09**	

0.10 Convergence Range
0.20 Convergence Limit

It is more difficult to find a cheaper approach, as the one presented in section *8-4.3 Business Drivers and Customer Needs* is already quite optimal; however, our team might come up with a strictly technical solution approach, trying to fill in a minimum number of cells only. This might look like Figure 8-19.

The fact that the team used cell values to a maximum of 3 instead of 9 only indicates less freedom in choosing cells; what matters is the smaller total business impact of 54 instead of 81 in Figure 8-16. Since business impact is calculated on the base of the normalized scale 0 to 9, it does not matter whether the team uses 3 or 9 as its highest score.

Figure 8-19 shows a focus on the two technical user stories *Q003: Enrollment Story* and *Q005: Get SIM Certificate*, yielding a solution strategy that targets at customers enchanted by technical functionality. The solution profile is an excellent tool for identifying strategic targets. Moreover, it will be used for prioritizing backlog user stories for the sprints. Thus, visualizing solution strategy to the development team is yet another advantage of the approach.

This solution can be implemented in 7 instead of 8 sprints; the cost savings are probably not as important compared to the loss of focus to *Q001: Customer Story*. Nevertheless, the solution is valid in view of the business drivers and cost constraints are of high importance to the sponsor of this project, it could be considered as a base for starting development.

8-5 Conclusion

Software project cost estimation for the 21st century cannot consist of macro estimation only – predicting cost based on benchmark data and suitable parameters. However, the old way of micro estimation – build a Gantt chart containing all activities, assign resources, sum up effort, and add risk contingencies – no longer works.

The idea for software project cost estimation presented in section 8-4 uses modern development paradigms and is novel insofar as it takes the customer viewpoint into consideration, paying respect at the customer's needs and embracing change. Traditionally, change requests had been labeled *Scope Creep* and considered as enemy of due diligence and project management. Today's micro estimation approaches must be able to predict scope creep.

Figure 8-20 shows how NFR evolve eventually into FUR during the lifetime of a project. It originates in the *COSMIC Guideline on Non-Functional & Project Requirements* (COSMIC Consortium, 2015, p. 6).

Figure 8-20: NFR may Evolve, Wholly or Partly, into FUR as a Project Progresses, adapted for Cost Engineering by Vogelezang (Vogelezang, 2015, p. 40)

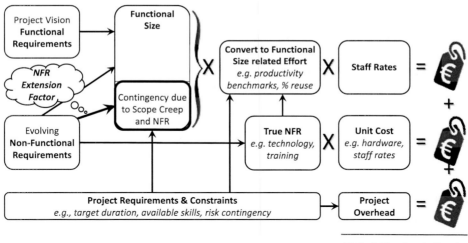

Newly introduced FUR in turn evolve eventually into additional project cost. This is the model for micro estimation in software projects, and for this century, a candidate model for most projects undertakings.

The challenge is to predict the NFR extension factor, and thus predict scope creep, which reflects the process of uncovering the true requirements for a product.

CHAPTER 9: REQUIREMENTS ELICITATION

What happens when you omit valuable information? Selection of topics. Functional, non-functional, quality, architecture, social status, recommendations, etc. Why remote possibilities start playing a role. Link to chaos theory.

Linking requirements among chain topics by transfer functions. The Arrow Model: additional to the matrix combination you can combine profile vectors.

9-1 THE NEW ECONOMICS BY DEMING

In the early 20th century, people started to understand that economics is not just a single-silo undertaking, and that it is not true that simply the fittest wins, but that in contrary one value provider depends from the other. It did not help avoiding the deep recession of the 1930ies but it could have done so in the financial crisis of this century, when people started understanding that the failure of one economy is likely to become a failure of every other economy, since they are interdependent.

Deming (Deming, 1986) introduced quality control on the intra-company level. Instead of delivering on time and on budget only, the functions or production steps preceding had to deliver at agreed quality levels. That made production far more efficient. Prices decreased and overall quality increased.

The move to *Internet of Things* (IoT) creates similar effects. Many goods that previously were of low to medium quality now add intelligence that makes them more useful and add value. Prices might now fall in the first instance; however, the effect of scale and the ability to replace high manufacturing cost by low-cost chips and pecking motors, adapting to the environment instead of withstanding rough usage, will ultimately have similar effects.

9-1.1 THE COSMIC INFLATION IN THE ICT OF THE 21ST CENTURY

The world of *Information & Communication Technology* (ICT) is currently undergoing a cosmic inflation. Like in the early universe, quantum fluctuations in a microscopic inflationary agile region become the seeds for the growth of structure in the ICT universe, and the universe becomes transparent. This phenomenon, familiar to cosmologists, happens right now to ICT. One of the reasons is that all "things" become intelligent, receive IP addresses and connect to the Internet. The possibility to create new

ICT-based products seems unlimited; however, there must be sponsors to fuel the inflation.

The other reason is the move from static product definition and waterfall models n product management agile products that adapt themselves to changing customer's needs. Delivery changes from one-upon-a-time to continuous delivery – not only for software, but also for all software-rich products such as cars, planes and trains; houses and household appliances, clothes, and even for food products.

It is the actual process of acquiring knowledge, translating them into requirements and implementing into software. It changes slightly from project to project, or with technology change. Although maturity models and guidance exist for the software development process, it is never completely repeatable except for a few process steps such as gate reviews or other key deliverables.

Imagine for example a system that connects cooking recipes with the inventory of kitchen fridges and cupboards, with food information, home delivery grocery stores, and finally with stoves and cooking plates to help people cooking at home. If cooking plates know what they heat up, cooking pots will not overspill. If the fridge knows, what the cook needs, it can tell the grocery shops what their home delivery shall bring this very afternoon. For such a case, the sponsors are grocery stores, cooking recipe portals and kitchen appliance suppliers. The challenge is to understand the needs of all stakeholders, especially sponsors, and continuously deliver what they need to keep the project going, profitable and successful. Software that implements for instance the *Kitchen Helper* project, as discussed in section 9-5 of this chapter, is a learning system that continuously adapts to customer's needs; else, it will prevail not for long. Thus, continuous delivery and creating a community is part of the project and an absolute must.

9-1.2 ARDUINO TECHNOLOGY

Arduino is some open-source electronics prototyping platform based on flexible, easy-to-use hardware and software, targeted at artists, designers, hobbyist and anyone interested in creating interactive objects or environments. Compatible platforms for professional and commercial use are available by now. Arduinos can sense the environment by receiving input from a variety of sensors and can affect its surroundings by controlling lights, motors, and other actuators. The microcontroller on the board understands the Arduino programming language and embraces the Arduino development environment; both are open source distributions. Arduino projects can be stand-alone or they can communicate with software running on a computer (Arduino Community, 2005).

Arduinos – and other related initiatives such as Raspberry PI – change the characteristics of embedded software. Arduinos come with a development platform and environment that is so easy to handle that every digital native is capable to program own devices. Arduino devices affect the physical world, as they sense the environment and actuate on it, following the Arduino program. It is very simple to build a device that opens and close windows depending on the weather report received from Internet, or pours water to plants before they are parched, or switch illumination on and off upon sensing a smartphone with near-field technology or Bluetooth switched on. Arduinos sense the human, based on their most basic appliances, and act. Such a technology on a shop window certainly creates attention among visitors and shoppers, making shopping fun.

However, Arduinos is a prototyping platform. Before a significant number of actual shops use such technology, the prototype must undergo some basic product development steps. It is less the question of reliability; the Arduino platform is so cheap that replacing a defective component never is an issue. More of an issue is maintainability, as not every user is keen on replacing failed components often; safety, because the physical environment is affected; and security, since hacking an Arduino is more attractive than using brute force against physical locks.

9-1.3 THREE-DIMENSIONAL PRINTING

3D printing or *Additive Manufacturing* (AM) is one of various processes of making a three-dimensional object from a 3D model or other electronic data source primarily through additive processes in which successive layers of material are laid down under computer control. A 3D printer is a type of industrial robot. By the early 2010s, the terms 3D printing and additive manufacturing became umbrella terms for all AM technologies. Although this was a departure from their earlier technically narrower meanings, it reflects the simple fact that the technologies all share the common theme of sequential-layer material addition/joining throughout a 3D work envelope under automated control. In the 2010s, metal parts such as engine brackets and large nuts start growing instead of being cut (or machined) from bar stock or plate in production.

Popular 3D printers use polymer products, industrial printers use metal mold. It lays down successive layers of liquid, powder, paper or sheet material. These joined or automatically fused layers, which correspond to the virtual cross sections from the CAD model, create the final shape. The primary advantage of this technique is its ability to create almost any shape or geometric feature, including sponge-like structures.

Together with Arduinos, the variety of feasible prototypes has increased at an enormous pace. With today's popularity of *FabLab's* – community-based 3D manufacturing laboratories – 3D printing has become within everybody's reach. Building model railroads for instance is no longer handcrafting; it consists of Arduino programming and 3D printing.

9-1.4 Creating new Gadgets

Primarily, these things look like gadgets for hobbyists, artists and individuals. However, for product managers, it is not only interesting to build prototypes but also to predict which prototypes open the way for future products and a business model. Certainly, it is possible to do experiments and try out whether there is interest for some of these gadgets. The problem is that gadgets often create attention even if no one ever will pay for them. If the customer sees no value in it or expects things free, there is no better way than use it as a possible carrier for advertising. This is not an exciting business model.

The Internet has created content accessible for free, which previously were linked to a business model such as newspapers. Today's Internet newspapers find it difficult to persuade readers to pay for content. In the beginnings, newspapers were keen to produce everything online for free to create better revenue from cheaper advertising, still hoping the print business will not break down. For many publishers, it broke, and especially local newspaper disappeared or merged to larger units, able to generate higher revenue for advertisement and lower cost for publishing and journalism.

When creating gadgets for household appliances or similar, how can the *"Newspaper Trap"* be avoided? For instance, when producing rain sensors, say for water pouring to plants, it might be cheaper relying on free weather forecast. If this produces similar results with far lower cost, your business model might fail.

9-2 Requirements Prioritization By Customer's Needs

Uncovering customer's needs is the approach that allows avoiding the newspaper trap. The need to know what happens in the nearer and wider neighborhood, the need to pour enough water on plants, the need for attracting customers to shop products is independent from available technology and did not change for centuries. The means to fulfill this needs changed. Thus, analyzing the needs, and looking for the best controls to fulfill these needs is the way to go. Therefore, we need analyzing transfer functions for requirements prioritization.

9-2.1 Identify Customer's Needs

Customer's needs never depend from technology. Switching on illumination at night, when entering a room, is not the need to find the switch. The switch is today's means, based on 19th and 20th century technology, to illuminate a room with electrical pulps. The need is to see something, not stumbling in darkness. In the 21st century, the Internet of Things with its sensors and actuators might provide new answers to such needs.

Thus, how do you distinguish customer's needs from solutions? When asking the customer about his needs at night, he or she will tell you the 19th and 20th century solution, and pretend this is the "need".

Often, time lining is an excellent means of distinguishing needs from solutions. A need seldom appears from nowhere; it was already there for some time but remained unfulfilled and unanswered. Constraints might block needs fulfillment. If someone is living alone on an island, without access to Internet, there is no way to satisfy his or her need to talk and to exchange social contacts. At Roman times, the need for speedy goods transportation existed as today: Romans liked eating fresh oysters even at places in mid-Europe what now is Switzerland. There were neither planes nor fast trucks; only horse riders could benefit from the excellent Roman streets from the sea to Switzerland. The need for speed was the same as today; constraints imposed by available technology dictated the solution. The oyster consumer is an interesting case because they mix basic requirements: ascertain that oysters a fresh enough such that they will not cause intestinal disease.

9-2.2 FUNCTIONAL AND NON-FUNCTIONAL REQUIREMENTS

Requirements can be functional (FUR) or non-functional (NFR). A sample functional requirement for oyster transportation is "As a conveyor, I need a transportation case such that I can protect oysters against dirt and dust, so that they arrive clean". Another functional requirement may read, as "As a conveyor, I need to keep seafood cool enough such that oysters do not overheat, so that they arrive alive". Both requirements are functional in nature, whereas the requirement to travel fast is non-functional. "Fast" means different things at Roman times than today; the requirement is still the same.

Besides time lining, another source for uncovering customer needs are the Voice of the Customer (VoC) techniques: for oysters, Going to the Gemba for inspecting customer's place, see *Chapter 4: Quality Function Deployment*. A typical Gemba is looking at oyster consumers how they try to find out whether an oyster still is alive by dropping lemon juice on them before eating.

Since eating oysters also is a social event, there are more requirements that are non-functional. First, they must origin from a reliable supplier. Consumers cannot control neither the oyster raising nor the transportation process, thus they must rely on trust. This too is a constraint. Trust is a customer's need; generating trust requires recommendation or certification by some third-party certification agency. Today, specialized organizations oversee controlling production and transportation; additionally, there are labels that guarantee organic raising conditions and/or exceptionally high quality levels. Consumers rely on this for their need to trust the stuff that they eat.

The separation between functional and non-functional is not always clear. Oysters die when staying too long in a transportation case, even if the temperature keeps at optimum level. To let living oysters arrive live, the transportation solution needs to be fast enough. While the original customer requirement is non-functional, it deploys into functional requirements as soon as one starts addressing it. The certificates and labels that guarantee fresh and gentle delivery in turn identify functional requirements that support the non-functional customer requirement regarding freshness.

9-2.3 Non-Functional Requirements in Software

The main problem with software requirements are that they are layered in Deming chains. On top, there are customer values, something people often confound with non-functional requirements, then these values deploy into functional requirements with a measurable functional size when going down the Deming chain. However, it is impossible to know all FUR at the beginning of a software project.

For instance, user friendliness is a value, sometimes even a business driver; privacy and security are values – security often needs lot of functionality otherwise not implemented. Reliability, Flexibility, etc., quality characteristics as listed in ISO 9126, now 25010:2011, are typical values that require specific functionality implementing them. Most quality characteristics that go as a true customer requirement imply some functionality. The often-cited example (Symons 2014): "we need the database implemented with some vendor's database" represents a constraint imposed by the business value "Realize economies of scale by focusing on one database license only" and the presence of other database solutions from the same vendor.

In 2015, the COSMIC Consortium published together IFPUG a glossary on FUR and NFR (COSMIC and IFPUG, 2015), and the COSMIC Consortium published a guideline (COSMIC Consortium, 2015) explaining how such FUR and NFR should be treated. It shows how NFR might "evolve, wholly or partially, into FUR as a project progresses." Ultimately, NFR disappear from a product when finished, while project constraints affect project cost.

9-2.4 EVOLUTION OF TICKET AUTOMATA IN LAST DECENNIALS

User-friendliness is a business driver, not a (functional or non-functional) requirement. We explain this with the example of ticket automata used for public transportation in the last thirty years or so.

How was it when you stood in the Nineties for the first time before a ticket machine, assuming you are old enough? Initially, you had one button per tariff zone, and it was up to you figuring out which ticket you need to reach your destination. With the second generation, it improved: you got one button per destination stop but still no means to change origin. In both the data movements map and the transaction map, this looks straightforward. An Entry, or EI, for selecting zone tariff and another for selecting the destination station. For any more complicated travel, you still had to go to a ticket shop manned with a human expert in travel ticket solutions.

Then, engineers detected that there are more possible destinations than fitting on a single automata panel; thus, they needed new functionality to implement an adaptable automata panel. Hardware button entries simply did no longer work. They invented a functional process allowing user type-ahead and proposing matching destination stations. Later, they found another functional process allowing entering address data, or GPS data, doing a database inquiry and getting the next bus stop as destination.

Today's ticket automata no longer have Entries for stations, but EQ, even EO, proposing suitable solutions from where the user can select. ILF help to recall personal, or public, preferences for providing quick selection button. All this is functional; nothing remains non-functional. Thus, the business driver making ticket automata and web ticket shops more user friendly has raised solid functional requirements, simply expected by the user and not even delighting them any longer. More functionality is required to better control user demand for mobility at rush-hour time. In turn, the number of traditional ticket shops has significantly decreased, and travel companies can afford offering much more complicated travel solutions, even if human no longer understand them, but machines do.

9-2.5 ASSESSING NON-FUNCTIONAL USER REQUIREMENTS (SNAP)

Despite the non-existence of non-functional user requirements in software, in the strict sense of the notion, assessing customer values and business drivers is still uttermost useful. Even more: the higher the complexity of software-based solutions, the higher the cost for understanding requirements. This is *Requirements Elicitation*, and this is today the most important, and most costly, part of any software project. Any means helping to predict cost of this part of a software project is highly appreciated.

In this respect, the approach taken by IFPUG is very interesting. In the *Software Non-functional Assessment Process* (SNAP), see (IFPUG Non-Functional Sizing Standards Committee, April 2013), Fehlmann and Kranich use the IFPUG transaction map based on the five transactional elementary processes, evaluating them according the importance for the customer, or user, of the software product. Even elements of the model that do not exhibit any functionality at all may yield value to the user. As an example, the so-called *Code Data*, build-in tables that translate e.g., airport three-letter codes into human readable airport names, provide value to the user, indeed. The ISO/IEC 20926 international standard does not identify code data as ILF, even if maintained by the application; however, they doubtless have an impact on user friendliness and thus count for SNAP. The weight of the count depends from application environment and user sensitiveness. This kind of "non-functional size" is very helpful for identifying cost, not depending from functionality alone, which will incur once the project gets started. Functional aspects and customer values together allow predicting project cost; in no way, you can expect good predictions from one or the other alone.

9-3 THE CAR DOOR EXAMPLE

Imagine that a group of engineers in automotive industry wants to design a new car door software. The mechanics of the door naturally matter as well for solution quality, but for simplicity we focus on the software only, to steer the car door. Assume a Kano analysis uncovered the following customer's needs for a car door; see Figure 9-1.

Figure 9-1: Customer's Needs for a Car Door

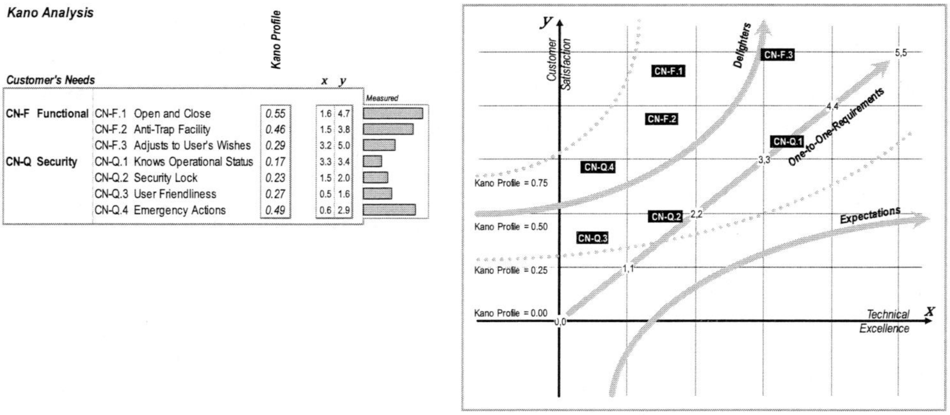

The following sample QFD process demonstrates the usefulness of the eigenvector approach in QFD. Stating technical requirements wrong results in an unbalanced QFD matrix and a bad convergence gap.

9-3.1 INITIAL WRONG SOLUTION APPROACH

Assume the engineers wanted initially to design a solution along the following solution qualities (Figure 9-2):

Figure 9-2: Initial List of User Stories for Car Door Control Software

	User Stories - Initial Topics	Attributes		
USt1 Normal Functionality	USt1.1 Open from Interior	Mechanical	Automatic	
	USt1.2 Open from Exterior	Mechanical	Remote	By Sensor
	USt1.3 Close from Interior	Automatic	With sound	
	USt1.4 Close from Exterior	Automatic	With sound	
USt2 Sensing	USt2.1 Engine Start	Door adjusts	to new status	
	USt2.2 Engine Stops	Door adjusts	to new status	
USt3 Emergency	USt3.1 Emergency: No Energy	Loss of power	Fail-Safe	
	USt3.2 Emergency: Collision Stop	Door open		

The QFD workshop predicted the transfer function as follow – with a dismal convergence gap. The comparison of response achieved with the goal profile shows major discrepancies, and the matrix is not well balanced.

The convergence gap of 0.44 in Figure 9-3 is off limit and thus the result is questionable – surprisingly, *UC-1.4: Close from Exterior* becomes the most prominent solution control.

Figure 9-3: Initial QFD Deployment for Car Door Control Software

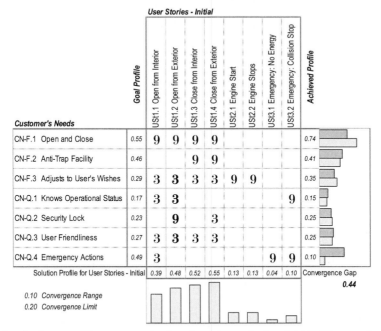

The initial solution profile represents the eigencontrols for this QFD transfer function. Apparently, the matrix is unbalanced, and the achieved profile is quite far away from the goal profile. Especially the emergency actions look unsatisfactory; the eigencontrols clearly cannot make much out of the two strong correlation factors with the emergency actions. Consequently, the (yellow) achieved response may point in a very different direction. The actual solution profile suggests a technical focus on the door open/close status. Since the convergence gap is bad, it is apparent that this conclusion is wrong.

9-3.2 IMPROVED CONTROLS

A simple modernization of the sensing status from of engine start/stop to car motion and power supply sensing yields a significant improvement:

Figure 9-4: Improved List of User Stories for Car Door Control Software

		User Stories - Final Topics	Attributes		
USt1	Normal Functionality	USt1.1 Open from Interior	Mechanical	Automatic	
		USt1.2 Open from Exterior	Mechanical	Remote	By Sensor
		USt1.3 Close from Interior	Automatic	With sound	
		USt1.4 Close from Exterior	Automatic	With sound	
USt2	Sensing	USt2.1 Motion Sensing	Door adjusts	to new status	
		USt2.2 Power Supply Sensing	Door adjusts	to new status	
USt3	Emergency	USt3.1 Emergency: No Energy	Loss of power	Fail-Safe	
		USt3.2 Emergency: Collision Stop	Door open		

It seems that sensing the engine status was the wrong requirement; it should extend to motion and power supply sensing yielding an expanded view on the car's status.

Figure 9-5: Improved QFD Deployment for Car Door Control Software

Customer's Needs	Goal Profile	USt1.1 Open from Interior	USt1.2 Open from Exterior	USt1.3 Close from Interior	USt1.4 Close from Exterior	USt2.1 Motion Sensing	USt2.2 Power Supply Sensing	USt3.1 Emergency: No Energy	USt3.2 Emergency: Collision Stop	Achieved Profile
CN-F.1 Open and Close	0.55	9	9	9	9	3				0.52
CN-F.2 Anti-Trap Facility	0.46			9	9	9				0.41
CN-F.3 Adjusts to User's Wishes	0.29	3	3	3	3	9				0.34
CN-Q.1 Knows Operational Status	0.17	3	3				1	1	9	0.18
CN-Q.2 Security Lock	0.23		9			1	3	9		0.25
CN-Q.3 User Friendliness	0.27	3	1	3	1	9				0.29
CN-Q.4 Emergency Actions	0.49	9				9	9	9	9	0.52
Solution Profile for User Stories - Final		0.41	0.31	0.36	0.35	0.58	0.25	0.17	0.22	Convergence Gap
										0.09

0.10 Convergence Range
0.20 Convergence Limit

Thus, the QFD workshop can validate results and estimate its measurement error and variability of estimate when looking at the convergence gap.

The resulting solution profile now shows much better what really matters for the customer; see Figure 9-6. The change towards Figure 9-2 is relatively small but reflects the wider view on the technical capabilities that are needed to meet customers' needs regarding functionality and safety of car doors.

Figure 9-6 yields a much better view on technical requirements; it becomes apparent that motion sensing is what serves customers best.

Figure 9-6: Resulting Solution Profile

User Stories - Final

		User Stories - Final Topics	Priority Weight	Priority Profile	
USt1	Normal Functionality	USt1.1 Open from Interior	16%	0.41	
		USt1.2 Open from Exterior	12%	0.31	
		USt1.3 Close from Interior	14%	0.36	
		USt1.4 Close from Exterior	13%	0.35	
USt2	Sensing	USt2.1 Motion Sensing	22%	0.58	
		USt2.2 Power Supply Sensing	9%	0.25	
USt3	Emergency	USt3.1 Emergency: No Energy	6%	0.17	
		USt3.2 Emergency: Collision Stop	8%	0.22	

This workshop in the car supplying industry happened long before everybody talked about self-driving cars. QFD, with the eigenvector method, could predict the need for intelligence in cars well in advance.

9-3.3 MEASURING THE TRANSFER FUNCTION

Since we are developing software, the impact of one control quality upon a response quality is easily measurable: the cost of developing the respective impact can be measured. This yields more details per cell and allows replacing the 0–1–3–9 values by more specific numbers proportional to the effort spent per cell. Instead of effort, other metrics for the impact can be used as well, if effort is not spread equally among the software development project; e.g., because of reuse of existing components.

9-4 UNCOVERING HIDDEN REQUIREMENTS

This case study originates from a software house supplying software solutions who does regular NPS surveys. Its customers are businesses doing customer communications, with personalized marketing, but also with transactional messages such as credit card or phone bills. Voice of the Customer (VoC) means listening to the organizations using the product; surveying end user consumers is less rewarding as they often are not aware of the communication software product behind their personalized communication with the organization.

Table 9-7. Sample NPS Profile According Customer Segments

		NPS	Profile	
EP1	Enterprise Decider	40%	0.42	
EP2	Enterprise Influencer	38%	0.40	
EP3	Enterprise User	50%	0.53	
SP1	Specialist Decider	31%	0.33	
SP2	Specialist Influencer	33%	0.34	
SP3	Specialist User	38%	0.39	

An NPS survey addresses the products' user community by surveys of the willingness to recommend this product by asking managers, influencers, and product users. Typical customers include banks and insurances, providing statements; telecoms and utilities, providing bills; postal services, providing online mail, bill presentment and payment service.

The NPS profile relies on three customer types: deciders, influencers, and users; in this case from two different industries: enterprises where ICT integrates all aspects of customer communication, and specialized service providers that serve many smaller organizations with dedicated customer communication. This yields six customer segments. The results of the NPS survey are shown in Table 9-7, and the transfer function analysis is based on seven business drivers shown in Table 9-8. The claim is that these seven business drivers explain the observed NPS response for the six customer segments.

Table 9-8: Business Drivers for High NPS for Customer Communication Software

	Topics	Attributes
C1	Technical Usability	Performs the intended tasks, ease of use, human interface design
C2	Service Integration	Ease of integration, interoperability, servitization, installation, cloud services
C3	Mobile Platforms	Support of mobile platforms, automation, flexible media formats
P1	Deployment & Licensing	Servitization; pricing & licensing schemes meet the needs of the customer
P2	Process Excellence	Ease of doing business; process excellence, Six Sigma
Q1	Time to Market	Service Quality - Always be first to make the product work
Q2	Fitness for Purpose	Product Quality - No bugs, no recalls, just working fine

An NPS of 40% – 50% is very high for a supplier of software. The company enjoys overall healthy growth and leaves competition behind. Servitization is the innovation of organization's capabilities and processes to create better mutual value through a shift from selling product to selling Product-Service Systems (Vandermerwe & Rada,

1988). Servitization describes the trend in product management that leads to the *Internet of Things* (López, et al., 2013). Normalization as explained in section 2-4.2.4 calculates the NPS profile in Table 9-7. The suspected business drivers are those listed in Table 9-8.

9-4.1 IMPORTANCE

The transfer function looks as shown in Figure 9-9. In terms of statistical indicators, the convergence gap of 0.28 is the vector difference between the measured NPS profile and the validation profile based on measurement with a confidence interval of 93% is just within limits. The convergence gap is less than one third of the unit vector. Measured NPS profile and validation profile almost coincide – see graph to the right of Table 9-7.

Figure 9-9: Importance Transfer Function   *Figure 9-10: Satisfaction Transfer Function*

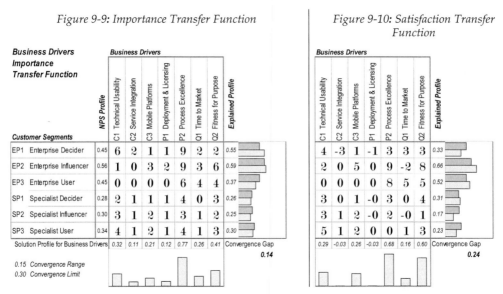

The calculated priority profile of the seven business drivers provides guidance both for managers and for developers with its strong focus on P2: Process Excellence and Q2: Fitness for Purpose. The signal detected is clear enough. Section 2-4 in *Chapter 2: Transfer Functions* explains how to calculate solutions to transfer functions. The numbers in the matrix cells indicate the frequency of mentioning the respective business driver. The results have been normalized to a scale from 0 to 9.

9-4.2 SATISFACTION

In this sample case, negative satisfaction is not as abundant to violate the condition to calculate the solution with Eigenvectors. The matrix AA^{T} still is positive definite.

Moreover, the satisfaction analysis shows an interesting result; see Figure 9-10. The convergence gap of 0.24 is at near to the limit; it seems these business drivers do not explain satisfaction well enough. Satisfaction is not an equally good explanation for the observed NPS per customer segment. However, it can be seen that customers have an issue with P1: Deployment & Licensing, and with C2: Service Integration.

9-4.3 COMBINING IMPORTANCE AND SATISFACTION

As usual, neither importance nor satisfaction alone provides a valid profile for the business drivers; see section 5.5 in *Chapter 5: Voice of the Customer by Net Promoter*®. In this case, the team decided to weight importance and the satisfaction gap equally, this yields a profile for the business drivers shown in Figure 9-11. The weak satisfaction profile affect importance especially hard in case of the drivers P1: Deployment & Licensing and C2: Service Integration, where dissatisfaction is overwhelming.

Figure 9-11. Combining Importance and Satisfaction

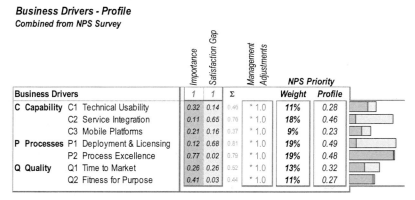

Here again it is apparent why importance alone does not explain all reasons why customers recommend. Dissatisfaction with relevant business drivers reduces the weight of importance for recommendation.

9-4.4 AGILE PRODUCT DEVELOPMENT

Our software house developed a new tool allowing organizations connecting interactively to its customers by providing two-way personalized business communication messages between the organization and its customers, integrated with *Customer Relationship Management* (CRM). Target users of this product may include banks doing wealth management allowing private investors to initiate and approve stock transactions and other investments through a network of mobile apps or slower, more traditional communication means such as phone calls. Other users of the software may use the same product to connect with their regular patrons to book vacations and wellness weekends, or opera houses may use it for their registered visitors when booking seats in performances. The product has a wide range of applicability and serves very different industries. It seems to extend the product portfolio of the software house in a very sensible way.

The business drivers that have been the controls of the NPS analysis become now the intended response for the second transfer function in the Deming chain. This transfer function looks like the well-known House of Quality (HoQ) in QFD – however, mapping user requirements to business drivers' profile rather than VoE to VoC.

9-4.5 USER REQUIREMENTS AS CONTROLS

Table 9-12 shows the initial capabilities required by these stakeholders:

Table 9-12. Selected User Requirements (Capabilities) for Interactive Customer Communication

Capability Requestor		User Requirements
RC1	C-Level Executive	Support brand recognition consistently in the product
		Increase customer loyalty
		Differentiate company through personalized service
RC2	Financial Manager	Leverage best channels to control delivery costs
		Benefit from new revenue streams
		On demand production replaces stock inventory
RC3	ICT Operator	Only one deployment for all platforms
		Flexible open interface
		Automated installation and maintenance processes
RC4	Marketing Professional	Exploit opportunities to sell similar products
		Include e-Marketing, mobile marketing
		Central marketing assets management
RC5	Compliance Officer	Approval workflow for marketing messages
		Auditable approval cycle
		Monitor and track editing access

Unfortunately, the first version of that product, meeting requirements RC1 to RC5, did not fulfill the expectations of customers, and flopped. Why, becomes apparent when looking at the business drivers' profiles detected by the NPS survey. Trying to match the product features with the targets set by the business drivers, as shown in the HoQ in Figure 9-14, did not help. The convergence gap of 0.38 shows limited support only for the required business driver.

The matrix cell are measurable by the functionality that they require; functional size measures how much a particular user requirement supports one of the business drivers when planning to create or improve the software product, compare with *Chapter*

6: Functional Sizing. Sometimes, experts try predicting the transfer function in a QFD workshop instead of measuring.

This example from software product management shall explain why Six Sigma professionals use transfer functions to help product managers making the right decisions. Others can be found in (Fehlmann & Kranich, 2012-3); (Fehlmann & Kranich, 2011-2), (Fehlmann, 2006-1), (Denney, 2005), and (Akao, 1990), and obviously in this book.

9-4.6 DETECTING MISSING REQUIREMENTS

The missing required capabilities originate from missing stakeholders. The missing requirements are from *RC6: Technology Officer* and *RC7: Information Officer* (Table 9-13):

Table 9-13. Missing Capabilities for Interactive Customer Communication

Capability Requestor		User Requirements
RC6	Technology Officer	Service Oriented Architecture - SaaS
		Ability to plug-in Bring-Your-Own-Device
		Industry Standards - focus on HTML 5
RC7	Information Officer	Customer Identity Management
		Recognizing customers in all media
		Track communications in all media

Adding required capabilities from those additional stakeholders to the House of Quality resolves the problem, as shown below. Transfer functions cannot detect missing requirements be itself, they only validate whether the analysis is correct or not. Once the gap becomes apparent, filling it is rather straightforward for people knowing the domain subject.

While Figure 9-14 shows impact measured by its functional size in each cell, the two new rows in Figure 9-15 reflect the product improvement design. Their cell values show the QFD workshop ratio scale according ISO 16355 (ISO 16355-1:2015, 2015).

Requirements focus shifts from *RC4: Marketing Professional* to *RC7: Information Officer*. The QFD matrix provides work instructions what the software developers shall do to improve the product. Cell entries show the impact of work on required capabilities per business driver. The control profiles in Figure 9-14 and Figure 9-15 consequently indicate the total impact of product development, or product improvement, relatively to each capability requestor.

Figure 9-14. Initial HoQ Detecting Missing Capabilities   *Figure 9-15. Final HoQ Completed*

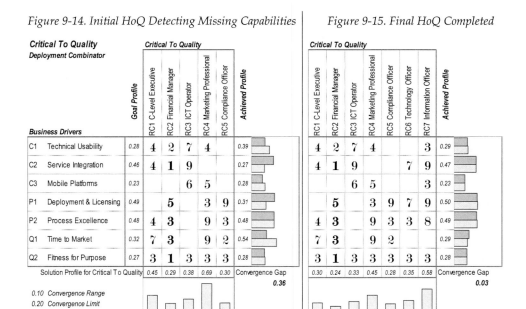

Adding requirements from the two new requestors in Figure 9-15 does not simply increase the weight of the *C3: Mobile Platforms* business driver. The product capability focus gains on *C2: Service Integration* and loses on *Q1: Time to Market*. Indeed, this is a novel insight on what makes the new product succeed. It means: the intuitive approach, namely simply adding support for *C3: Mobile Platforms* is not the right strategy. In order to expand successfully to mobile platforms, *C2: Service Integration* is the key.

9-4.7 COMPLETING USER REQUIREMENTS

For this case study we concentrate on the interesting point, how requirements from *RC7: Information Officer* support *C2: Service Integration*. The user requirements in Table 9-13 mention *Customer Identity Management*, *Recognizing customers in all media*, and *Track communications in all media*. Which user stories implement these requirements?

There are several. Among them, the most important might be stated as follows (in the Grant Rule format for User Stories, following (Fagg & Rule, 2010)): "Me as an Information Officer, I want to recognize customers who contact us by phone, e-mail, Facebook, Twitter or else such that I can retrieve them and record their activity on the Corporate CRM". Obviously, such a requirement implies quite a few FUR and NFR, among them the service interfaces needed to access phone numbers, SIM card information, e-Mail addresses of senders, credentials in Facebook and Twitter, and more. These service interfaces are FUR that require data movements and can easily be counted and accessed by the ISO/IEC 19761 COSMIC functional sizing method (COSMIC Measurement Practices Committee, 2015). However, none of those FUR has impact on business drivers. Functional size; i.e., the number of data movements needed, drives these story cards.

The more complicated aspect of this user story is that customers need to agree to such information management. The implied NFR and FUR might include a customer cockpit that allows users to connect corporate customer accounts with social media, e-Mail addresses and phone numbers, if they agree to let the corporation providing services know their identity. Such functionality has impact on the business driver *C2: Service Integration*. Customers will benefit from integration, but need to agree on it. Functionality does not drive these work tickets, rather usability, although the user cockpit also requires some functionality.

The main concern is making complicated service integration topics understandable and acceptable for end customers. This is not only a matter of software development and ergonomic design but as well of psychology and legal user rights, and not captured in functional size measurements. This might have been the reason – complexity – why these requirements had not been included in the initial product setting.

9-5 THE KITCHEN HELPER CASE

The authors created a fictitious *Kitchen Helper* project, connecting cooking recipes with the inventory of kitchen fridges and cupboards; with food information, home delivery grocery stores, and finally with stoves and cooking plates to help people cooking at home. If cooking plates know what they heat up, cooking pots will not overspill. The sponsors are grocery stores, cooking recipe portals and kitchen appliance suppliers. This is an agile system integration project; its intension is not restricted to software development. The result of the project is an integrated service product involving suppliers of kitchen appliances, grocery stores, and home delivery.

The challenge is to understand the needs of all stakeholders and continuously deliver what they need to keep the project going, profitable and successful. The Kitchen Helper is a learning system that continuously adapts to customer's needs; else, it will prevail not for long. Continuous delivery and creating a community is an absolute must; transfer functions supporting the teams.

9-5.1 STAKEHOLDER'S PRIORITIES USING AHP

The *Analytic Hierarchy Process* (AHP) is the best tool choice for understanding the variety of stakeholder requirements, listed in Figure 9-16. The four stakeholders look at different business drivers. On the other hand, the customer, the kitchen helper user, decides who wins.

The project team must compare them against each other in view of their impact on project success. Success means that users flock to this new service product; only then, it affects grocery store sales in relevant quantity to produce effects on consumer prices, as well as on kitchen constructors and the other stakeholders.

Figure 9-17 illustrates the AHP method to determine the top business drivers for the kitchen helper case. Each of the four stakeholders has to determine its own priority ranking by looking at what attracts the kitchen helper user most. As a result of the comparison, and taking the top level priorities into account, the four pair-wise comparisons add up into top priorities. From those, the first six account as goal profile for developing the kitchen helper app.

The *Kitchen User* with *D04: Get High Esteem* wins because her or his priorities govern the overall attractiveness of the *Kitchen Helper* application; closely followed by the *Cooking Community* with *B01: Collect New Recipes* because of this topic's influence on users' perceptions and beliefs. The other top business drivers find due consideration; thus, the project can start, based on the above goal profile.

Figure 9-16: Business Drivers for Kitchen Helper

AHP Summary
Kitchen Helper

A Grocery Shop

A01 Increase Turnover
A02 Promote High-end priced Food
A03 Customer Loyalty

B Cooking Community

B01 Collect New Recipes
B02 Promote Specialities

C Kitchen Manufacturer

C01 Extra Appliance Value
C02 Willing to Recommend
C03 Low Maintenance Cost

D Kitchen User

D01 Surprise Friends
D02 Cook without Stress
D03 Everything in Stock
D04 Get High Esteem
D05 Stay in Control

Figure 9-17: The AHP for Determining Top Business Drivers for the Kitchen Helper Case

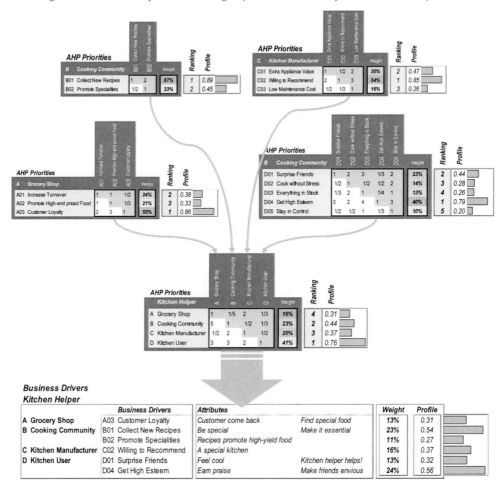

To keep things simple, we present the user stories for the kitchen user only. Other user stories that talk about the expectations and needs of the other stakeholders to optimize their food logistics, profits, and marketing messages are not included in Table 9-18. The following covers only user stories affecting the kitchen user. She or he will run apps on the tablet or smartphone, and almost not be aware of the infrastructure and additional functionality needed to make all this happen. Moreover, there are no such difficulties addressed in this case study such as home delivery of items missing in the kitchen fridge while people are absent. It's only about cooking recipes.

9-5.2 User Requirements and Stakeholder's Business Drivers

Table 9-18: Functional User Requirements recorded as User Stories in the Grant Rule Format

	Label	As a ... [Functional User]	I want to ... [get something done]	Such that ... [quality characteristic]	So that ... [value or benefit].
1)	Q001: Collect Recipes	Kitchen User	collect recipes	I can select one that interests me	my family and guests are impressed
2)	Q002: Identify Food	Kitchen User	identify food components	My shopping list is accurate	the recipes use correct components
3)	Q003: Search Recipes	Kitchen User	find new recipes	I can select one that interests me	my family and guests are impressed
4)	Q004: Manage Inventory	Kitchen User	know what's in my kitchen	I can get rid of obsolescent food	before it decays
5)	Q005: Shopping List	Kitchen User	get a shopping list	I buy everything that's needed	I can cook my recipe
6)	Q006: Cooking Process	Kitchen User	start cooking	my appliances know what I'm doing	the can help me doing it right
7)	Q007: Process Control	Kitchen User	execute the cooking process	heat and treatment is correct	boiling pans don't overspill
8)	Q008: Remember	Kitchen User	remember what I cooked last time	for my family or guests	they won't get weary of my recipes

Four functional objects execute the eight functional processes; see Figure 9-19, Figure 9-20 and Figure 9-21. Based on the user's view, it is considered as one application.

There are other applications needed that interact with the kitchen helper application. Among them are the *Recipe Portal*, the *Grocery Shop*, and the Internet of Things components such as the *Kitchen Controls*, the *Boiling Plates*, and the *Oven & Steamer*. More "things" such as intelligent pots already exist and can be added based on the emerging customer's needs.

9-5.3 The Data Movement Map for the Kitchen Helper

The overview shows four functional objects within the application - *Collect Recipes*, *Identify Food*, *Kitchen Inventory* and *Execute Recipe* – with a total of eight functional processes. Each functional process has a trigger; each actuated by the kitchen user.

Figure 9-19: Application Overview for the Kitchen Helper

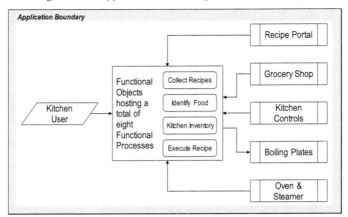

The *Collect Recipes* functional object executes the functional process *Enter* a new recipe. It allows the kitchen user enter own recipes, modifying existing recipes, and delete recipes from her or his personal recipe database. The same functional process executes the functional process *Search* for existing recipes in the recipe database.

The *Identify Food* functional object executes the functional process *Identify* food and food components. This functionality can resolve different spelling or language variants used retrieving the identical food ID, offering language independency. It also hosts the functional process *Manage* inventory, maintaining the food database.

The *Kitchen* Inventory object case for the functional process *Shop* for a recipe, issuing shopping lists automatically or with manual amendments, maintaining the inventory database.

The *Execute Recipe* objects is the most prominent: it executes the functional process *Start* the cooking process, starting the cooking process and monitoring it, instructing the kitchen user what she or he has to prepare. The functional process *Control* the cooking process reminds the kitchen user when the next process steps are due or overdue. Finally, the functional process *Remember* previous recipes cooked avoids loss of variety in the kitchen's menu. Especially, guests should not get the same recipe cooked too many times, even if they like it.

Figure 9-20: The Kitchen Helper Data Movement Map – Preparing Recipes

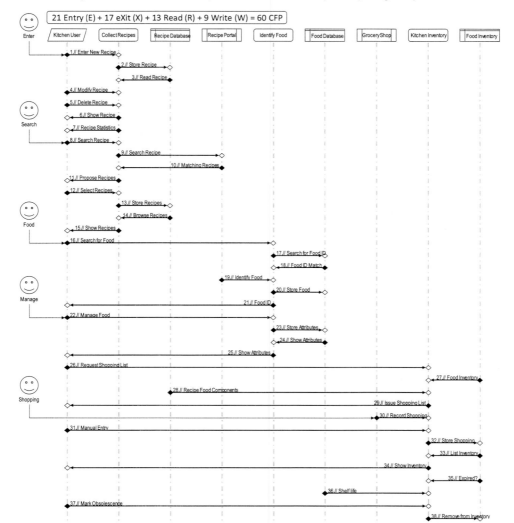

The total is 60 CFP. Figure 9-20 displays the first five functional processes, represented by their triggers. The data movement map continues with the functional object *Execute Recipe* part, executing the remaining three functional processes in Figure 9-21.

Figure 9-21: The Kitchen Helper Data Movement Map – Executing Recipes

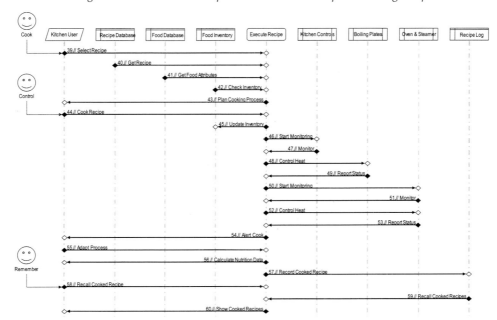

Thus, the kitchen user can use the helper anytime from work or while commuting home; the recipe selected governs the shopping list and thus determines what she or he finds when returning home, either still in the fridge or delivered home by the grocery shop's delivery service. By adapting the cooking process, spontaneous changes are possible, for instance when passing by at an outlet featuring fresh vegetables or fruits to be included as amendments in the modified cooking recipe.

9-5.4 THE BUGLIONE-TRUDEL MATRIX FOR THE KITCHEN HELPER

When this project starts, functionality definition is on a visionary level only. It is almost impossible to write an exact specification because there is little previous knowledge helping to separate the necessary from volatile ideas.

The development team used the functional effectiveness concept to assess whether the functionality devised so far covers the business drivers identified by the AHP with the stakeholder in Figure 9-17. The cellar of the BT-matrix looks satisfactory, the development teams and the stakeholders detect neither missing nor excess functionality; see Figure 9-24.

The initial BT-matrix shows clear deficiencies (Figure 9-22), despite it covers all functionality planned in the data movement map and the functional effectiveness is quite satisfactory with a convergence gap of 0.05.

Figure 9-22: Initial Buglione-Trudel Matrix for the Kitchen Helper Project

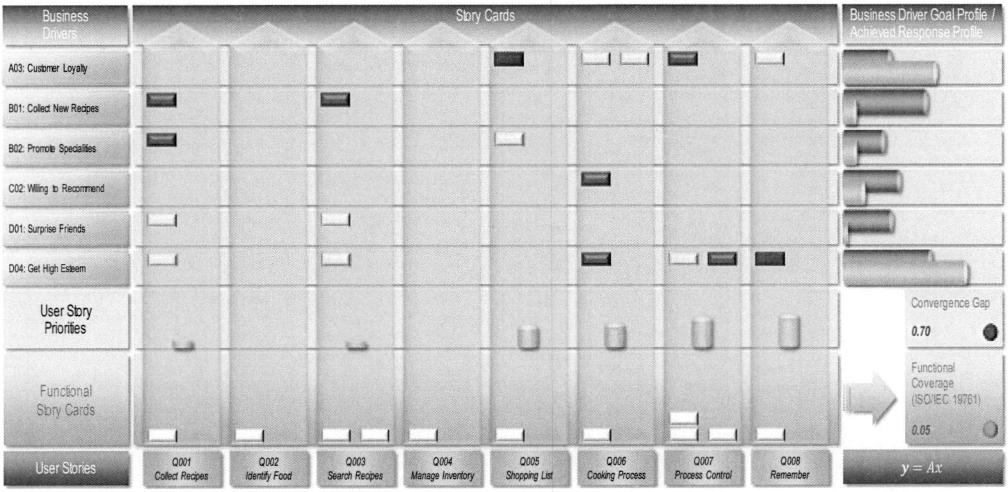

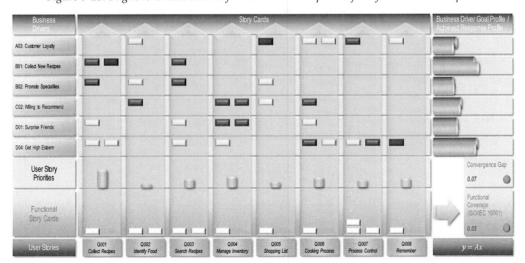

For instance, the second and the fourth user story are not contributing any value to the business drivers. The development team needs becoming creative and add value.

Typically, agile development teams detect such situations at early stages in the project; in this case, the team came up with a much-improved solution after two sprints only; see Figure 9-23. Compare with Figure 9-22; the deficiencies are removed by adding story cards to almost all user stories.

Most of the story cards remain in the backlog for the first few sprints. Nevertheless, the BT-matrix looks much better now (Figure 9-23).

Figure 9-23: Buglione-Trudel Matrix for the Kitchen Helper Project after the Second Sprint

The added story cards are mainly non-functional, in the sense that added functionality does not yet appear in the functional sizing model represented by the data movement maps (Figure 9-20 and Figure 9-21), and the additions are guided by the visual appearance of the BT-matrix.

Figure 9-24: Functional Effectiveness of the Kitchen Helper Project

Business Drivers	Goal Profile	Q001 Collect Recipes	Q002 Identify Food	Q003 Search Recipes	Q004 Manage Inventory	Q005 Shopping List	Q006 Cooking Process	Q007 Process Control	Q008 Remember	Achieved Profile
A03 Customer Loyalty	0.31	8	3		4				5	0.28
B01 Collect New Recipes	0.54	5	5	9	4	6	4		6	0.57
B02 Promote Specialities	0.27	6	3	4			2		2	0.27
C02 Willing to Recommend	0.37		6	5	2		2	5		0.34
D01 Surprise Friends	0.32	3		2		4	3	8		0.31
D04 Get High Esteem	0.56	2	6	4	5	2	6	10	1	0.56
Solution Profile for User Stories		0.37	0.42	0.45	0.29	0.24	0.33	0.41	0.25	Convergence Gap

Convergence Gap **0.05**

0.10 Convergence Range
0.20 Convergence Limit

A typical added story card is shown in Figure 9-25. It impacts both the business drivers *C02: Willing to Recommend* and *D01: Surprise Friends*; both valued by the team with an impact value of 3 (blue story cards). Otherwise, not much is yet known about how such a webcam shall be implemented, whether a software is added that can recognize barcode-labeled items, or RFID chips. Thus, it is yet unknown what kind of data movement is needed to help the kitchen user inspecting his cupboard from abroad.

It is likely that more functionality than represented with the 60 CFP from the initial data movement map will finally be implemented when the kitchen helper product hits the market.

Whatever the implementation details are: the effect when demonstrating the cupboard view on a smartphone is certainly both surprising and makes the user of such a kitchen helper application attractive; thus, it will lead to more recommendations among friends and colleagues, and the Net Promoter® Score will raise.

Story Card for Q004: Manage Inventory	Story Points:		Name:			Test is Ready	Draft is Ready	Review Done	Final-ized	Appro-ved	Func-tional
Q004-02Q: **Inspect Cupboard**	Functional Size:	0	Sprint:								
	Business Impact:				C02: 3	D01: 3					

As a Kitchen User, I want to know what's in my kitchen, such that I can get rid of obsolescent food, so that it doesn't decay

Use a webcam to inspect the cupboard

Another typical non-functional story card is the one needed for setting up a database, for instance, the food database. Figure 9-26 has neither functionality nor specific business impact; however, it started in the first sprint and already reported some progress.

Figure 9-26: Setting up a Database has neither Functionality nor Specific Business Impact

Story Card for Q002: Identify Food	Story Points:	13	Name:	Jean		Test is Ready	Draft is Ready	Review Done	Final-ized	Appro-ved	Func-tional
Q002-02F: **Setup Food DB**	Functional Size:	0	Sprint:	#1 - Overture		9	2				
	Business Impact:										

As a Kitchen User, I want to identify food components, such that My shopping list is accurate, so that the recipes use correct components

Create database model for food DB

Several such cards always exist; they are part of the cellar if they do not have any business impact.

9-5.5 DETECTING FATAL-BLOW REQUIREMENTS

Fatal-blow requirements are A-defects that completely change the game – see the upcoming section 10-1.2 in *Chapter 10: Software Testing and Defect Density Prediction*. For instance, a non-anticipated change of legislation after some unexpected incident can block a project and its related product idea completely, irrespective of how much money and effort the development team already has spent on the matter. This can happen for instance after a change of legislation against money laundering in the finance sector. It can block projects to ease access to money transfer services. With the Internet of Things, unsafe communications also might block bright product ideas for safety reasons.

To the best of knowledge of the author, transfer functions can hardly predict such fatal-blow requirements, as the controls are out of reach for measurement.

9-6 CONCLUSIONS

The development certainly will not stop with this initial adjustments and improvements; more will be uncovered when the development team advances and identifies new requirements and expectations. The development team together with sponsors and stakeholders will proceed adding value if time and budget permits and the convergence gap remains small.

Requirements elicitation happens in real-time and new features will continuously enhance the product upon each new sprint. This is *Continuous Delivery*. The team learns from the kitchen users via feedback in NPS surveys and via user communities in social networks and assesses new requirements based on the business driver's profile. This works even when business drivers change themselves; adjusting to such change continuously is less difficult than upon a break in customer experience.

The Buglione-Trudel matrix is an interactive, visual tool that supports requirements elicitation during development. For further discussion of how COSMIC supports requirement elicitation, consult the PhD thesis of Sylvie Trudel (Trudel, 2012).

CHAPTER 10: SOFTWARE TESTING AND DEFECT DENSITY PREDICTION

When undertaking ICT projects, defects are common. Since writing software is knowledge acquisition, not civil engineering, and operating software involves human interaction in many cases, missing knowledge might lead to defects, even to failures, whose effects are hard to predict. However, the approach allows defect density prediction based on fault detection effectiveness and efficiency during development.

For prediction, understanding the origin of defects, and the ability to measure defects, are essential. Defects originate from missing requirements, from missed requirements and inadequate testing, while others are simply due to laxness, oversights and careless-ness. We propose measuring defects by functional sizing rather than by cost of defect removal, or by number of entries in a defect-tracking tool. With this new approach, transfer functions can help creating meaningful statistics and test strategies.

10-1 INTRODUCTION

When software is newly written or existing software amended, it is important to know how many defects in operations to expect. Many decisions and foreseeable cost depend from the amount of work needed for fixing defects after release.

In the last few years, a tendency became dominant to cope with defects by issuing new releases in regular intervals. This is an excellent strategy to avoid huge cost of testing, meet time to market, and customers accept it for new software. However, when new releases affect existing software by introducing new failures, things look different. Customers might already rely on the software for their business and do not like introducing new releases when new defects are susceptible to hit operations. Moreover, with the move to Servitization (Vandermerwe & Rada, 1988) and service-oriented software architectures, it is no longer the case that most software services originate from own code and thus can be judged for reliability. It is vital for any software or application provider to know defect density before releasing it to general customers, and most often, there is not much time for conducting pilots and asking sensing groups to find out.

However, there are some intrinsic difficulties when predicting the future: how do you know how many defects you are about to release before you have found them? In addition, this is only the most apparent difficulty when speaking about defects.

10-1.1 WHAT ARE DEFECTS?

When undertaking ICT projects, defects are common. Since writing software is knowledge acquisition, not civil engineering, and operating software involves human interaction in many cases, the effects of faults are hard to predict. Defects are the most common manifestation of faulty software. However, not all faults lead to defects; some remain undetected and do not harm the value of software. Others manifest themselves, increasing the efforts needed and thus cost of detecting faults. Removing defects is a consequence of defect density and thus is predictable from the fault distribution in software.

For prediction, it is important to understand the origin of defects. Most defects originate from missing requirements, some from missed requirements and inadequate testing, and others are simply due to laxness, oversights and carelessness. It is possible to analyze and understand the lifecycle of all these three kind of defects and estimate their density, based on observing the organizations and stakeholders involved into software creation and operation.

Figure 10-1: Mistakes, Defects, and Failures to Perform

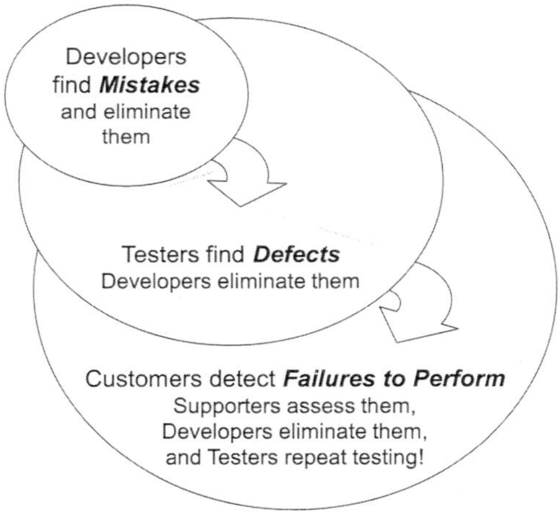

Failures to perform hit the user, or customer. When quality assessment techniques before release avoid such failures, the test team records defects found and defects removed in defect tracking tools such as Bugzilla. The term "bug" reflects the very early days of computing, when living insects walked through circuits, nibbled at electrical isolation materials affecting computing logic. Static quality assessment methods such as reviews and walkthroughs, and dynamic software tests, are the methods of choice for finding defects.

Before a fault becomes apparent as a defect, many mistakes were detected and removed at the stage of creation; e.g., by developers when walking through code, doing peer reviews or trial runs. Trial runs are not testing, although sometimes referred to as self-tests, because the expected outcome is unknown beforehand. Trial runs are what most developers use to validate their expectations towards the code they have written, and they are very effective for eliminating failures early – namely now when developers' mind keep a fresh picture of what the code is supposed to do.

There is one inherent difficulty: How to size mistakes? They are difficult to count and probably their number is not indicative. Developers do not count the mistakes that they detect and eliminate before anybody else sees them. It would look like waste. Defects in contrary require communication and therefore leave auditable trails. Nevertheless, just counting entries in a defect-tracking tool might be misleading. Many defects manifest themselves differently in different contexts, and are hard to consolidate. They do not come alone; they might pop up under different viewpoints but still share one common root cause. On the other hand, one observation refers to several, not necessarily dependent defects.

However, every defect removed before release means the risk decreases that customers experience failures to perform. Thus, the number of defects removed might be indicative for the remaining defect density in the released product. It is therefore undeniably constructive to track defects and use them as an indicator for the amount and effectiveness of quality assurance activities.

10-1.2 A-defects, B-defects, and C-defects

The aim of this chapter is to draft a model how to predict defect density after release of a piece of software or a software service. To achieve this, we first must understand how to identify and count defects in software and then to investigate into the transfer functions that explain how solutions map into *User Stories* (USt), and consequently how defects are injected, and into the transfer function that map defect detection methods – called tests – onto the same user story.

Note that some failures to perform do not relate to user stories, rather to non-functional requirements such as usability and responsiveness.

10-1.2.1 Terminology

There is a common perception what faults are. However, it is difficult to distinguish faults from features. The United Kingdom Software Metrics Association (UKSMA) defines in its new *Software Defect Measurement and Analysis Handbook* (UKSMA, 2015) a software defect as a functional or non-functional construct that leads to system failure or incorrect behavior including being the root cause for human error and misunderstandings (mistakes). This can be due to a departure in a software product from its expected properties as identified in the software contract or as realized after delivery, thus suggesting incorrect requirement specification or changes in the requirements.

10-1.2.2 MISTAKES, DEFECTS, AND FAILURE TO PERFORM:

A *Requirement Defect* is an *incorrectly specified requirement leading to defective behavior*. This may be due to unrealized requirements, changes, or misunderstanding of requirements (mistakes).

In addition to the UKSMA handbook, we use the notion *Mistake* for faults uncovered and corrected immediately, within the creation process. Spelling mistakes are typical samples. *Defects* are mistakes that escape the creation process, detected by an immediately related verification and validation process, by either peer review, tests, or any other kind of quality inspection suitable for detecting defects for removal.

Failures to Perform are what the UKSMA Handbook calls *Residual Defects* that escaped the assessment process; they ship with the released product and hit the user of the product, be it the next instance in the value chain or the end-user. Such failures to perform block people relying on the correctness of the product; they cannot continue their normal way of operation but must take some precaution to address the issue caused by the defect that escaped timely detection. *Faults* in turn refer to all three; mistakes becoming defects if not detected; and defects shipped in turn cause failures to perform.

The expectation is that if the quality assurance process detects more defects, and developers can remove them before release, fewer failures will show up later. The underlying assumption is that the total number of opportunities to introduce faults is limited. This is the usual assumption in the Six Sigma methodology – however, it deeply affects counting of defects.

Note that not only customer detect defects after release. Defects shipped with the initial product release might as well have been found and removed internally within the organization; e.g., during maintenance work (Damm, et al., 2008).

10-1.2.3 TEST STRATEGY AND SLIP-THROUGH:

A *Test Strategy* (Damm, et al., 2006) focuses on detecting certain kind of faults most effectively. A well-known test strategy is to rely on the testing phase for identifying and removing faults. A different test strategy is to remove all faults in the phase, or right after the phase, where they were injected (Fehlmann, 1999), requiring techniques like peer review and pair programming. A third one is to rely on beta testers and customer feedback. Various combinations of such approaches are conceivable, depending from the market. *Fault Slip-Through* refers to the percentage of faults remaining undetected when they should according test strategy, but slip through the quality assessment barriers that have been set in place to find them.

10-1.2.4 A-DEFECTS, B-DEFECTS AND C-DEFECTS:

Defects can be of different natures. We distinguish *A-Defects*, *B-Defects* and *C-Defects*. "A" stands for missing requirement, "Anforderung" in German; "B" stands for nonconformity with specified requirements, or a missed requirement that should have been detected as a defect, or "*Bug*"; "C" is short for both *Consistency* and *Contradiction*.

The impact of A-defects can be disastrous. For a software product, it can be that the market, or the customer, will not use the product at all; it becomes a commercial failure and disappears from the market. For a software project, missing requirements seriously diminish the value created and the project becomes a failure. This can happen, even if the responsible people manage the project technically perfect and professionally.

To detect A-defect early enough; i.e., before release, is difficult; however, *Quality Function Deployment* (QFD) is the method of choice for uncovering unspoken customers' needs, for instance by analyzing social media content (Fehlmann & Kranich, 2012-2).

Figure 10-2: Model for Test Strategy

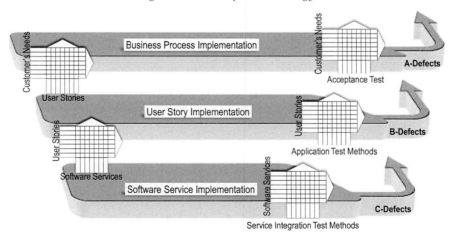

B-defects usually are easier to detect because stated requirements and even specifications exist that make existing gaps clear to everybody. Fixing such defects is – although more expensive the removing a fault before release – relatively easy and is often done when preparing a new product or software release.

There are also C-defects, cursory defects that sometimes are hard to detect, causing unacceptable behavior of software but not violating an explicit requirement. C is short for both *Consistency* and *Contradiction* – most of the C-defects arise from inconsistent specifications or requirements, such as legal regulations contradicting each other. They typically result from laxness in both development and testing. Examples of such C-defects are user input not trimming trailing blanks that cause blocking access or losing data, or changes to data selection while processing without notifying the user, or also the famous 64-bit floating-point number to 16-bit signed integer conversion that failed and caused the Ariane 5 flight 501 rocket to crash (June 4, 1996). Such C-defects can have severe economic impact, are hard to measure and detect since not covered by explicit requirements or specifications. The only means is maintaining good communication and high quality standards within software development. Thus, quality managers have developed many answers to the problem of C-defects in software development, for instance by conducting daily stand-ups in Scrum (Schwaber & Beedle, 2002).

While severe B-defects and C-defects might cost quite a bit, A-defects could account for losing customers and credibility, and that can fatally hurt business, on the long term much more than B- or C-defects.

Figure 10-2 shows a model for a test strategy with three levels, focusing on A-defects, B-defects and C-defects respectively. The transfer functions to the right show test effectiveness; small convergence gaps are indications for optimum effectiveness; the two transfer functions to the right show how user stories support customer needs (CN), as well as software services (SS) that in turn support user stories. Testers normally uncover C-defects level with unit and integration tests; the B-defect level with application tests and the A-defect level with user acceptance tests.

10-1.3 FAULT SLIP-THROUGH MODEL

This fault slip-through model as proposed in Figure 10-3 allows tracing the origins of defects. The graph shows how mistakes accumulate when detection and removal are classified according the eight phases shown in Figure 10-3, and finally become defects.

In only one case, the final defects originate from a faulty contractual statement; ten defects are coding mistakes that remained undetected by tests or other applicable means of quality assessment, and two defects newly introduced during the testing phase, most likely by faulty fixes. The total fault density is build up until a maximum during coding, and reduced, but not completely, by tests.

The cost of removing those defects after release might be substantial; however, trying to totally avoiding them by removing all defects within the following development stage can also become quite costly, even if defect costs would become significantly lower when reducing the final defect count to the two test defects only. So where is the trade-off?

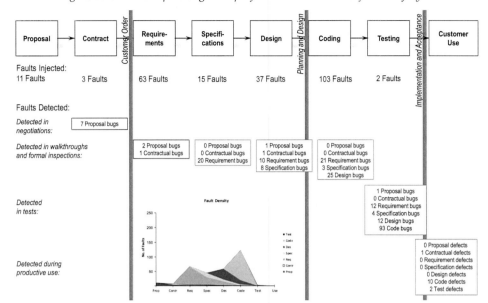

One of the intrinsic problems is who pays for the quality assessments needed to un-cover defects, and the cost for removing them. The eight stages PROP-CONTR-REQ-HLD-LLD-CODE-TEST-USE[2] do not designate a waterfall model; they characterize tasks while developing software or services. Some of them can happen simultane-ously, as for instance proposal and contracting activities may continue while an agile team already started development. However, usually both supplier and sponsor share cost of proposal and contract, while usually the sponsor pays the supplier for require-ments elicitation, solution specification and design, coding and some testing, while the customer carries all cost for using the product. Some separation might even exist between requirements elicitation & design, and implementation & testing activities.

Fagan (Fagan, 1974) points out, that faults introduced at some earlier stage that remain undetected cost much more to fix when they become defects at some later stage. How much cost increases for fixing a defect when detecting it at a later stage may very much depend from the specific situation. Fagan postulates cost increase by a factor of ten – 10:100:1000 – per stage, making a business case for early inspections. Today's agile teams consider faults that survive one single day as inacceptable; see e.g., the Pop-pendiecks (Poppendieck & Poppendieck, 2007).

However, faults become defects only by quality assurance activities, and they are costly at least. Especially the early stages such as proposal and contract are crucial

---

[2] HLD/LLD refers to High-Level/Low-Level Design

steps: faults injected there but not detected until very late normally are typical A-defects and may become fatal for the product. In addition, REQ faults can cause disastrous high costs at late stages if not addressed early enough.

10-1.4 TECHNICAL DEBT

Today, developers consider such costs issues as *Technical Debt*. Technical debt is work that developers must complete before properly finish a job, or a development stage. If debt remains unpaid, it will keep on accumulating interest, making it hard to implement changes later.

Agile literature usually uses the term technical debt for coding work only. In Six Sigma, software & services development is an integral undertaking and it makes no sense to look after technical debt in coding only. Most such work depends intrinsically from business needs, as shown elsewhere in this book, and cannot be separated without creating muda from its context, be it business, scientific or production. Looking again at Figure 10-3, the matrix in Figure 10-4 shows when these faults originated and how many remained undetected at later stages.

Figure 10-4: Fault Density in a well-controlled Development Environment

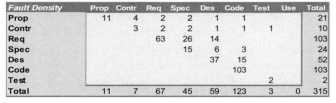

Fault Density	Prop	Contr	Req	Spec	Des	Code	Test	Use	Total
Prop	11	4	2	2	1	1			21
Contr		3	2	2	1	1	1		10
Req			63	26	14				103
Spec				15	6	3			24
Des					37	15			52
Code						103			103
Test							2		2
Total	11	7	67	45	59	123	3	0	315

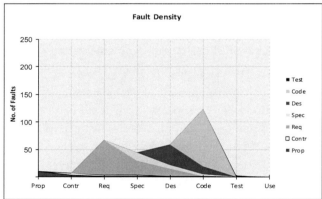

Since we lack data, we do a *Gedankenexperiment* (thought experiment). Assuming cost of € 5 for fixing a mistake found at its inception stage, and an error propagation factor of 2.5 – which is likely higher in reality – the fault density shown in Figure 10-3 leads to the technical debt shown in Figure 10-5:

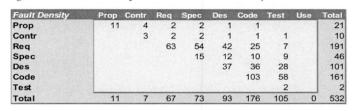

Figure 10-5: Technical Debt in a well-controlled Development Environment

Technical Debt	Prop	Contr	Req	Spec	Des	Code	Test
Prop	€ 55	€ 50	€ 63	€ 156	€ 195	€ 488	
Contr		€ 15	€ 63	€ 156	€ 195	€ 488	€ 1'221
Req			€ 315	€ 4'219	€ 8'203	€ 12'207	€ 8'545
Spec				€ 75	€ 2'344	€ 4'883	€ 10'986
Des					€ 185	€ 17'578	€ 34'180
Code						€ 515	€ 70'801
Test							€ 10
Total	€ 55	€ 65	€ 440	€ 4'606	€ 11'123	€ 36'160	€ 125'742

While technical debt peaks after writing code, testing detects almost all defects and removing them leaves marginal debt only to the "*Use*" stage.

In some less well-controlled development environment, defects remain undetected for a longer period, significantly increasing fault density; see Figure 10-6.

Figure 10-6: Fault Density in a less controlled Development Environment

Fault Density	Prop	Contr	Req	Spec	Des	Code	Test	Use	Total
Prop	11	4	2	2	1	1			21
Contr		3	2	2	1	1	1		10
Req			63	54	42	25	7		191
Spec				15	12	10	9		46
Des					37	36	28		101
Code						103	58		161
Test							2		2
Total	11	7	67	73	93	176	105	0	532

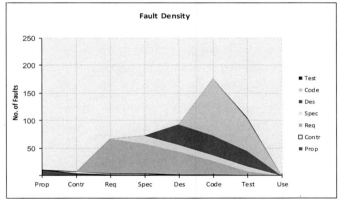

Consequently, technical debt also increases and even might increase to an unacceptable level. In this sample case, the product hits the user with no less than eight defects that originate from the contractual (one) and the requirements elicitation stage (seven). It is quite unlikely that the user will be happy with the product.

Technical debt shows a significant higher level as well when leaving the TEST stage:

Figure 10-7: *Technical Debt in a less controlled Development Environment*

Technical Debt	Prop	Contr	Req	Spec	Des	Code	Test
Prop	€ 55	€ 50	€ 63	€ 156	€ 195	€ 488	
Contr		€ 15	€ 63	€ 156	€ 195	€ 488	€ 1'221
Req			€ 315	€ 4'219	€ 8'203	€ 12'207	€ 8'545
Spec				€ 75	€ 2'344	€ 4'883	€ 10'986
Des					€ 185	€ 17'578	€ 34'180
Code						€ 515	€ 70'801
Test							€ 10
Total	€ 55	€ 65	€ 440	€ 4'606	€ 11'123	€ 36'160	€ 125'742

The fact that some faults slip through quality assurance and remain undetected yields high technical debt and might become a threat for the product.

The aim of software & services testing and defect density prediction is avoiding situation like in Figure 10-7. For this, we need prediction by Lean Six Sigma transfer functions.

10-2 KEY FACTORS FOR DEFECT DENSITY

The Six Sigma approach requires establishing size metrics, including software. As already seen in *Part 1 – Chapter 6*, the ISO/IEC 19761 COSMIC approach for functional sizing also sizes tests and thus allows calculating defect density using the same measurement unit. Data movements are something the developers understand as well as ICT people who configure services, combining various data sources with applications over the web, and data movement maps are common practice in serious software development. Moreover, data movement mapping is also helpful in understanding, configuring and testing application services.

Thus, the natural choice of counting defects is in relation to functional size measured with ISO/IEC 19761 COSMIC. Earlier approaches related defects to the effort needed to remove them, see (Fehlmann, 2009-1).

10-2.1 STAKEHOLDERS

Stakeholders are the organizations and groups that are involved into creating and using the software under question. They are typically very interested in specifications and supply the constraints needed for the software to work together and support the business processes. Stakeholders have a predominant role in injecting, and in removing, defects, including A-defects that arise from misunderstanding constraints and environmental settings.

10-2.1.1 SPONSORS

Sponsors pay for the development, the operation, and for all cost of removing defects, and receive the benefit from developing and running the software product.

Sponsors are involved into requirements elicitation and prioritization. They identify Functional User Requirements. Especially with an agile development team, they interact on an operational level with developers and sometimes with users. For internally used and developed customer software, sponsors and users even can coincide. Most often, sponsors are not writing specifications but leave the explicit definition of user stories and functionality to other stakeholders.

Sponsors are typically responsible for injecting most of the A-defects.

10-2.1.2 DEVELOPERS

Developers do also create requirements, but more on a technical level. They are responsible for creating a sound solution and thus meeting technical standards and avoid technical debt as much as possible.

Developers do not create many A-defects, but are responsible for most B- and almost all C-defects – with the noteworthy exception of inconsistencies injected by regulations and conflicting requirements. Developers often detect these kinds of C-defects but they cannot remove them completely.

10-2.1.3 USERS

Users detect most defects; for new software, faults typically will not hit users. However, users often participate as stakeholders in stating requirements and even in writing specifications, thus users often share with stakeholders the same role and the same influence for products.

10-2.2 TRANSFER FUNCTIONS AND QFD

QFD is the method of choice for selecting controls and validating solutions in requirements engineering. QFD constitutes *Transfer Functions* that map process controls onto process responses. For instance, QFD maps *User Stories* (USt) onto *Business Drivers* (BD) with transfer function A. In each mapping, the convergence gap is an indication whether that mapping completely explains – or enforces – the expected process response, as outlined in Part 1 of this book.

$$A: USt \to BD \tag{10-1}$$

Compare (10-1) with the top left transfer function in Figure 10-2. Business drivers come as a priority profile; i.e., as a vector $y = \langle y_1, y_2, \ldots, y_n \rangle$ in the space of customer's needs of unit length $\|y\| = 1$. The generic term for the vector components y_i is *Topic*. Thus, a priority profile defines the relative weight for each topic, here in customer's needs.

Using transfer function in this way is standard practice in Design for Six Sigma and QFD; see (Fehlmann & Kranich, 2011-2). If the transfer function A is a linear mapping and measurable as a matrix, a solution x can be determined solving the equation $Ax = y$ for the transfer function (10-1), see *Chapter 2: Transfer Functions*. The vector x is an *Eigenvector* of AA^T where A^T is the transpose of the matrix A in the vector space of Functional User Requirements RC. Then x is the priority profile for the Functional User Requirements RC, and the solution z of the transfer function. Quality management prioritization of the test strategy requires these priority profiles.

10-3 A MODEL FOR DEFECT DENSITY CALCULATION

Counting software defects in defects recording database is no suitable measurement method in Six Sigma because the number of defect opportunities remains unknown. There is no upper limit for the count, set by the size of the software. Defects per Million Opportunities (DPMO) remains unclear as well, and the defect distribution cannot be determined. Thus, there is a wide research gap, how to measure defects in software such that Six Sigma tools are applicable. It is incomprehensible why such widely recognized authorities, for instance the SEI, overlooked such a straightforward difficulty (Siviy, et al., 2008).

Defects measurement must follow the same rules as functional size, and test metrics as well. Then defect measurements become independent from implementation techniques, and statistical evaluation methods and Six Sigma tools become applicable.

10-3.1 MEASURING DEFECTS

A common misconception is to count defects based on several entries into a defect-tracking tool. This is not possible. There is no way to distinguish defects from each other as they form a complicated, interrelated network of misconceptions, mistakes, errors and defects. You need a reference model before you can start measuring faults.

A defect refers to some identifiable entity in the model for the software. The model must fit not only bespoke software, written on purpose, but include software services acquired from some provider, as custom coding becomes less and less necessary to provide software services.

Suitable models exists; the ISO/IEC 19761:2011 COSMIC international standard for sizing software (COSMIC Measurement Practices Committee, 2015) suitably sizes such models – see *Chapter 7: The Modern Art of Developing Software* – and sizing can be automated (Jenner, 2011). The COSMIC standard counts *Data Movements*; an entity in software that moves a unique *Data Group* and is easily identifiable, independent of coding or of service implementation details. Data groups capture the business logic of the software or service. In such a setting, *Defect Size* is the number of data movements affected by one or more defects.

Every defect count depends from the number of tests applied in the software development, or service customizing process. When measuring defects, we do not only measure software products but the development process as well. Part of that process are tests, and a software product is only as good as its tests. Software or services without tests are unpredictable and not in the Lean Six Sigma focus. They might produce waste and therefore might not be lean at all. Thus, we must state rules how to count tests as well.

10-3.2 COUNTING RULES FOR DEFECTS COUNTING

The following counting rule holds: a data movement might account for one defect for each test story that demonstrates a failure to perform. A *Test Story* or test scenario describes a (qualitative or quantitative) feature linked to a user story by some *Test Cases*. A test case consists of test data together with preconditions and post conditions that has the capability to fail or succeed. Every test story consists of several test cases that demonstrate capabilities required by the user story, and thus are logically connected. If one test case fails, the test story fails, and one or more of the data movements executed by the test cases adds a defect to its count. As an example, we use the *Fast Ticket App* from section 7-5.3 in *Chapter 7: The Modern Art of Developing Software*.

Defect counts not only depend from the software structure but from the testing process as well. A single data movement might demonstrate as many defects as can be identified by a distinct test story. The total functional size, times the number of test stories executing a data movement, constitutes the total number of defect opportunities in our piece of software, making it possible to derive Six Sigma indicators such as DPMO and use the related defect counting tools. However, when testing software, the total number of tests will increase if testers detect still more defects.

This measurement method has all characteristics needed for a measuring both functional size and defects. The most important is linearity. The functional size, as well as the defect size, of any piece of software is the sum of its components' sizes.

10-3.3 A SAMPLE DEFECT COUNT

In the extract from a data movement map shown in Figure 10-8, a simple app for a mobile device had been sized to a total of 38 *COSMIC Function Points* (CFP) allowing the user to buy a personal ticket for a train using the smartphone while boarding. Other timetable-based ticket purchases are not possible with this App. Standard destinations stations can be prepared; e.g., near home; that part is shown in Figure 10-8.

All defects affect some data movements, and thus relate to functional size. The *Defect Rate* is the ratio between total defect counts divided by functional size. If defects affect more than one data movement, they count once per data movement affected. Since more than one defect might affect a data movement, defect rate can be superior to 100%.

Figure 10-8: Ticket App for Mobiles with Seven Data Movements and One Defect Shown

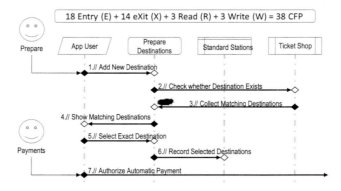

10-3.3.1 WHAT ARE LEARNING OPPORTUNITIES?

The number of defects found and corrected before release allows for defect density prediction. If a significant number of defects found and removed during development never hit the testing phase, the number of refactoring runs is a reliable indicator for such defect removed; this number too refers to data movements. It is essential that developers can identify data movements during development, or while configuring a service. Refactoring, even for other reasons than fixing an identified defect, is an indication for avoiding future defects. The same data movement undergoes refactoring many times, and each run counts. If the total refactoring count exceeds the total functional size, it means that several consecutive fixes were necessary and thus the *Learning Opportunity Ratio* (LeOR) becomes larger than 100%.

$$LeOR = \frac{Learning\ Size}{Functional\ Size} \tag{10-2}$$

The *Learning Size* in equation (10-2) is the total functional size of all refactored functionality. The refactoring count refers to the number of times that the progress indicator within the six steps to completion method has undergone a setback. This is counted automatically when managing software development or setting up a software service by *Story Cards*, as explained in *Chapter 7: The Modern Art of Developing Software*. The higher the refactoring count, the fewer defects remain undetected until release.

Figure 10-9: Sample Story Card with Three Defects Removed

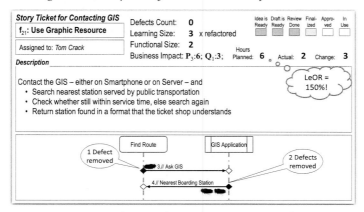

LeOR measurements are part of normal project progress tracking and very transparent to all stakeholders involved; see Figure 10-9. COSMIC-based counts fit agile approaches well thanks to its transparency and linearity; see (Cohn, 2005) and (Buglione, et al., 2011). Refactoring avoids defects. The team should be encouraged to drive the LeOR as high as possible, and refactor them as often as possible. Refactoring runs count when tests stories having identified additional defects case fixes in some data movement.

According agile customs, and lean principles, the defect count per story card should be zero when final acceptance testing starts. If testing still uncovers defects, it is possible to trace them back to the story card that implemented them. Comparing with refactoring count might yield interesting learnings again.

As a side note, story cards refer to business impact. This is essential for Lean; it allows inferring priorities based on business drivers, or customer's needs. Whatever drives the software development can be used to avoiding waste; see *Chapter 9: Requirements Elicitation.*

10-4 DEFECT DENSITY PREDICTION BY MEASUREMENT

By measuring and tracking defects, for a software project it is possible adding as many tests as needed to reach the previously agreed defect density after release.

10-4.1 BUSINESS IMPACT

The key to defect metrics is an understanding of the defects' impact on business goals and targets. A defect is a failure only because it hurts. Anomalies that go unnoticed are not defects, probably not even mistakes; fixing them would be a waste of time and energy.

User stories need a certain amount of data movements executed to fulfil requirements. Non-functional requirements typically attach to user stories by the Grant Rule format. Thus, the data movement count determines the size of test stories for functionality tests.

10-4.2 ASSESSING TEST EFFECTIVENESS

Test stories cover a distinct number of data movements, and data movements belong to some specific feature. Thus, a transfer function T exists that maps the test stories TSt onto user stories USt. The number of specific data movements covered by the test cases that constitute the test story defines the test story size. Note that data movements count as many times as they execute in test cases.

The transfer function T describes *Test Effectiveness*:

$$T: TSt \rightarrow USt \qquad (10\text{-}3)$$

To calculate this transfer function, the goal profile y for the user stories USt represents customer's priorities, as defined by the story cards; see *Chapter 7: The Modern Art of Developing Software*. If the test stories TSt cover the USt, the test story profile $x = T^{\mathsf{T}}(y)$ is near to a solution of the equation (10-4)

$$T(x) = T\left(T^{\mathsf{T}}(y)\right) \cong y \qquad (10\text{-}4)$$

The difference between $T\left(T^{\mathsf{T}}(y)\right)$ and y is the well-known convergence gap. If effectiveness is sufficient, the convergence gap of T approaches zero. It is indicative for the ability to predict test effectiveness of the test stories.

Thus, the convergence gap of the transfer function (10-3) indicates test effectiveness.

10-4.3 DEPENDENCY FROM TEST INTENSITY

When sizing tests with ISO/IEC 19761 COSMIC, there is a maximum number of defects, depending from the *Test Intensity*; i.e., how many times a data movement executes in tests on average. If tests uncover defects, test intensity must increase because the test strategy requires more tests. A lean test strategy attempts to increase test intensity already with LeOR; i.e., the number of refactoring runs during development. The higher LeOR, the less defects remain uncovered by testing, and the less tests are necessary to reach a sufficient level of test density.

Measuring actual LeOR during development or installation of software services is part of project progress tracking and control. The story card in Figure 10-9 shows how to achieve this in practice, by counting refactoring runs with respect to data movements.

10-5 THE TICKET APP EXAMPLE

We use the "Ticket App" application already presented in section 7-5.3 in *Chapter 7: The Modern Art of Developing Software*.

10-5.1 BUSINESS DRIVERS FOR THE TICKET APP

Suitable methods; e.g., Net Promoter Score (NPS) as explained in *Chapter 5: Voice of the Customer by Net Promoter®*, transfer raw customer voices backed by an NPS score into a goal profile for the business drivers that refers to a variety of customer segments.

For the business drivers, we received the following goal profile as shown in Figure 10-10:

Figure 10-10: Business Driver's profile for the Ticket App

Topics	NPS Priority	
	Weight	Profile
CN1 Have Ticket to Ride	27%	0.47
CN2 Use Standard Settings	39%	0.66
CN3 Buy with Few Clicks	34%	0.59

The user stories connected to these business drivers are listed in Grant Rule format in Table 7-17, *Chapter 7: The Modern Art of Developing Software*. Functional effectiveness yields the following profile (Figure 10-11):

Figure 10-11: Solution Profile for Ticket App User Stories, representing Customers' Priority

User Stories

User Stories Topics	Priority	
	Weight	Profile
1) Q001 Prepare Destinations	28%	0.53
2) Q002 Find Route	24%	0.47
3) Q003 Authentication	33%	0.64
4) Q004 Issue Ticket	15%	0.29

Compare with Figure 8-13: Deming Chain for Business Driver's Coverage and Functional Effectiveness.

10-5.2 FUNCTIONAL EFFECTIVENESS

Functional effectiveness of business drivers is as shown in Figure 10-12. The numbers in the matrix cells in Figure 10-12 represent the number of data movements per user story that have an impact on some business driver.

Figure 10-12: Functional Effectiveness in the Ticket App Example

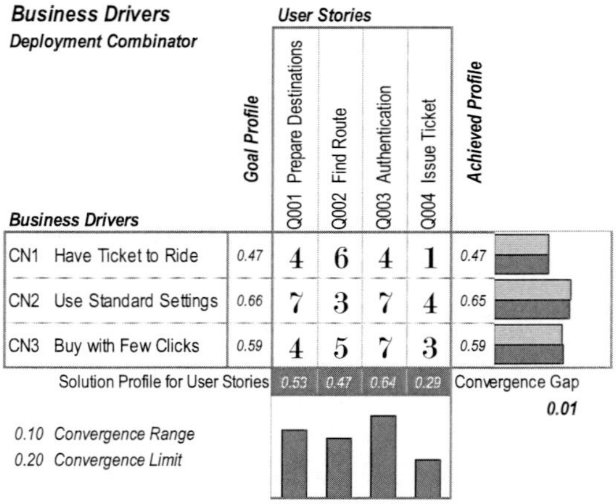

The total of the cell content numbers can well exceed the functional size of the App; usually, data movements needed to, say, find a route can easily affect e.g., the two business drivers *CN2: Use Standard Settings* and *CN3: Buy with Few Clicks*. See section 8-2.5 in *Chapter 8: Lean & Agile Software Development*.

Effectiveness thus refers to the statement that the data movements implemented in the ticket app do exactly what is compulsory to run the app, and meet business goals. There is nothing meaningless; the solution is a lean solution, avoiding 無駄 (= muda, waste). The bottom of Figure 10-12 shows the resulting solution profile for user stories, or turned counterclockwise Figure 10-11 makes the same profile better readable.

Effectiveness does not necessarily imply completeness. It is possible that the ticket app has data movements implemented that never count for any of the matrix cells in the transfer function $USt \rightarrow BD$ shown in Figure 10-12.

That would be the case for data movements, which are not part of the implementation for any user story. Although this might sound strange, it typically happens if something should remain hidden from the user. Such data movements are therefore of some interest and should be investigated, whether the implement some unknown user story of some other functional user, or should not be there at all.

10-5.3 TEST EFFECTIVENESS

Based on this priority profile for user stories, testers can select test stories such that they cover customer's needs. The sample of nine test stories in cover the functionality of the user stories.

The test stories in Figure 10-13 represent customers' priorities. This is a consequence of tests covering functionality according customers' priorities with a convergence gap of only 0.04; see Figure 10-14.

Figure 10-13: Test Stories for the Ticket App with Customers' Priority

	Test Story		Priority		Test Size	Measured Defect Profile
			Weight	Profile		
CT-A Prepare	CT-A.1	Find Nearest Station	0.21	0.30	23	
	CT-A.2	Served Stations only	0.13	0.18	14	
	CT-A.3	Enter New Destination	0.16	0.23	14	
CT-B Ticketing	CT-B.1	Select Destination	0.21	0.31	22	
	CT-B.2	Get Ticket	0.25	0.37	24	
	CT-B.3	Price Calculation	0.29	0.43	30	
	CT-B.4	Issue Ticket	0.27	0.39	28	
CT-C Payment	CT-C.1	Contract Validation	0.24	0.34	22	
	CT-C.2	Payment Tests	0.26	0.38	25	

Test stories consist of several test cases. Test cases execute with some predefined data with known test results in advance; else, it is not a test but rather a behavioral exploration. In other words, counting data movements measures the transfer function shown in formula (*10-3*). The measurement rule is that a data movement in a test case counts for the user story if the user story also executes it. Counting test stories is therefore a consequence of the functional count of the user stories supported.

The transfer function thus is again defined by measurement and simply measure per matrix cell how many data movements the user story and the test case have in common when executed; see Figure 10-14. The test stories should achieve exactly the effectiveness required for the user stories; however, that this type of effectiveness does not necessarily imply that all data movements of the ticket app are covered by these test stories (*Test Coverage* – the percentage of data movements executed in at least one test case). Full test coverage might not be necessary when some of the data movements belong to an already existing tested service. The numbers in the matrix cell represent the number of data movements executed for some test case with a test story that tests the respective user story.

Obviously, one single data movement can contribute for several different counts in matrix cells. *Test Intensity* – the average number of test cases touching one data movement – might be well in the range of ten or more test cases per data movement. Test coverage and test intensity are indicators for the quality of tests; together with the *Test Effectiveness* – the convergence gap in the transfer function $USt \rightarrow BD$, see Figure 10-14 – they characterize the density of residual defects after tests.

Figure 10-14: Test Effectiveness of the Ticket App User Stories

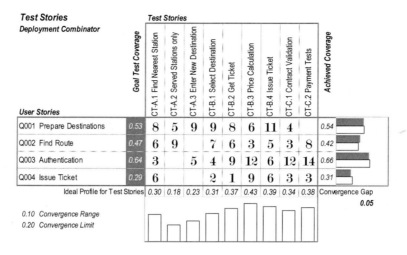

User Stories	Goal Test Coverage	CT-A.1 Find Nearest Station	CT-A.2 Served Stations only	CT-A.3 Enter New Destination	CT-B.1 Select Destination	CT-B.2 Get Ticket	CT-B.3 Price Calculation	CT-B.4 Issue Ticket	CT-C.1 Contract Validation	CT-C.2 Payment Tests	Achieved Coverage
Q001 Prepare Destinations	0.53	8	5	9	9	8	6	11	4		0.54
Q002 Find Route	0.47	6	9		7	6	3	5	3	8	0.42
Q003 Authentication	0.64	3		5	4	9	12	6	12	14	0.66
Q004 Issue Ticket	0.29	6		2	1	9	6	3	3		0.31
Ideal Profile for Test Stories		0.30	0.18	0.23	0.31	0.37	0.43	0.39	0.34	0.38	Convergence Gap 0.05

0.10 Convergence Range
0.20 Convergence Limit

10-5.4 DEFECT REPORTING

The following Figure 10-15 visualizes the defect reporting process.

Figure 10-15: Defect Reporting Process in Effect for the Ticket App

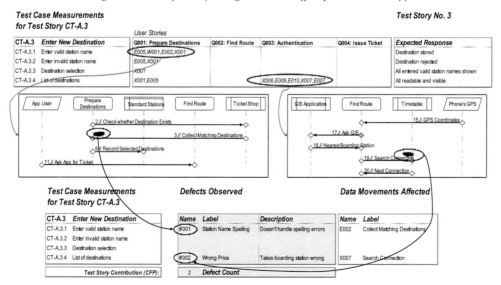

As a rule, one test story can uncover only one defect in a data movement. It can uncover several defects in different data movements, but the definition of one defect refers to test stories. The reason is the following: it is always possible to add as many test cases to a test story by slightly modify test data only. Even if defects exist that manifest themselves for some crucial test data only, and for other data, the function responds as expected, and even though other reasons or bad code might be responsible for different test data, we restrict counting to only one defective data movement per test story. This avoids the limitless growth in defect counts otherwise observed with more traditional defect counting methods.

Within test story No. 3, two test cases uncover two different defects:

1) Defect #001 – Station Name Spelling: Doesn't handle spelling errors
2) Defect #002 – Wrong Price: Takes boarding station wrong

Test story No. 3 cannot uncover the other two defects seen in the data movement map clips, as the four test cases shown do not execute the related data movements. This test story can place only one defect indicator (the "bug") on the data movement in the data movement map. To place a second bug on the same data movement, another test story identifying another defect is needed.

10-5.5 TEST STATUS REPORTING

Test status report for the ticket app looks as shown in Figure 10-16. We already explained the meaning of test intensity and test coverage; *Defect Density* in turn means the percentage of defective data movements among all data movements in the ticket app. The *Test Size* simply is the sum of all test case sizes; i.e., the total number of data movements covered by tests when executed. In all these counts, the COSMIC rules apply.

With these six key performance indicators, the test process becomes transparent and comparable between different testing organizations and products.

Looking at the test summary for our ticket app, it is obvious that neither a test intensity of 11% nor a test coverage of 50% is enough in practice. This sample case serves for explaining the principle only. In real life, the test manager for the ticket app project probably should do more.

Figure 10-16: Sample Test Status Report for the Ticket App

Test Status Summary

Total CFP:	38	Test Size in CFP:	202
		Test Intensity in CFP:	5.3
Defects Found in Total:	4	Defect Density:	11%
Defects Pending for Removal:	4	Test Coverage:	39%

The defect density definition presented here explicitly refers to the functional sizing method ISO/IEC 19761 COSMIC and cannot easily convert to other sizing methods except if they meet the criteria of the VIM (ISO/IEC Guide 99:2007, 2007).

10-6 DEFECT DENSITY PREDICTION BY Q CONTROL CHARTS

Once it is clear how to measure defects, it becomes possible predicting defects. The authors present a method for defect density prediction based on Q control charts (Quesenberry, 1997).

10-6.1 PRECONDITION

As already mentioned, expecting that Lean Six Sigma statistics provides realistic results is possible only because we used the COSMIC defect measurement approach for both functional size and defect counting and tracking.

10-6.2 STATISTICAL PROCESS CONTROL FOR DEFECT TRACKING

It is common practice in manufacturing industries that Statistical Process Control (SPC) methods are applied for monitoring, controlling and improving processes over time using classical Shewhart control chart; see e.g., (Montgomery, 2009). Shewhart control charts are well suited for long-run processes. However, software testing usually is a once-upon-a-time, or *Short-Run*, process; thus, lacking the historical data otherwise needed to construct classical Shewhart control charts.

10-6.3 PREDICTING DEFECT DENSITY

Self-starting control charts enable monitoring and controlling of a process without a large amount of historical data at hand. They center at the median of an actual observation series and update the upper and lower control limits with each new process run.

Hence self-starting control charts such as Q control charts (Quesenberry, 1997) are more appropriate for software development processes, especially when the software testing process is monitored by the fault-slip-through process (Damm, et al., 2006), (Fehlmann, 2009-1).

Q control charts monitor LeOR metrics. The basis of Q control charts is a sequence of iterative LeOR measurement x_r, which transforms by the following Q statistics

$$Q_r(x_r) = \Phi^{-1}\left\{G_{r-2}\left[\sqrt{\frac{r-1}{r}}\left(\frac{x_r - \bar{x}_{r-1}}{S_{r-1}}\right)\right]\right\}, r = 3,4,\ldots \qquad (10\text{-}5)$$

with

$$S_r^2 = \frac{1}{r-1} \sum_{j=1}^{r} (x_j - \bar{x}_r)^2 \qquad (10\text{-}6)$$

into a normally distributed random variable where the mean equals zero and variance equals one. In (10-5) and (10-6), $G_r(.)$ denotes the Student t distribution with r degrees of freedom and $\Phi^{-1}(.)$ the inverse of the standard normal distribution. Furthermore, at iteration $r+1$ the corresponding observation x_{r+1} contributes to the mean $\bar{x}_{r+1}$ and the sample variance S_{r+1}^2 as shown in (10-7) and (10-8):

$$\bar{x}_{r+1} = \frac{r}{r+1} \left(\bar{x}_r + \frac{x_{r+1}}{r} \right), \ r = \ 2, 3, \dots \qquad (10\text{-}7)$$

and

$$S_{r+1}^2 = \left(\frac{r-1}{r}\right) S_r^2 + \frac{1}{r+1}(x_{r+1} - \bar{x}_r)^2, \ r = 3,4, \dots \qquad (10\text{-}8)$$

respectively. The equations (10-7) and (10-8) describe defect prediction; the short-run control charts adapt to the sequential observations. The difference $|\bar{x}_{r+1} - \bar{x}_r| >$ *threshold* indicates that the actual observation x_{r+1} may cause $|Q_{r+1}(x_{r+1})| > 3$; i.e., x_{r+1} may be an outlier. Whether x_{r+1} is in fact an outlier depends on the difference $|S_{r+1}^2 - S_r^2|$. If this difference is small, then x_{r+1} has a minor impact on the observations series and we can remove it from further calculations.

The approach proofed to produce reproducible results in software testing. Fehlmann and Kranich have published details about the experiences in (Fehlmann & Kranich, 2013-2).

It is tempting to apply it for agile developments, where the r's refer to sprints.

10-7 CONCLUSIONS

Counting defects by means of functional size leaves non-functional defects out. Nevertheless, if combined with the Lean approach proposed in *Chapter 7: The Modern Art of Developing Software*, the concept of defect counting extends to non-functional story cards, using story points to calibrate non-functional story cards with functional story cards. In addition, the non-functional story cards become functional when it comes to implementation; thus, the concept of counting functional defects carries over to non-functional requirements.

While functional sizing for effort estimation is a long-standing practice, using functional sizing for defect density measurements and prediction in software services, especially in a service-oriented architecture, is a new approach that was not possible with first generation functional sizing methods. Moreover, test effectiveness and consequently test planning and test strategy becomes quantitatively manageable when based on the data movements' count with the ISO/IEC 19761 COSMIC method.

Using COSMIC data movements as defect measurement rules allows managing tests quantitatively, and since COSMIC defines a linear measurement topology on software, Lean Six Sigma and the related statistical methods, including transfer functions, are applicable. Furthermore, the same measuring rules apply to LeOR, and for defect density predictions.

As a summary, defect density prediction becomes operationally much simpler, more reliable, and better reflects observable reality when based on

- Defect sizing based on ISO/IEC 19761;
- Identifying and measuring LeOR;
- Validating transfer functions by convergence gaps;
- Monitoring the SW process with Q charts.

Functional sizing connects with perception of developers thanks to data movement mapping, making measurements of LeOR and defects transparent and acceptable for all stakeholders.

It is also remarkable that non-functional user requirements need no special treatment in this framework. As every non-functional requirements links as a quality characteristics to a user story, expressing how some functional user requirements shall be executed, there is no need to expressively state non-functional test stories. The IFPUG Software Non-functional Assessment Process (SNAP) relies exactly upon the fact that non-functional requirements always refer to some specific functionality. SNAP assigns quantifiable quality characteristics to the IFPUG functional model (IFPUG Non-Functional Sizing Standards Committee, April 2013), quite similar as we do.

PART THREE:

LEAN SIX SIGMA APPLICATIONS

CHAPTER 11:   APPLICATION TO PRODUCT MANAGEMENT

Six Sigma Transfer Functions for Analyzing Market Preferences; using New Lanchester Strategy, and some industry experiences. A general framework for product management with Lean Six Sigma.

How can product managers avoid 無駄 *Muda?*

11-1 TRANSFER FUNCTIONS IN PRODUCT MANAGEMENT

Competitors excel at different levels with respect to market preferences; we call this a *Competitive Profile*. In fact, profiles characterize all topics in Deming processes. The profile represents the priorities set for the various aspects of customer needs or market preferences and normalized as unit vectors. Profiles compare with each other; see *Chapter 2: Transfer Functions*. The convergence gap measures the difference between market preferences and the competitive profile. The vector difference between the market preference vector and the competitive profile vector is measured in the linear vector space of market preferences.

11-1.1 WHAT ARE MARKET'S PREFERENCES?

This chapter explains why market's preferences – essential for software products – are on a different level than customer's needs in a project QFD. Market's needs are different – the market decides based on preferences rather than on analyzing individual customer's needs, formulating requirements and assessing whether the software product meets these requirements. Many voices shape such preferences – professional journals, trade magazines, experiences made, assumptions and prejudices play their roles. In ICT, market preferences are expressed as a selection of product features – those features that are relevant when evaluation products and doing buying decisions. Sometimes, what the market selects comes quite as a surprise for the supplier of the product – for instance, when adopting mobile telephones, the market decided for a preference for the *Short Message Service* (SMS text messages, originally designed for internal use only when servicing network hubs), it was not expected for inventors and providers of that service.

However, when no direct customer is involved into development, Voice of the Customer is not as readily available as for customer software projects, project sponsors cannot formulate requirements by analyzing customer's needs; thus, product management must guess somehow what the potential customers want. Nevertheless, it is possible to predict market precedence using the power of Six Sigma and of Deming Value Chains.

11-1.2 DEMING VALUE CHAINS

The *Deming Value Chain* – see *Chapter 2: Transfer Functions* – consists of a series of transfers from one topic level of the value chain into another. It depends on the domain; for instance, with software projects it looks as shown in Figure 11-1. Advice how to set up a Deming Value Chain for various deployments is available in the Quality Function Deployment Best Practices (Herzwurm & Pietsch, 2009). W. Edwards Deming published organizational deployment schemes for production chains in the early 1930'ies already (Deming, 1986). Prof. Akao used similar schemes for "QFD in the Broad Sense" (Akao, 1990) and (Mizuno & Akao, 1994).

Deming Value Chains link Deming Value Creation Processes, value-added production steps that transform resources into business value.

Figure 11-1: Deming Value Chain for Software Project Deployment

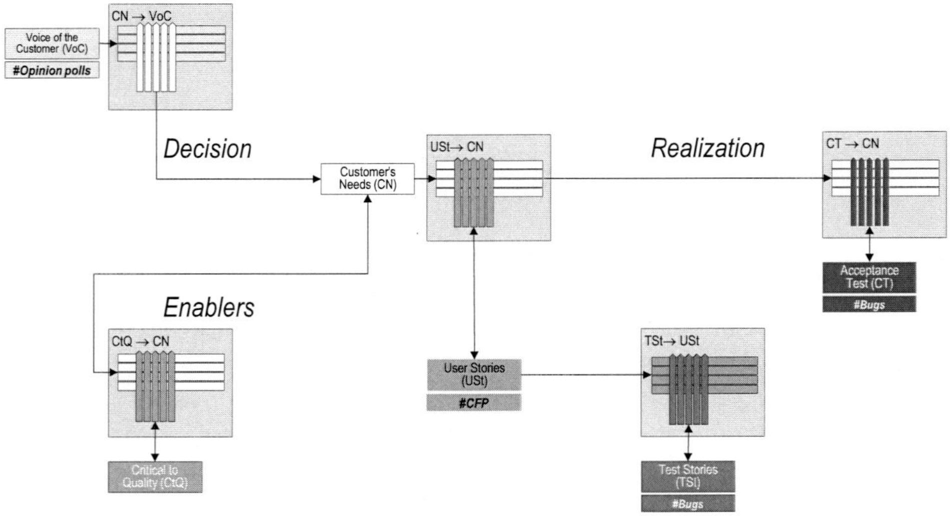

We write $B \rightarrow A$ when B is a set of controls that yields business value A as a response. For instance, *Customer's Needs (CN)* result in *Voice of the Customer (VoC)* that reflects those needs $CN \rightarrow VoC$. Customer's needs (CN) also cause buying decisions $CN \rightarrow LT$; LT stands for "Lanchester Theory" (Taoka, 1997), reflecting the same needs. In turn, *User Stories (USt)* must meet *Customer's Needs (USt $\rightarrow$ CN)*; i.e., they fulfill those needs what we also understand as sort of production process. In this sense, we adopt and re-use Deming's value chains for software and service purposes.

Knowing about *VoC* and *LT*, the software supplier should understand *CN*, and, if successful, he should be able to collect (*USt*) and transpose them into *Test Stories* (*TSt*).

There is also an upward branch: *Test Stories* (*TS*) and *Acceptance Tests* (*CT*) support the respective topic according their level; and there is a second downward branch in the value chain dealing with non-functional requirements against the software development process such as the *Critical-to-Quality* (*CtQ*) characteristics. Both quality and functionality control software project success.

11-1.3 DEMING VALUE CHAIN FOR SOFTWARE PRODUCTS

When setting up the Deming value chain for software products, there is a significant difference (Figure 11-2). Not *Customer's Needs* (*CN*) are pivotal but *Business Drivers* (*BD*). Market preferences develop around business objectives; the selection and valuation of business objectives changes when market preferences evolve.

Figure 11-2: Deming Value Chain for Software Product Deployment

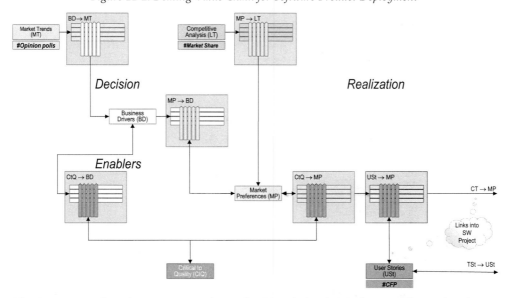

The Deming value chain starts with *Market Trends* (*MT*). Market trends are harder to assess, they are the result of decisions made by market players about their *Business Drivers* (*BD*), which in turn depend from *Market Preferences* (*MP*) rather than from software engineering artifacts directly. Both the functionality offered for use cases in the software product, as *Critical-to-Quality* (*CtQ*) characteristics affect market preferences. This makes the Deming value chain for software products deployment significantly more complicated than for software projects.

A software project is at the origins for creating a software product, but the software engineering artifacts such as user stories or test stories link not to requirements or customer's needs, but to market preferences. Moreover, market preferences depend both from critical-to-quality characteristics and from the functionality provided by the product. Thus, both quality characteristics and functionality are more difficult to align to software product development than in a software project. Moreover, no formal acceptance testing is possible; the market decided probably based on pilot experiences but not on tests.

11-1.4 CONTROL BY MEASUREMENTS

Six Sigma offers metrics and tools needed to provide *Decision Metrics* to this Deming Value Chain, by linking the controls to observable facts. A decision metric is an evaluation function $A \to A$ that yields stable results, where A is some business value topic. This means, repeating the decision method always yields the same results.

Market preferences are not directly measurable; but surveys, market research provide observable measurements, and market share is measurable. Using Six Sigma *Transfer Functions*, we can measure market preferences indirectly; however, we need to know how reliable such indirect measurements are. A transfer function evaluates a Deming process $B \to A$.

Competitors excel at different levels with respect to market preferences; this is called a *Competitive Profile*. The difference between market preferences and the competitive profile can be measured using the *Convergence Gap*, which is the length vector difference between the market preference vector and the competitive profile vector, measured in the linear vector space of market preferences.

11-2 A CALL CENTER EXAMPLE FROM LONG TIME AGO

The following simple example may illustrate the problem; although the technology constraints of this QFD are now completely outdated. It originates from a QFD Workshop analyzing the market share trends for a call center whose primary service consisted of answering inquiries for customer phone numbers. It goes back to a time when phone numbers where not yet readily searchable in the Internet, and people had to ask call centers if they did not want carrying voluminous phone directories.

11-2.1 CALL CENTER MARKET SHARE

The market share was measurable. Following the deregulation at the time, other emerging phone companies started their own call center operations, advertising them and trying to get market share from the former monopolists.

Figure 11-3: Market Share of Competing Call Centers

		Customer's Needs Topics	Attributes		Weight	Profile	
We	**Group 1**	Our Market Share (0%)	Solid Brand	Tradition	26%	0.51	
They	**Group 2**	C1 Competitor 1 (0%)	Quick. Fast.	Unreliable	20%	0.39	
		C2 Competitor 2 (0%)	No Waits	Responsive	31%	0.61	
		C3 Competitor 3 (0%)	Rock Solid	Me Too	23%	0.45	

(Priority)

However, customer segmentation was unknown – the assumption was that people using the service had limited access to Internet.

A transfer function maps the competitive profile of each competitor onto market share. The competitive profile consists of the *Solution Profile for Market Preferences*, inferred from the market share and the transfer function. It describes customer experience that otherwise is difficult to measure directly. Multiplying it with the matrix and normalizing it yields the *Achieved Profile*. It compares with the *Goal Profile*. In theory, the expectation is that competitors share the market according their specific competitive advantages. This depends from the selection of controls.

Figure 11-4: Transfer Function Not Explaining Observed Market Share

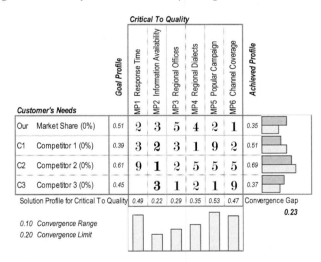

The matrix cells in Figure 11-4 resulted from a *Quality Function Deployment* (QFD) workshop; they reflect in each column the relative performance of competitors against one market preference topic. The competitive ranking of the *MP1: Response Time* was measurable; Data for *MP2: Information Availability* was gained with some test calls; *MP3: Regional Offices* and *MP4: Regional Dialects* have been measured as well, while

MP5: Popular Campaign and *MP6: Campaign Channel Coverage* were assessed by workshop team agreement. Nine is highest, zero or empty is least. Equal settings are admissible.

Figure 11-5 shows the initial list of market preferences that went as critical-to-quality controls in the transfer function represented by Figure 11-4. The solution profile now turns clockwise.

Figure 11-5: Initial List of Critical-to-Quality Controls for Market Preference

Critical To Quality

	Critical To Quality Topics	Attributes	Weight	Profile	
SC1 Group 1 MP1 Response Time	No waiting loops	21%	0.49	
	MP2 Information Availability	Requested Info is available	10%	0.22	
SC2 Group 2 MP3 Regional Offices	Employees are working from distributed regional offices	12%	0.29	
	MP4 Regional Dialects	Employees use regional dialects if appropriate	15%	0.35	
SC3 Group 3 MP5 Popular Campaign	How well campaign for help desk number is known	22%	0.53	
	MP6 Channel Coverage	Campaign on TV, billboards and paper media	20%	0.47	

(Priority: Weight, Profile)

In Figure 11-4, the observed *Our Market Share (26%)* in the achieved profile (in green) is higher than the predicted market share (in yellow). Why this? What is the cause?

11-2.2 MARKET PREFERENCES

The missing market preference factor seemed not related to any technical or organizational feature. It rather related to the fact that our product was well known and trusted for its long existence. We did call that the *Trust Factor*.

Figure 11-6: Transfer Function Better Explaining Observed Market Share

Customer's Needs	Goal Profile	MP1 Response Time	MP2 Information Availability	MP3 Regional Offices	MP4 Regional Dialects	MP5 Popular Campaign	MP6 Channel Coverage	MPT Trust Factor	Achieved Profile
Our Market Share (0%)	0.51	2	3	5	4	2	1	9	0.51
C1 Competitor 1 (0%)	0.39	3	2	3	1	9	2	3	0.49
C2 Competitor 2 (0%)	0.61	9	1	2	5	5	5	3	0.63
C3 Competitor 3 (0%)	0.45		3	1	2	1	9	1	0.32
Solution Profile for Critical To Quality		0.43	0.22	0.29	0.33	0.47	0.40	0.43	Convergence Gap

Convergence Gap **0.16**

0.10 Convergence Range
0.20 Convergence Limit

Clearly, the added control improved the convergence gap in Figure 11-6, there is reason to believe that this extended set of customer preferences with the trust factor added better reflects market reality. The transfer function makes it possible to quantify the relative importance of the trust factor versus other factors.

However, the convergence gap of 0.16 is near the limit of the stated *Convergence Range* – the range of trust defined in view of the data quality – and thus unsatisfactory.

11-2.3 IMPROVING THE ASSUMPTIONS

Since market preferences are assumptions not stated requirements, it is tempting to experiment with removing some of the supposed market preferences to see what happens.

As candidates for removal, we considered two groups with three topics:

1. *MP2: Information Availability* is not relevant for market preference, because the users of the call center will not notice if information is not available.

2. Although the two regional product features (*MP3* and *MP4*) were of high political interest and much discussed in the news, typically people are not ready to pay any price tag for such product features and therefore they are good candidate for removal from the list of market preferences.

Figure 11-7: Dismal Result when Removing some of the Supposed Market Preferences

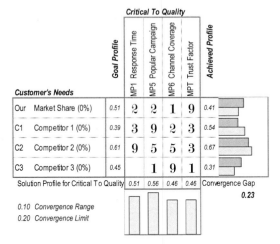

The result shown in Figure 11-7 is interesting: the market share predictions in the *Achieved Profile* for both *Our Market Share (26%)* and *C3: Competitor 3 (23%)* again fell below the observed value, while the prediction for *C1: Competitor 1 (20%)* became much too high. That gave a hint that some controlling factor is missing that increases market share for both *Our Market Share (26%)* and *C3: Competitor 3 (23%)*.

In fact, both *Our Company* and *C3: Competitor 3 (23%)* made heavy use of the evolving Internet search engines and provided full integration of the customer call center within their web pages. A quick assessment of the team looking at their Internet portals brought a competitive ranking for *MPN: New Technology* that best reflected the actual market share, see Figure 11-8, and brought the convergence gap down to an excellent 0.02.

Figure 11-8: Result of Adjusting the set of Market Preferences

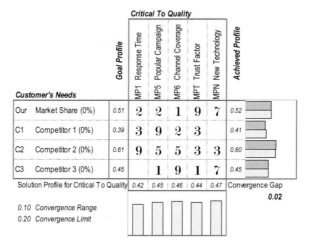

In this case, it turns out that with a modified set – the trust factor and the technology factor added – the improved convergence gap may better reflect market reality.

Figure 11-9: Final Critical-to-Quality Profile for the Market Preferences

Critical To Quality

	Critical To Quality Topics	Attributes	Priority	
			Weight	Profile
SC1 Group 1	MP1 Response Time	No waiting loops	19%	0.42
SC3 Group 3	MP5 Popular Campaign	How well campaign for help desk number is known	20%	0.45
	MP6 Channel Coverage	Campaign on TV, billboards and paper media	21%	0.46
SC4 Trust	MPT Trust Factor	Long-time player - well known number	20%	0.44
SC5 Tech	MPN New Technology	Integrated with Internet Query	21%	0.47

Figure 11-9 shows the final profile for market preferences. That the priorities do no longer differ much is probably a consequence of removing *MP2, MP3* and *MP4* and makes the marketing tasks not easier; however, it reflected reality at the time when googling the Internet was still novel.

11-2.4 LEARNING FROM INCOMPLETE KNOWLEDGE

Where customers are not readily available for answering questions about their needs, Six Sigma offers with transfer functions a nice way of analyzing incomplete information. The convergence gap offers a quick means to assess consistency of that analysis; however, Six Sigma is no automated problem solver. It requires sound data, domain knowledge and excellent QFD practices when predicting market share with Six Sigma.

Clearly, the trust factor is limited in time; as the new competitors consolidate their offerings, its effect will vane and new strong arguments must replace trust for continuing to increase market preference. Recommendations are therefore to watch closely what the competitors do with new technology.

11-3 THE IDEAL SITUATION

Frederick William Lanchester was born 1868 in London and educated as engineer. He created aircrafts, gas motors, disk brakes, four-wheel drives, and servo¬ drive, and received 236 patents. In 1916, while he devised the operational strategy for the Royal Air Force, he formulated the 1st and the 2nd Law of Lanchester, which describe forces needed for winning military battles. This was particularly helpful for the US in the Pacific campaign against the Japanese fleet. Until the late 20th century, the Lanchester Laws successfully predicted the outcome of military battles. Today, with massive computing power, more elaborate simulation approaches have taken the lead.

The *New Lanchester Approach* originated in Japan, where Nobuo Taoka and Shinichi Yano adapted the two laws of Lanchester for strategic and tactical marketing (Taoka, 1997). Competition in the market is sort of a battle, where a sales organization only can win when they can convince customers to buy their products. Instead of counting won battles, the measurement for marketing is the market share for the product or service; instead of defeated warriors, convinced customer minds count.

11-3.1 THE LANCHESTER THEORY

If the market is in equilibrium, the buying decisions of customers follow a stable pattern. Given a vector of market preferences in the vector space of relevant product features, the transfer function is the matrix of all comparisons buyers made between competitive offerings. This transfer function acts on the market preferences and the results are market shares, again a vector but this time in the vector space of competitive product offerings.

The New Lanchester Strategy distinguishes *Strategy of the Strong* for those dominating the market, and *Strategy of the Weak* for those that try to find or create market niches. It deducts a formula drawn from a military model that explains when to use which strategy. Ideally, the competitors are distinct only by product features – not by other influence factors such as branding and market image. The target market segment is homogeneous and product features are stable.

The Lanchester *Strategy for the Weak* requires that weak competitors identify areas where they can locally beat the market leader. "Locally" means that they need to focus on some limited market segment.

A famous formula, the *Lanchester Formula,* allows comparing two competitive profile vectors. The so-called "Weapon Strength" E, the *Exchange Rate*, is the ratio between two normalized profile weights (i.e., their vector lengths). If E exceeds $\sqrt{3}$ then this competitor is dominant, wins all competitive comparisons and reaches market dominance, see (Taoka, 1997). Under such circumstances, the New Lanchester Strategy creates excellent products, including software products.

11-3.2 PREDICTING MARKET SHARE BATTLES

The two laws of Lanchester allow predicting, who wins a market share contest.

First, we look after the traditional one-to-one contest as described so exquisitely in Homers Trojan War. Success depends from the individual performance only. Whoever has better weapons and more strength, wins. The following formula describes the likely outcome of the battle:

$$m_0 - m = E(n_0 - n) \tag{11-1}$$

where m_0 counts the initial number of fighters from one parts, n_0 the number from the controversial site, and m, n the undefeated fighters left after the fight, on both sides. E is the *Exchange Rate*, the relation imposed by the respective weapon strength. This is the 1st law of Lanchester. Balance is reached when the weapon strength E corresponds to the number of casualties:

$$E = \frac{(m_0 - m)}{(n_0 - n)} \tag{11-2}$$

The 2nd law in turn takes mass victory effects into account, such as neuronal activity of a brain cell that influences a multitude of other neighbor neurons; like modern mass communication media.

In this case, the law holds in square form:

$$m_0^2 - m^2 = E(n_0^2 - n^2) \tag{11-3}$$

Equilibrium is reached when the exchange rate E equals the ratio of the square differences:

$$E = \frac{(m_0^2 - m^2)}{(n_0^2 - n^2)} \tag{11-4}$$

As a proof idea for this formula, imagine two groups of brain cells (Figure 11-10): Group A with three, group B with two neuronal bundles. Each of them fires electrical pulses and thus influencing its neighbor neuron cells.

Each cell of group A is exposed to neuronal fire from two active neighbor cells that hit them with probability $1/3$.

On the other hand, a member of group B is "under fire" by three active neurons. It is hit with probability $1/2$.

The exchange rate becomes

Figure 11-10: The Neuronal "Battlefield"

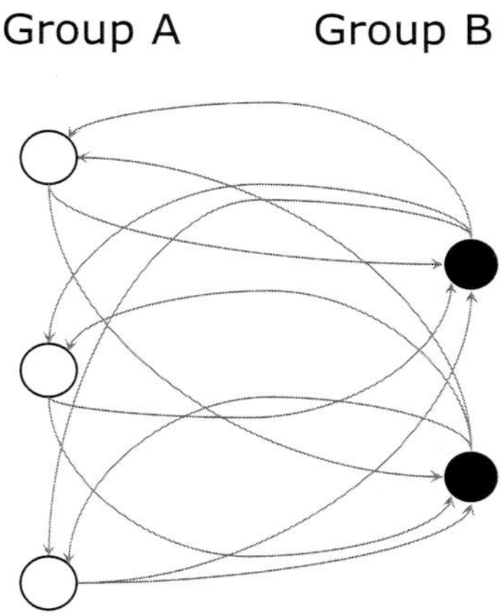

$$E = \frac{2/3}{3/2} = \frac{2*2}{3*3} = \frac{4}{9} = \frac{(m_0^2 - m^2)}{(n_0^2 - n^2)} \tag{11-5}$$

Thus, the group A benefits from the double amount of exchange rate effectiveness compared to group B, with only 50% more members. The 2nd law of Lanchester allows to optimize scarce resources significantly.

11-3.3 THE SHOOTING RANGE OF A COMPETITOR

The following is excerpt from Taoka (Taoka, 1997, p. 44).

When the exchange rate is 3 or less, its square root is the *Shooting Range*. When comparing two competitors, the shooting range decides who has the advantage. There are three values of importance for the shooting range:

- *Relative Safety Value* 41.7%. This is the primary target in a competitive battle among multiple companies. It is enough to keep the two immediate followers off shooting range.

- The *Upper-Limit Target* 73.88%. This so-called monopoly condition does not depend on the number of competitors. Companies try to defend this market share as the game play changes when they fall below the upper-limit target.

- The *Lower-Limit Target* 26.12%. This is the upper-limit of the inferior ranking, and is not stable even though the company may share the top ranking.

Obviously, lower-limit target plus upper-limit target equals 100%; in case of a duopoly – only two companies competing in one market – the inferior company has no way of attacking market share of the dominant player.

11-3.4 „THE STRATEGY OF THE WEAK "

If competing product features are weak, one still can outperform competition locally by creating a niche; e.g., near the customer, knowing its business in depth, or speaking local languages and knowing local habits. This corresponds to additional customer's needs criteria that you can explore from a weak position. Custom software development often is the *Strategy of the Weak.*

The weak competitor tries to outperform the stronger in so-called local battles, playing on an advantage that the competitor failed to notice in time.

11-3.5 „THE STRATEGY OF THE STRONG "

On the contrary, if holding a share above 41.7% in a market with more than two competitors, the *Strategy of the Strong* applies. The goal is to block smaller competitors from market niches. This works typically by announcing new developments, even without target dates and without a real investment. It blocks competitors from gaining market share. However, as soon as the market share falls below the "magic barrier" of 41.7%, business goes into trouble and margins fall. Especially in ICT, there were many examples (IBM, Nokia, Microsoft,) of the strategy of the strong; most of them failed when the threshold was hit.

Another popular strategy is to let smaller competitors explore new market niches and then buy them, including their skills and market shares.

11-3.6 NEW LANCHESTER APPLIED TO CALL CENTER EXAMPLE

It is tempting to look at the previous example to see what increases market share. To do this, we calculate the competitive profiles – normally, sum of the competition's profiles against our own competitive profile.

Figure 11-11: Market Share of Competing Call Centers (recall of Figure 11-3)

		Customer's Needs Topics	Attributes		Weight	Profile	
We	Group 1	Our Market Share (0%)	Solid Brand	Tradition	26%	0.51	
They	Group 2 C1	Competitor 1 (0%)	Quick. Fast.	Unreliable	20%	0.39	
	C2	Competitor 2 (0%)	No Waits	Responsive	31%	0.61	
	C3	Competitor 3 (0%)	Rock Solid	Me Too	23%	0.45	

For the above call center example, the comparison of competitive profiles shows that the current market is almost in equilibrium with $E^2 = 0.91$; nobody has a significant competitive advantage and therefore nobody is able to gain market share significantly.

Figure 11-12: Actual Competitive Profiles are in Equilibrium

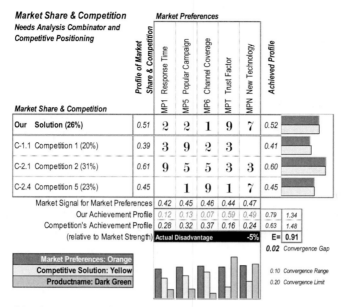

Market Share & Competition
Needs Analysis Combinator and
Competitive Positioning

Market Share & Competition	Profile of Market Share & Competition	MP1 Response Time	MP5 Popular Campaign	MP6 Channel Coverage	MPT Trust Factor	MPN New Technology	Achieved Profile	
Our Solution (26%)	0.51	2	2	1	9	7	0.52	
C-1.1 Competition 1 (20%)	0.39	3	9	2	3		0.41	
C-2.1 Competition 2 (31%)	0.61	9	5	5	3	3	0.60	
C-2.4 Competition 5 (23%)	0.45		1	9	1	7	0.45	
Market Signal for Market Preferences		0.42	0.45	0.46	0.44	0.47		
Our Achievement Profile		0.12	0.13	0.07	0.59	0.49	0.79	1.34
Competition's Achievement Profile		0.28	0.32	0.37	0.16	0.24	0.63	1.48
(relative to Market Strength)	Actual Disadvantage					-5%	E=	0.91

0.02 Convergence Gap

Market Preferences: Orange
Competitive Solution: Yellow
Productname: Dark Green

0.10 Convergence Range
0.20 Convergence Limit

The visual profile shows areas for improvement:

Figure 11-13: What to do for Winning Market Share

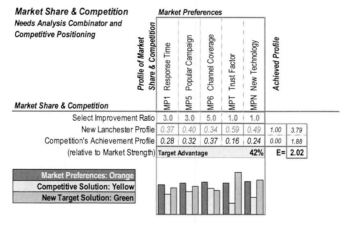

Market Share & Competition
Needs Analysis Combinator and
Competitive Positioning

Market Share & Competition	Profile of Market Share & Competition	MP1 Response Time	MP5 Popular Campaign	MP6 Channel Coverage	MPT Trust Factor	MPN New Technology	Achieved Profile	
Select Improvement Ratio		3.0	3.0	5.0	1.0	1.0		
New Lanchester Profile		0.37	0.40	0.34	0.59	0.49	1.00	3.79
Competition's Achievement Profile		0.28	0.32	0.37	0.16	0.24	0.00	1.88
(relative to Market Strength)	Target Advantage					42%	E=	2.02

Market Preferences: Orange
Competitive Solution: Yellow
New Target Solution: Green

With the New Lanchester Strategy, it is possible to select topics for product improvement and calculate how well the improved competitive profile of our product, called the *New Lanchester Profile*, matches market preferences. The *New Lanchester Profile* is obtained by multiplying *Our Achievement Profile* with the manually selected *Improvement Ratio*. Normally, imrovement factors are increased until the *Target Advantage* increases competitivity well enough.

Some of the proposed improvements might be costly – especially the increase by an improvement ratio of five (5.0) of the *MP6: Channel Coverage,* and other investments needed as well, but an increase of competitive advantage to 42% (corresponding to $E^2 = 2.02$) is quite rewarding. According the New Lanchester Strategy, 41.7% would guarantee market dominance.

11-3.7 CREATING A WINNING PRODUCT

The other relevant transfer functions, according Figure 11-2, are

- $UC \rightarrow MP$: From use case functionality into market preferences – how well does the product meet market preferences? The transfer function TUC→MP identifies the features needed to meet market preferences.

- $CtQ \rightarrow MP$: From quality characteristics into market preferences – how well does the product quality meet market preferences, sometimes called "House of Quality"? The transfer function CtQ→MP explains the impact of quality characteristics onto market preferences.

By combining the two transfer functions with the New Lanchester Strategy, it is possible to predict which added or removed features gain market share. This prediction is needed for successful release planning. The first transfer function defines the functionality needed to identify the added or superfluous functionality; with the second, quality characteristics adjust to meet market preferences better.

11-4 THE WEB PORTAL EXAMPLE

The next example covers a web portal. Comparing with other competing portals, measures become apparent that increase the attractiveness of some web portal against competition.

11-4.1 MARKET SHARE

Assume the following market distribution, shown in Figure 11-14:

Figure 11-14: Market Share for Our Web Portal

Profile of Market Share & Competition			Market Share as of today (Total: 100%)	Measured	
Our Product		Our Solution	19%	0.44	
They Competitors	C1	Competition 1 - easy to use	16%	0.37	
	C2	Competition 2 - all nice	19%	0.44	
	C3	Competition 3 - best in class	26%	0.60	
	C4	Competition 4 - struggling	9%	0.21	
	C5	Competition 5 - niche player	11%	0.26	

A previous Voice of the Customer (VoC) analysis (Kano, NPS, or else) suggests the following customer's needs to be decisive for the customer experience with competing web portals.

Figure 11-15: Customer's Needs for Our Web Portal

		Customer's Needs Topics	Attributes		
F Functionality	F1	Information Content	Access to relevant information	Availability	Reliability
	F2	Trustworthiness	High Reputation	Partners	
Q Quality	Q1	Up-to-date Functionality	Can adapt to changes	Continuity in performance	Actuality in content
	Q2	Ease of Use	Simple and straightforward to use	Ergonomically tested	
	Q3	Compelling Service	Clear in language & formulation	Adapts to cultural background	Personalized

There are many other choices for assessing customer preferences such as questionnaires, following clicks, assessing responses and call center complaints and wishes; see e.g., (Oudrhiri, 2005), (Pande, et al., 2002).

11-4.2 THE NEW LANCHESTER TRANSFER FUNCTION

According the New Lanchester Strategy, the customer's needs are instrumental for the customer's decision to buy a product or use a service.

In case of our Web Portal, it is possible to trace the frequency, how many times the portal had been visited and used by customers. The two functional needs – *F1: Information Content* and *F2: Trustworthiness* – are easy assessable by counting clicks on relevant information on the web portal. The three quality characteristics are harder to assess – either by an NPS survey asking for the likeliness of recommendation, or by an expert's assessment in a QFD workshop. Normalizing the frequency counts allows comparison with values originating from QFD workshops. Therefore, we can use the New Lanchester strategy to find the true customer's needs profile as seen by the market.

Figure 11-16: Competitive Comparison

Market Share & Competition
Needs Analysis Combinator and Competitive Positioning — Customer's Needs

Market Share & Competition		Profile of Market Share & Competition	F1 Information Content	F2 Trustworthiness	Q1 Up-to-date Functionality	Q2 Ease of Use	Q3 Compelling Service
Our	Solution (19%)	0.44	5	9	8	1	5
C1	Competition 1 - easy to use (16%)	0.37	5	3	6	9	3
C2	Competition 2 - all nice (19%)	0.44	6	7	6	5	7
C3	Competition 3 - best in class (26%)	0.60	9	9	8	9	9
C4	Competition 4 - struggling (9%)	0.21	2	3	3	4	5
C5	Competition 5 - niche player (11%)	0.26	3	6	3	6	5

The caveat is that market share in general is the result of a long period of competition and may not always reflect the current attractiveness of a service. For our Web Portal, we assume that the market share reflects the attractiveness for our target customers, and not some heritage characteristics like traditional name or well-established brand.

Figure 11-17: Competitive Advantage

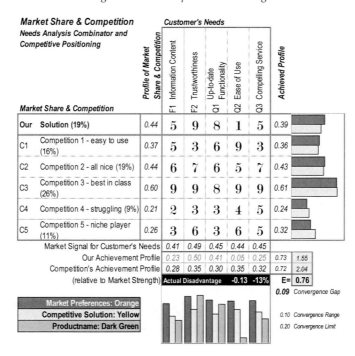

The cells in this cause/effect transfer function compare competitor's performance on a ratio scale. Thus, a "9" indicates three times better performance, or three times more clicks, than a "3", and indicates the highest rating. Least performance is recorded as a "1", which unfortunately holds for *Our Service (19%)* for the criteria *Q2: Ease of Use*.

For the matrix in Figure 11-17 (the same as in Figure 11-16), the small convergence gap of 0.09 means that the selection of customer's needs is a valid explanation for the observed attractiveness of our service, expressed by its market share. Thus, the competitive analysis for the web portal example looks sound.

Therefore, it is possible to compare the customer's needs profile with the market signal obtained from the web portal. First, the overall distribution of market share yields the *Market Signal for Customer's Needs* – the solution profile obtained as the eigencontrols for the full transfer function. *Our Achievement Profile* is a normalization of the first row in the matrix, holding the competitive standing of *Our Service (19%)*. *Competition's Achievement Profile* contains the eigencontrols for the matrix limited to the competition only, and this is the benchmark for comparing with *Our Service (19%)*. The exchange rate $E^2 = 0.76$ refers to it; in this case, this corresponds to a sensible competitive disadvantage of -13% – this is the negative square root of exchange rate $100\% - E$.

11-4.3 IMPROVING THE WEB PORTAL

There is significant room for improvement. With such performance, the web portal is likely to lose visitors and users, due largely to the bad rating in criteria *Q2: Ease of Use*.

Since with a market share of 19%, market dominance is far away, the choices are:

- Go out of business, or
- Re-define our business, or
- Significantly improve our profile.

If deciding for the third choice, the New Lanchester method effectively selects the development direction for improving the market share. Adding feature strength to selected customer's needs categories optimizes performance against competition. The transfer function defined by the matrix shown in Figure 11-17 predicts market strength.

As before, by selecting improvement ratios, the competitive profile can be turned in a *New Lanchester Profile* that serves as a target for improving the product, eventually increasing market share.

Figure 11-18: Improvement Ratio

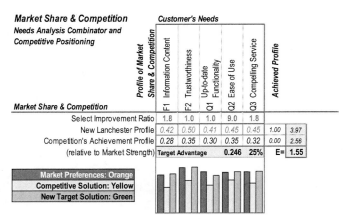

Market Share & Competition
Needs Analysis Combinator and
Competitive Positioning

Market Share & Competition	F1 Information Content	F2 Trustworthiness	Q1 Up-to-date Functionality	Q2 Ease of Use	Q3 Compelling Service		Achieved Profile
Select Improvement Ratio	1.8	1.0	1.0	9.0	1.8		
New Lanchester Profile	0.42	0.50	0.41	0.45	0.45	1.00	3.97
Competition's Achievement Profile	0.28	0.35	0.30	0.35	0.32	0.00	2.56
(relative to Market Strength)	Target Advantage			0.246	25%	E=	1.55

Market Preferences: Orange
Competitive Solution: Yellow
New Target Solution: Green

The improvement ratio of 9.0 for *Q2: Ease of Use* corresponds to the settings of 1 versus 9 when comparing competitive strength in the transfer function.

11-4.4 VALIDATION WITH KANO

It is advisable for better understanding the New Lanchester Strategy and its results, to validate the market preferences and customer's needs using Kano. For the web portal example, a Kano analysis matching the market signal obtained from market share shows again the high esteem customers have for the criteria *Q2: Ease of Use*.

Figure 11-19: Kano Analysis for Web Portal

Kano Analysis

Customer's Needs		Kano Profile	Goal Profile	x	y
F Functionality	F1 Information Content	0.43	0.42	0.3	1.4
	F2 Trustworthiness	0.50	0.50	1.1	2.1
Q Quality	Q1 Up-to-date Functionality	0.40	0.41	2.1	2.5
	Q2 Ease of Use	0.46	0.45	2.8	3.9
	Q3 Compelling Service	0.44	0.45	2.5	3.2

0.02   Convergence Gap

Although Kano allows placing the topic *Q2: Ease of Use* anywhere on the hyperbolic curve matching the profile, indications exist for its importance as perceived by customers, notwithstanding the fact that it provides the highest technical challenge for implementing the web portal. The Kano method is the fastest way to prioritize requirements; the method depends whether we find a knowledgeable team that can agree on positioning requirements in the Kano grid.

11-5 OVERCOMING DIFFICULTIES

Product development is not a science; it is related to creativity and innovation. Nevertheless, supporting creativity and innovation with a sound theory and some mathematical background allows making better decisions.

11-5.1 IDENTIFYING MARKET PREFERENCES

QFD offers many techniques for analyzing information available on the market – such as verbatim analysis, Google search, and many more. The best reference is the QFD Best Practices (Herzwurm & Schockert, 2006) drafted by practitioners of the German QFD Institute.

In the author's experience, the following approaches provide good results:

- Going to the Gemba – observe potential customers how they use the product – is useful for software products even more than for anything else. Software products are more easily traceable than hardware, thus the supplier knows what users do with it.

- One source of information is an ergonomic test – market preferences are more easily detectable when observing what software users are trying to do, than by asking them.

- Another very valuable source of information is building trackers into the software that count how often certain functions or use cases have been activated – care must be taken to get the user's agreement before collecting such information. If doing this, the recommended way is to ask the user explicitly for her or his collaboration before sending collected usage statistics back to the software supplier.

- Asking the product user has the additional advantage that the software product can collect qualified feedback – obviously, such build-in surveys are additional product features and use cases that enhance the product.

- Watching evolving technology is a very valuable source for potential market preferences as well – they might point into the future but when correlated with business value they allow reliable predictions.

- Another approach is with classical marketing. If markets are well defined and some customers willing to provide information, a sensing group or a market survey can help.

The preferred selection depends from the Deming Value Chain used for building the software product.

11-5.2 USING THE IT PRODUCT COMPASS

An easy way of detecting unknown market preferences is using some standard model; e.g., the IT Product Compass published by Herzwurm and Pietsch (Herzwurm & Pietsch, 2009, p. 22). A basic set for selecting market preference topics for investigation in transfer functions consists in this example by the following:

- Development
  - Product Life Cycle status (new product, or refurbished…)
  - Target Definitions (target market, target customers, target problems)
  - Training Needs (depends on roles; e.g., user, super-user, administrator, …)
  - Installation Ease (interoperability, interfaces, standards, constraints, …)
- Operations
  - Data Security (privacy, redundancy, disaster recovery, …)
  - Operational Ease (service dependency, autonomy, availability, …)
  - Administration (account management, login management, …)
- Application
  - Target Users (office-base, home-based, mobile, …)
  - Consultancy Needs (skills availability, …)
  - Helpdesk Support (channels, competency, response time, …)
- Service and Support
  - Incidence Management (non-conformities, maintenance, …)
  - Defect Prevention (release cycles, regression testing, …)
  - Preventive Maintenance (new feature management, …)
  - Corrective Maintenance (hotfixes, support interventions, …)

This set of standard market preferences topics serve for analyzing market share observations, if none more specific are available. It allows benchmarking software products that do not share common functionality.

11-5.3 FILLING THE MATRIX

Having a set of potential market preferences is one problem solved; however, the transfer function remains unknown. Usually, the transfer function is represented as a correlation matrix; i.e., a matrix whose cells indicate the strength of coupling between the respective market preferences as controls, and the desired market share response. It has previously been shown that methods exist to measure the matrix components, be it with NPS score (*Chapter 5: Voice of the Customer by Net Promoter®*), or data movements (*Chapter 8: Lean & Agile Software Development*).

Once candidates are selected for the market preference topics MP, product support tickets are the most valuable source of information for the two transfer functions $USt \rightarrow MP$ and $CtQ \rightarrow MP$, see below in 11-5.4 . Trackers, as explained in section 11-5.1 , can also be used to fill in the correlation matrix, since USt and CtQ are stable at least within one product release.

See (Herzwurm & Schockert, 2006) for suggestions how to measure Deming Chains such as $USt \rightarrow MP$ when developing software. Note that usually there is functionality in the product that does not contribute to any market preferences topic at all – such as housekeeping functions, security checks, etc., however, market preferences change fast.

11-5.4 ANALYZING SUPPORT CASES

A wealth of information is available for most of the companies producing software: support calls. Support cases tell more about changing market preferences than anything else. Customers tell the support people what they are trying to do and cannot achieve, for some reason.

Support calls most often address the $USt \rightarrow MP$ and $CtQ \rightarrow MP$ Transfer function, sometimes $MP \rightarrow BO$ as well, when users try to explain their business objectives to the call center. Support data primarily fills the matrix. Collecting support case data successfully involves classifying them into the applicable Deming Process, train support personnel to ask the right questions, and set up the respective data collectors.

For most suppliers of software products, this is most rewarding data source and probably the best investment into market research. With such data, trends towards new market preferences are detectable using techniques shown in section 11-2 .

11-5.5 Using Traditional QFD Workshop Techniques

If measuring the matrix components is difficult, traditional QFD workshop techniques can help to "guess" the correlations factors if they are not measurable. The convergence gap is helpful for such workshops as well since the accuracy of matrix correlation is uncertain, and therefore it must be measured.

The crucial thing about QFD workshops is that the QFD team pools together knowledgeable experts from different fields join to agree on the correlation matrix and finally on a solution. The QFD moderator is responsible to make this work. *Chapter 4: Quality Function Deployment* provides guidance on how to conduct a QFD workshop.

11-5.6 Closing the Convergence Gap

If the New Lanchester transfer function shows a large convergence gap, it hints at the possibility that relevant Market Preferences remain undetected. As in the section 11-2 example, additional topics can be constructed from the matrix correlation values observed.

Sometimes it helps reformulating the preference topic statement; sometimes it should be split into two for better modeling the buyer's behavior. The criterion is orthogonality; i.e., the preference topic statements should not be dependent from each other. In the competitive assessment, it should be possible to change one topic's evaluation without affecting others.

However, if this trial-and-error approach does not lead to tangible results, another check can be performed. Let $\mathbf{y}$ denote the observation; a rectangular AHP matrix must exists that has $\mathbf{y}$ as its result. To construct this matrix in practice is not difficult; however, the pairwise comparison involved in its construction will point the team towards the relevant market preferences.

11-6 Conclusions

Several software houses used New Lanchester Strategy to steer their 30% – 50% yearly growth and outperform its competition. The New Lanchester Strategy selects new features for inclusion in new releases. For the software start-up mentioned below, it took a 12 x 200+ matrix. This is easy to handle; it means that you must assess every new feature request against the 12 customer's needs only, based on your competition's achievement profile. Using a controlled approach, by closely observing and tracking the critical-to-quality aspects of the software development aspects, it did not only allow to win market share, but also to avoid winning too much market share, in selecting new features such that these companies remained able to manage its growth. With the New Lanchester Strategy, growth becomes predictable. The yearly growth corresponds roughly with the competitive advantage derived from the exchange rate E. Thus, you can manage growth for instance such that after three year of full growth at, say, 40%, you plan one year with growth limited to, say, 15%, to regain breath and consolidate the development facilities. All you must do is selecting new features to control the exchange rate.

Naturally, the question arises what happens if the competition also goes for the New Lanchester Strategy to compete for market share. However, the risk is slim; the hurdles to overcome when adopting transfer functions, AHP, QFD, and trust in these business models are important. It means that the corporate culture must change away from the usual pride of humans – or that leaders must have a physical or mathematical background strong enough to bear with eigenvector theory. This is most difficult to achieve, and in fact, it happened only rarely. One was the marketing department of a Swiss airline, the other a software start-up led by a physicist with education at the Swiss Federal Institute of Technology. The former disappeared because of the "hunter strategy" of its executive leaders who did not understand anything like QFD or New Lanchester; the latter became world leader in its area.

In addition, turning the "Strategy of the Weak" to the "Strategy of the Strong" works as predicted by the theory; thus, the need to be more innovative than the competition and block possible market niches for new competitors guaranties continuous market dominance.

Fehlmann published some details of these approaches in (Fehlmann, 2000) and (Fehlmann, 2008). Other experience reports are difficult to find, as the adoption of New Lanchester strategy, if ever used in corporate strategic decisions, is not widely communicated.

Statistical methods open a wide range of application possibilities to product management and product improvement. Finding the best combination of traditional product marketing and Six Sigma approaches is not easy, and experiences are not widely shared; however, prospects are very promising.

CHAPTER 12: EFFORT ESTIMATION FOR ICT PROJECTS

Software engineering is not civil engineering, where you first create a plan then execute the plan, and all you need to do is making sure the plan takes all eventualities into due consideration. Software must explain complex tasks in a language simple enough such that ICT systems can understand and execute it correctly. It is a translation process starting with humans, some actual processes, some explicit and many more implicit requirements, involves social behavior, organizational capability maturity, ability to communicate, to formulate in different industry-specific languages, of keeping trust and continual engagement that eventually ends in an integrated men-machine system creating value.

How do transfer functions master this complexity?

12-1 WHY IS SOFTWARE PROJECT ESTIMATION SO DIFFICULT?

Project estimation is a difficult task. It requires predicting the future, and this is much more difficult than explaining the past – although this might be difficult as well, sometimes.

Software project estimation is even more difficult. You must predict a future that the envisioned project is going to change. The prediction affects the prediction – like measurement affecting measurement, as with quantum physics. It is self-referencing and recursive.

In the past, software project managers used Gantt charts to predict cost. These charts proved going out of date more rapidly than computers could print them. Today's lean and agile software developers have been accustomed to use various kind of Delphi techniques, assigning story points to developing tasks, and never commit to anything except to the next sprint. For the next sprint, this is ok. For project cost estimation, or for knowing how many sprints it will take, this is not very helpful.

Today's society and economy needs software by the large. ICT has become the essential core competency for all but very few industries. Thus, getting reliable predictions what an ICT project is likely to cost at the end is mission-critical.

12-1.1 WHAT ARE PROJECT ESTIMATIONS?

An estimation is the transfer function mapping cost drivers into projects. Assume a range of projects with known costs from history. The question, which cost drivers apply, is the same question as in any QFD or other Six Sigma problem when it is unclear, what are the controls that drive the response. Given that the range of known projects is finite, and the number of cost drivers is limited as well, we might even expect that our transfer function is representable as a linear matrix.

Thus, the principle is clear and straightforward. The practice is not. We face several problems. Firstly, the response of our estimation transfer function must be known and measurable. Measurement standards for measuring software projects exist by ISBSG, for other projects consult the Basis of Estimates by the AACE; see (American Association of Cost Estimators, 2014).

Secondly, how to measure the correlation matrix? Some of the 27 productivity impact factors (PIF) that a workgroup of GUFPI-ISMA identified (see section *3-5 Productivity Impact Factor Determination by AHP*) are very difficult to measure. The usual workaround is asking experts for their opinion. However, this induces yet another problem. Experts must be able to discuss when assessing PIF impact. A good means of enabling and controlling discussion is concentrating on only three possible opinions, namely low, medium or high impact. Any other scale serves as well; however, when matching 27 PIF with a few dozen projects, the size of such a matrix is large and does not allow for a larger scale. QFD knows the same issues with large matrices.

The problem with the three-size scale is the same as with the IFPUG approach. It is not a ratio scale. High impact is not just three times low impact. As with traditional QFD, exponentiation could better model reality.

Boehm used exponentiation in his COCOMO models for transforming low-medium-high scales into effort. Effort is preferred against cost for services such as software writing, to avoid dependencies from local salary levels.

The idea of Boehm is, given m known projects that rely on n cost factors or PIF; it is possible to predict the cost of some $m + 1^{st}$ project after selecting the impact factors for the n PIF. This method is known as *Parameterization.* The idea was nucleus for very successful project estimation tools like QSM and Galorath. The m known projects are called *Estimation Stack,* providing the base of estimation for new projects. This is a successful approach, if the predicted new project exhibits similar characteristics (*Parameters*) as those already existing in the estimation stack.

If the convergence gap of the extended matrix remains equally good as before, this means that the cost of the new project becomes predictable when based on the same assumptions as for those already in the estimation stack. The eigencontrols remain the same; there is no jump in the eigenvector solution. This ensures quality of estimation.

12-1.2 FUNCTIONAL SIZE

In the past, functional size was the most important cost driver. For certain application areas, it still is. In any case, be it mobile apps or large, software development or software service operations, single application or networked IT, all cost predictions start with a functional size count. Everything else is as unsafe as building a bridge without measuring its span. There are international standards, ISO/IEC 20926, 19761, and more, which define how to measure functional size. *Chapter 6: Functional Sizing* introduces the most popular measurement standards in detail and gives examples.

However, although functional size is easily measurable, quality requirements and non-functional requirements are not, and they contribute quite a bit to drive cost up- or downwards. The term is *Cost Drivers*. Thus, it is necessary to complement functional size with other project parameters when doing estimates.

12-1.3 MACRO ESTIMATIONS

Macro estimations try to make a comparison, based on functional size, with other, similar projects, either by benchmarking or by using parameterized cost drivers. This gives a ballpoint number for cost, depending from parameters that model unknowns such as how many skilled people will be available for the chosen technology. Usually, macro estimations provide a likeliness, and an upper and lower bound for cost.

Estimations based on cost drivers are macro estimations that this chapter will expand upon.

12-1.4 MICRO ESTIMATIONS

Micro estimations try to predict the development process. Although they too start with a functional size model, this model serves for breaking down the project into work items. On the Buglione-Trudel matrix, introduced in *Chapter 7: The Modern Art of Developing Software*, the story cards represent the work items. Summing up the estimates for each work item yields a cost prediction for the software development part.

Thus, macro estimations should always be complemented by micro estimations. In section *8-4 Early Project Estimation the Six Sigma Way*, it is explained how to micro estimate today's agile software development projects. Older approaches did rely on a work breakdown.

12-1.5 Benefits

Organizations need software project estimations before evaluating solution approaches and eventually suppliers. This allows to base decisions and software acquisition on a realistic budget. Based on sound estimation techniques, solutions can be chosen that pay off and match the financial capabilities. Moreover, it is easy to identify suppliers whose offers are too high or too low.

12-2 The Traditional Approach to Parametrization

Finance managers need reliable estimates to be able to fund software and ICT projects without running risks. Estimates are usually readily available – for instance based on functional size and benchmarking. However, the question how reliable these estimations are, is often left out, or answered in a purely statistical manner that gives no clue to practitioners what these overall statistical variations means for them.

This chapter explains how to make use of Six Sigma's transfer functions that map cost drivers onto project cost. Transfer functions reverse the process of estimation: they show how much a project costs under suitable assumptions for the cost drivers. If cost drivers can be measured, and transfer functions can be determined with known accuracy, not only project cost can be predicted but also the range and probability for such cost to occur.

12-2.1 Types of Software Project Cost Estimation

Since the early days, when manual coding was still the main task of developers, software project managers had attempted to predict software project costs by detail analyzing tasks and duration, using sophisticated methodologies based on detailed *Work Breakdown Structures* (WBS). While a WBS helps management understanding complexity of software development, it turned out to be unreliable in predicting the actual work needed. Nevertheless, effort predictions based on work breakdown structures were not all bad – even if they tended to predict other work than that needed to complete the project. The major problem with these approaches is that they do not reflect the nature of software development – namely to acquire the knowledge needed to complete the project.

A recent reaction to WBS-based project estimations is agile development – not planning for the work details but for allocating the time slots needed to complete the project, and do the work in fixed-time increments; e.g., sprints.

However, the most popular approach to cost estimation is benchmarking – based on own experiences or comparisons with industry. Because benchmarking always suffered from the difficulties of collecting reliable and comparable data, we also consider the so-called *Expert Estimation* approach as kind of benchmarking – instead of numerical database using the memories of experienced developers. Expert estimation is particularly successful in agile – collecting *Story Points* for sizing software projects and allocating enough sprints (Cohn, 2005).

12-2.2 The ISBSG Benchmarking Database

Among the numerical database collections, the ISBSG database is certainly the most popular one (Hill, 2010). It is an open collection of software development and maintenance projects collected all over the world and across all kind of industries. While other such collections exist; e.g., the proprietary QSM collection, none has the advantage of open, standardized and controlled collection practices. It is relatively easy to estimate a project – given its functional size; that is, when it is clear what needs to be build. Unfortunately, this happens relatively late in the project, namely when requirements are known up to a certain degree and at a defined granularity level. At this point of time, the development team has already spent a large part of the project budget to find out what these functional user requirements eventually actually are. Nevertheless, for solution design, model-driven development, scope management of projects, and defect prediction and planning of software operations: functional sizing is the method of choice; see also (Fenton, et al., 2008).

Figure 12-1: ISBSG MIS Projects: Function Points vs. Actual Effort

Figure 12-2: Same Projects with Cost Driven Estimations vs. Actual Effort

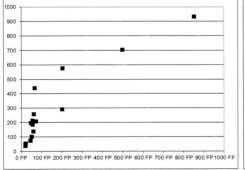
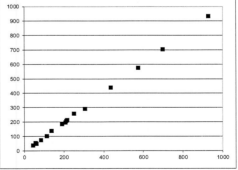

The ISBSG database comes with a large list of project attribute parameters for comparing different projects: industry, choice of modeling and coding language, team size, usage characteristics, methodology approach, architecture, target platform. Users can filter the database for large, medium or small functional size, development platform, and development type (new, enhancement, re-development, or customization).

Nevertheless, variations between functional size count and effective effort needed are significant. A random sample of 17 MIS Projects in the ISBSG R10 Database of 2009 with similar project attributes shows almost no correlation between functional size and actual effort reported, see Figure 12-1.

On the contrary, assuming the project parameters used by ISBSG control the cost drivers of these projects, the profile of these cost drivers explains the actual cost observed, see Figure 12-2.

12-3 PARAMETRIC APPROACHES TO COST PREDICTION

Parametric approaches are the most promising for predicting software project cost. Barry Boehm made this idea famous by 1981 when he started publishing the range of COCOMO prediction models (Boehm & et.al., 2000). He based project estimation on several cost drivers.

12-3.1 COST DRIVERS

Each of such cost driver functions has a different slope that models how the cost driver influences overall effort (Figure 12-3).

Figure 12-3: Cost Drivers with Different Slopes

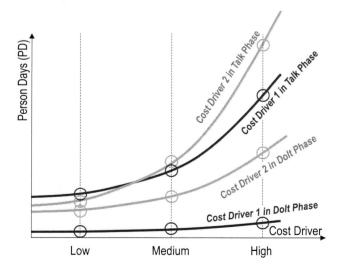

The *a-Parameter* defines slope; the parameter *x* denotes the selected cost driver impact. Boehm used general system characteristics such as requirements volatility, functional sizing, technical complexity, impact on current application, communication needs, and so one. These cost drivers are intuitively easy to understand but behave differently in the requirements elicitation, in the design & development phases, or for application testing or documentation. Moreover, the cost drivers may behave differently among different products. Total effort prediction is a function of these cost drivers.

These cost drivers are good candidates for predicting the effort needed to implement the project tasks. Boehm characterizes project cost by a *Cost Driver Profile*. Users of the model use a discrete scale marked with "Low"– "Medium"– "High" characterizing the cost driving force. The medium data point should be defined in such a way that profiles remain comparable. Thus, there is a need to state measurable standard value ranges for medium profile values; e.g., for team size or for people's skills against which comparison is possible. Ideally, cost drivers should be measurable; however, only quite rough assessments are usually available for soft factors such as "skills level", or "need for communication".

The impact of cost drivers varies among the development stages. Thus, as Figure 12-3 shows, we may have different impact even of the same cost driver, depending on the product (Fehlmann & Kranich, 2012-1).

12-3.2 MEASURING THE RESPONSE OF THE SOFTWARE PROJECT PROCESS

We need to analyze the response of our process in a way that allows distinguishing the various contributions from the cost drivers. An obvious choice is looking at cost per phase: for instance, distinguishing cost of requirements elicitation, analysis & design, technical implementation, solution integration, and start of operation phases already allow for analyzing impact of various cost drivers that relate to people and requirements volatility.

Another approach is based on the five CMMI process areas *Requirements Development* (RD), *Technical Solution* (TS), *Quality Assurance* (QA), *Product Integration* (PI), and *Project Management* (PM); however, as the experience of ISBSG shows, it is very difficult to get reliable results for phases across organizations, since different, internally developed and applied methodologies are common. In view of practicality, effort data should be collected as closely to roles and physical evidence as possible, to allow for comparisons among different organizations and methodologies. Automatic measurements based on meeting calendars, connection time to code repositories, integrated development environments and test automation tools must be possible.

We therefore propose to distinguish effort spent for requirements elicitation in team and stakeholder communications (*Talk*), work and rework needed (*DoIt*), reviews and tests conducted (*Test*), time needed for technical and financial project administration (*Adm*); e.g., documentation, configuration management, time and records keeping, and project management (*PM*). Since this kind of effort data is effort spent by the roles sponsor, developer, tester, administrators, and project managers, it is easier to collect and allows getting more reliable effort data. Note that cost drivers affect all effort types in the same way; e.g., high.

Efforts spent for a few relevant effort types measure the project process response sufficiently well. Table 12-4 shows which they are.

Table 12-4 Measurable Effort Types

Roles\Effort Types	Talk	DoIt	Test	Adm	PM
Sponsor	Meetings Decisions				
Developer	Meetings, Chat	Design, Code			
Tester	Meetings, Chat		Integrate, QA		
Admin	Meetings, Chat			Enable, Track	
Project Manager	Meetings, Chat				Manage Decisions

The profile vector for the project effort response thus runs over two levels: over the number of projects estimated or effort-measured, and for each project over the five effort types: *Talk, DoIt, Test, Adm,* and *PM.* These points of measurement are *Estimation Items.* Total dimension of the project effort response vector is five times the number of projects considered for process measurement; possibly a few dimensions less if not all projects cover all five effort types.

12-3.3 THE ESTIMATION FORMULA

Boehm (Boehm & et.al., 2000) uses exponential functions for the impact of cost factors[3]:

$$g_i(a_i) = e^{x_i a_i} \tag{12-1}$$

where a_i represents the slope of the ith cost driver and x_i defines the impact of the ith cost driver a_i, see Figure 12-3. The products $a_i x_i$ refer to the cross-point values for the impact function $g_i(a_i)$. The practical reason for taking an exponential function is that you do not have to care for dimensions nor for any static minimum cost; experience shows, on the other hand, that cost factors tend to soaring when they start growing. The *a*-parameter takes care of all that. The difference between low impact and medium impact is much less than between medium to high impact. High impact has

---

[3] Note that Barry Boehm uses x_i and a_i the other way around, compare (Boehm & et.al., 2000)

almost no upper limit, whereas low impact always has a limit: a minimal cost associated to it.

Boehm combines the effects of those individual cost drivers by multiplication with an exponential factor:

$$PI(\boldsymbol{a}) = \prod_{i=1}^{n} g_i(a_i) = \prod_{i=1}^{n} e^{x_i a_i} \tag{12-2}$$

where n is the number of cost drivers, and $PI(\boldsymbol{a}) = PI(\langle a_1, a_2, \dots, a_n \rangle)$ is the total cost influence profile per estimation item for the cost driver profile $\boldsymbol{a} = \langle a_1, a_2, \dots, a_n \rangle$. Note that the a_i represent low/medium/high and can be set without loss of generality to some equally distanced values around 1.0: $a_i = 0.5, 1.0$, and 1.5 respectively. No impact means $a_i = 0$, thus $e^0 = 1$. The *Impact Function* $PI(\boldsymbol{a})$ calculates the cost profile vector $\langle x_1, x_2, \dots, x_n \rangle$ for each cost driver vector $\boldsymbol{a}$ that represents an estimation item.

12-3.4 COMBINING WITH FUNCTIONAL SIZE

If the model contains more than just one cost driver that depends from functional size, we cannot use the above equation (12-2). However, since modules should not interfere with each other, an additive model is more appropriate than the multiplication of influential factors.

Let $f_i(a_i)$ denote the impact of functional size where the index $1 \le i \le n_0$. The functional contributions $f_i(a_i)$ sum up for the *Functional Cost Driver* FD:

$$FD(\boldsymbol{a}) = \sum_{i=1}^{n_0} f_i(a_i) \tag{12-3}$$

Exponentiation of the functional contributions $f_i(a_i)$ also yields excellent results, see (Fehlmann, 2009-2), but makes the model unnecessarily complex. With only one functional cost driver a_0,

$$FD(\boldsymbol{a}) = e^{x_0 a_0} \tag{12-4}$$

defines the logarithmic base for a_0.

For instance, a_0 can be selected such that, say, 512 COSMIC Function Points correspond to $x_0 = 1$; the exponential factor a_0 indicates the cost driving impact of functional size.

The *Calculated Effort in Person Days* (PD) for an estimation item with cost driver profile x is therefore

$$PD(a) = FD(a) * PI(a) = \prod_{i=0}^{n} e^{x_i a_i} \tag{12-5}$$

This is a simplification of equation (12-2).

12-3.5 CALIBRATION

The cost driver vector $x = \langle x_0, x_1, \dots, x_n \rangle$ represents by x_0 the impact of functional size and by $x_1, \dots, x_n$ the impact of non-functional cost drivers. If there are enough estimation items with cost driver profile x with known $PD(a)$, it is possible to calculate the a-parameters by multi-linear regression. The a-parameters hold for a series of similar estimation items. Taking the cost driver profiles into account, allowing no impact in the profile, at least $4n$ estimation items – with each cost driver once in no, low, medium and high profile state – are necessary for calibration. We use the term *Estimation Stack* for such as set of estimation items with known $PD(a)$. However, the more estimation items are available, the better for reducing measurement errors by redundancy.

So, if cost prediction for software projects is that easy, why is it not current successful standard practice?

12-3.6 QUALITY OF ESTIMATIONS

There are a few problems. The first is certainly the data collection used for calibration. Very few organizations are mature enough to collect their project data, know their cost drivers, and keep them under control for a long enough time to predict project cost successfully. Moreover, even if collectable data exists, how do we know how accurate the calibration data is? This is the second problem. Collecting actual data is significantly more difficult than collecting expert estimations. That is the reason why many organizations rely on expert estimations rather than on actual data when calibrating their estimation stacks. However, the third problem is probably the most intrinsic: since cost prediction is necessary for contracting ICT projects, it is not sufficient if some mature organization keeps collecting effort data and profiling their projects, its customer must have the possibility to compare and validate calibration data.

While high maturity organizations might address the first problem, the GUFPI-ISMA cost driver catalogue is a big step towards addressing the third issue. A detailed description is in section 3-5 : *Productivity Impact Factor Determination by AHP* in *Chapter*

3: What is AHP? The remaining part of this chapter focuses on the second problem: how to assess quality of estimations.

Figure 12-5: One Sigma and Six Sigma Estimations depend on Estimation Stack Variance

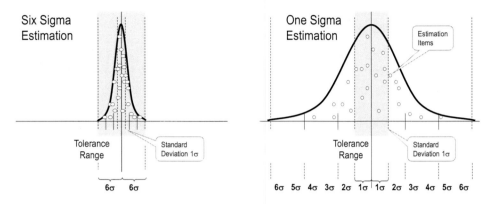

12-4 ESTIMATION STACKS AS TRANSFER FUNCTIONS

Transfer functions can combine with non-linear functions. An estimation stack represents a transfer function that maps the vector profile x of the cost driver onto $PD = \langle PD_1(a), PD_2(a), \ldots PD_m(a) \rangle$, an actual estimation item efforts vector of dimension m, using equation (12-2).

When creating an estimation stack for the m estimation items, this vector constitutes the response of the process. Each estimation item row depends from their cost driver profile $a_j = \langle a_{0,j}, a_{1,j}, \ldots, a_{n,j} \rangle$, for $j = 1, \ldots m$, see equation (12-6). This transfer function $F(x)$ predicts project cost, based on the settings for the cost driver profiles.

$$PD = F(x) = \langle \prod_{i=0}^{n} e^{a_{i,1}x_i} , \prod_{i=0}^{n} e^{a_{i,2}x_i} , \ldots, \prod_{i=0}^{n} e^{a_{i,m}x_i} \rangle \qquad (12\text{-}6)$$

If the cost driver profiles remain restricted to discrete values, such as $a_i = 0.0, 0.5, 1.0$, and 1.5, the number of estimations for a stack is limited to the permutation of all possible cost driver profiles, thus to 4^n possible response predictions – zero, low, medium, or high. Thus, every response of this cost prediction model comes with a known variation, with known accuracy. However, intermediate values for the a_i are allowed; it is a ratio scale. Thus, if the a_i are measurable, this is better than limiting them to discrete values.

12-4.1 Selecting Cost Drivers

Transfer functions map process controls into process responses – not the other way around. In general, we see responses first and can measure them before the relevant critical controls. Predicting the critical controls is a relevant issue for understanding transfer functions for processes.

Ideally, cost drivers should be *orthogonal* to each other, that is, one cost driver value must not depend from other cost driver values. The condition is that the value of one cost driver never depends from any combination of other cost drivers. However, cost drivers can compensate each other: if one cost driver has no impact, other cost drivers might provide the necessary impact to yield the observed process response.

Selecting the right cost drivers is in principle not different from selecting suitable sets of controls for any Six Sigma transfer function. The recommended practice is, to select as few controls as possible, since the amount of data needed for calibrating the controls increases exponentially with the number of controls.

12-4.2 Quality of Estimation Stacks

After calibration; i.e., calculation of the a-parameter by multi-linear regression, the prediction accuracy, as seen in Figure 12-5, characterizes the quality of an estimation stack. The prediction accuracy σ is mathematically the convergence gap for the transfer function $PD = F(x)$. The assumption is that project effort follows normal distribution. The American Association of Cost Engineers has recently put this into question, see (American Association of Cost Estimators, 2008-1), suggesting a double triangular distribution, skewed to allow for larger cost overruns than undercuts. In this case, both a left side lower limit and a right side upper limit become necessary.

With a sufficiently large number of cost drivers, it is always possible to find suitable a-parameter. Note that positive a-parameter increase cost; negative parameters would decrease cost. Sometimes this is not straightforward. For instance, if you add as a cost driver "Need for extensive documentation", it is not clear whether this increases or decreases cost. It might decrease cost of quality assurance and thus of effort type *Test* but increase effort spent for *DoIt*.

The prediction accuracy σ is also expressible in terms of confidence intervals. This metric gets the right kind of management attention. For instance, a variation of $\sigma = 3.5$ corresponds to 99.8% confidence based on the 95th percentile.

However, even if the estimation stack has high confidence, it only demonstrates that the selected cost drivers model the estimation items in the stack. It is unclear what happens when using the stack for predicting new projects. To understand how to assess quality on an estimation stack for prediction, we need to turn somewhat more into theory of transfer functions.

The cost driver profiles $a_j = \langle a_{0,j}, a_{1,j}, ..., a_{n,j} \rangle$ that define the cost for the jth estimation item using the estimation function (12-5) yield the matrix $A = (a_{i,j})$ of dimensions $(n+1) \times m$. n denotes the number of cost drivers as before; m is the size of the estimation stack. Let PD_M denote the vector obtained by actual measurement of all m estimation items. Obviously $PD_M \neq PD$ but, if the model is *capable*, $PD_M \cong PD$ up to prediction accuracy holds. The aim is to predict how capable the model based on the chosen cost driver profile x is.

12-4.3 ACTUAL ESTIMATION

The linearized transfer function looks at the cost driver. Let j be the index of the jth project in the estimation stack. Then the exponents x_i in equation (12-6) for this estimation stack define a linear mapping by the matrix $T = (x_{i,j})$

$$y = Tx = \langle y_1, ... y_n \rangle \text{ where } y_j = \sum_{i=0}^{m} a_{i,j} x_i \qquad (12\text{-}7)$$

The vector $y = \langle y_1, ..., y_n \rangle$ is a profile vector; however, it is not yet profiling effort. The matrix T, used for actual estimation, is not being confounded with the matrix A previously used for calibration of the estimation stack. Because of equation (12-4), the values of the profile vector need being converted to its logarithm.

$$PD = F(x) = ln(Tx) = \langle ln(y_1), ..., ln(y_n) \rangle, \text{ where } y_j = \sum_{i=0}^{m} a_{i,j} x_i \qquad (12\text{-}8)$$

The cost PD_j of the jth project is not a function of the organization's effort profile component y_j, as defined in (12-8), but of all components in cost driver coefficients $x_{i,j}$, where i is running over all cost drivers and j running over all projects in the estimation stack. Thus, the cost PD_j takes information from all cost drivers in all projects into consideration, according formula (12-5).

12-4.4 EIGENVECTORS AS QUALITY CRITERIA

In Figure 12-6, the cost driver profiles x represent an estimation stack. The analysis function is denoted by G; thus $x = G(PD_M)$ and $PD = F(x)$. Such an estimation stack meets the quality criteria for model estimations shown in Figure 12-6 and predicts other projects that rely on the cost drivers used for the same estimation stack.

Prediction accuracy is mathematically the same as the convergence gap; however, the transfer function must take equation (12-8) into account.

$$Prediction\ \ Accuracy = \left\| PD_M - F\big(G(PD_M)\big) \right\| = \left\| PD_M - PD \right\| \qquad (12\text{-}9)$$

For the convergence gap, and for how to use eigenvectors of transfer functions for validating cause and effect relations, consult *Chapter 2: Transfer Functions* or the literature; e.g., Hu & Antony (Hu & Antony, 2007).

Figure 12-6: Observed Response and Explained Response

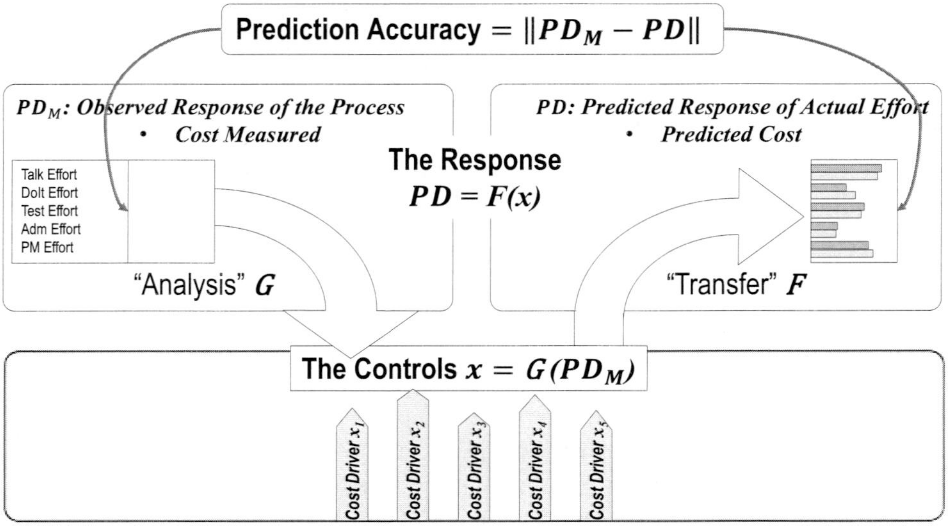

The consistency check for the matrix $A = (x_{i,j})$ cannot validate the semantics of the cost drivers x; it only limits the convergence gap. The labels must be determined by identifying the meaning of the cost driver in the real world. However, it is possible to find estimation stacks that allow for consistent selection of cost drivers.

12-5 QUALITY OF ESTIMATIONS IN PRACTICE

The following example uses a random sample of 17 projects from the *ISBSG Database Release 10* – replaced by *Release 13* as of early 2015 – for calculating an estimation stack and validating its consistency. Although the original data is realistic, some data is changed in order not reveal business secrets. The aim is to show how the principle works when assuring the quality of estimations based on transfer functions.

12-5.1 THE ISBSG SAMPLE ESTIMATION STACK

As explained before, effort data splits into the five categories of Table 12-4. Calibration yields the *a*-parameters as before for each of the categories. After calibration, the model was able to predict project effort with a 4 Sigma accuracy based on a target deviation of 10 Days (upper and lower limit). This is most probably better than average prediction models, due to the favorable selection of projects. Figure 12-7 shows the distribution of the estimation accuracy, in days. It shows the deviations per effort type; thus, the model was able predicting two (2) projects with better than -7.3 days negative deviation, and five (5) projects with less than +7.3 days positive deviation per effort type. Figure 12-7 shows this as a diagram.

Figure 12-7: Distribution of Estimation Accuracy

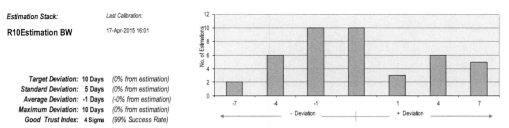

The histogram in Figure 12-7 shows the deviation per estimation item; that means, per effort type. This shows more detail than the convergence gap, which relies on the *a*-parameters only and levels inaccuracies out between effort types within a project. The cost driver selection in Table 12-8 is small compared to the *GUFPI-ISMA Productivity Impact Factors* (PIF); nevertheless, it already provides useful insights.

Table 12-8: Cost Drivers for the ISBSG Estimation Stack

		Cost Drivers Topics	Attributes			Weight	Profile	
DFP	COSMIC	Cfs Functional Size	COSMIC	Size		36%	0.86	
Cost	Drivers	M01 Requirements Volatility (#CR):	Number of	Change	Requests	8%	0.18	
		M02 Privacy Level (L-M-H):	Low	Medium	High	8%	0.20	
		M03 Technical Complexity (L-M-H):	Low	Medium	High	7%	0.18	
		M04 Communication Needs (#Sh):	Number of	Coordinations	Events	6%	0.15	
		M05 Team Size (#Tm):	Size of	Team		7%	0.16	
		M06 Duration Constraint (#day):	Schedule	Allowance		7%	0.17	
		M07 Availability Requirements (#h):	Tolerable	Downtime		7%	0.18	
		M08 Reliability Requirements (%):	Test	Coverage		7%	0.17	
		M09 Documentation Needs (L-M-H):	Low	Medium	High	6%	0.14	

The cost drivers *M01* to *M09* in Table 12-8 are used in the estimation practice because these cost drivers are easily measurable. The transfer function considers that the effort profile must be logarithmic; otherwise, it would not be comparable to the *a*-parameters.

The priority of functional size in Table 12-8 as the dominant cost driver reflects the effect of calculating the dominant controls for the observed project cost in this proof of concept.

Figure 12-9: Validation of Sample ISBSG Estimation Stack

Cost Drivers / Deployment Combinator / Measured Efforts	Goal Profile	Cfs Functional Size	M01 Requirements Volatility (#CR)	M02 Privacy Level (L-M-H)	M03 Technical Complexity (L-M-H)	M04 Communication Needs (#Sh)	M05 Team Size (#Tm)	M06 Duration Constraint (#day)	M07 Availability Requirements (fh)	M08 Reliability Requirements (%)	M09 Documentation Needs (L-M-H)	Achieved Profile
ISBSG-01 Telecommunications	0.24	8	2	2	3	3	2	1	2	1	2	0.22
ISBSG-02 Telecommunications	0.24	9	2	2	3	2	2	1	2	2	1	0.23
ISBSG-03 Banking	0.22	8	2	3	1	2	1	2	1	2	1	0.22
ISBSG-04 Telecommunications	0.27	8	2	2	1	1	2	2	3	3	2	0.23
ISBSG-05 Fine Enforcement	0.17	6	1	3	2	1	1	1	1	1	1	0.17
ISBSG-06 Banking	0.36	14	3	3	3	3	3	3	3	3	3	0.37
ISBSG-07 Engineering	0.19	8	1	1	2	1	1	1	1	2	1	0.19
ISBSG-08 Telecommunications	0.23	8	3	3	1	1	2	1	2	2	1	0.22
ISBSG-09 Banking	0.28	11	3	2	1	2	2	3	2	2	2	0.28
ISBSG-10 Telecommunications	0.25	11	3	3	2	2	2	1	2	1	1	0.27
ISBSG-11 Telecommunications	0.24	8	2	1	2	2	2	3	2	2	1	0.21
ISBSG-12 Telecommunications	0.23	9	1	2	2	1	1	2	3	2	1	0.22
ISBSG-13 Transport & Storage	0.29	12	1	1	2	1	2	2	3	2	3	0.30
ISBSG-14 Case Management	0.30	13	2	3	3	2	2	3	2	2	2	0.34
ISBSG-15 Fine Enforcement	0.16	6	1	1	1	1	1	1	1	1	1	0.15
ISBSG-16 Policy Enforcement	0.20	8	2	3	2	1	1	1	2	1	1	0.21
ISBSG-17 Telecommunications	0.17	6	2	1	1	1	1	1	1	1	1	0.16
Solution Profile for Cost Drivers		0.86	0.18	0.20	0.18	0.15	0.16	0.17	0.18	0.17	0.14	Convergence Gap 0.08

0.10 Convergence Range
0.20 Convergence Limit

Figure 12-9 lists the project sample from the ISGSB database used for this proof of concept. The sample is small, showing only a few traditional software development projects.

The left column in Figure 12-9 affects the solution profile much stronger than the rest; this reflects the domination of project costs by functionality in the ISBSG database R10 of 2009, as already seen in Table 12-8. The convergence gap of 0.08 is satisfactory small and suggests that the estimation stack is a sound base for predicting project costs of similar projects. Projects are similar if it is possible to add the new project to the estimation stack and the convergence gap does not deteriorate. This similarity requirement is easy to check. Adding the new project to the estimation stack and computing the convergence gap will do.

12-5.2 THE GUFPI-ISMA PRODUCTIVITY IMPACT FACTORS

The productivity impact factors (Table 3-15) defined by a working group of GUFPI-ISMA in 2011 (Beni, et al., 2011) have been introduced in section 3-5.2 to 3-5.5 of *Chapter 3: What is AHP?* For a discussion how the PIF serve as cost driver, we selected a number of typical projects and investigated their usefulness as estimation stack.

Figure 12-10: Validation of PIF Estimation Stack

Figure 12-10: Validation of PIF Estimation Stack

Productivity Impact Factors

Measured Efforts	Goal Profile	Cfs Functional Size	H1 Domain Knowhow	H2 Personnel Capability	H3 Technology Knowledge	H4 Team Turnover	H5 Management Capability	H6 Team Size	P1 Organization Maturity	P2 Schedule Constraints	P3 Requirement Completeness	P4 Reuse	P5 Project Type	P6 Methodology	P7 Stakeholder Cohesion	P8 Project/Program Integration	P9 Project Logistics	S1 Product Size	S2 Product Architecture	S3 Product Complexity	S4 Other Product Properties	S5 Required Documentation	S6 System Integration	S7 Required Reusability	T1 Programming Language	T2 Development Tools	T3 Technical Environment	T4 Technology Change	T5 Technical Constraints	Achieved Profile
PIf-001 First Project	0.19	1	3	5	5	8	5	5	5	5	5	5	5	3	3	5	5	3	5	8	8	5	5	5	5	8	5	5	8	0.21
PIf-002 Next Project	0.20	1	5	8	3	1	5	5	8	5	5	5	8	5	3	5	3	3	5	5	8	8	5	5	3	5	8	5	3	0.20
PIf-003 Third Project	0.22	8	3	5	3	8	5	5	8	5	3	5	5	3	5	5	5	8	5	8	5	3	5	3	5	3	8	5	5	0.20
PIf-004 Fourth Project	0.20	3	3	3	8	4	3	5	1	3	8	5	3	5	8	1	2	5	5	8	2	8	4	8	5	3	5	5	8	0.21
PIf-005 Fifth Project	0.20	2	2	6	6	4	8	5	1	8	2	6	5	2	3	8	2	3	6	8	2	6	4	3	4	6	2	2	4	0.23
PIf-006 Large Project	0.23	8	8	5	5	8	5	5	5	8	5	8	5	8	3	5	5	5	8	8	8	8	5	3	5	8	3	3	5	0.23
PIf-007 Important Project	0.20	2	5	5	5	8	3	5	5	5	3	8	5	3	3	5	5	8	8	8	8	8	3	5	5	8	5	5	5	0.21
PIf-008 Yet Another 1	0.20	1	3	5	3	8	8	8	5	8	5	5	5	5	3	5	5	3	5	5	5	5	3	5	5	5	5	5	3	0.18
PIf-009 Yet Another 2	0.20	1	3	5	5	8	5	3	5	5	3	5	5	3	5	5	5	3	5	8	8	8	3	5	5	8	5	5	5	0.19
PIf-010 Yet Another 3	0.21	3	3	5	5	8	5	8	5	8	5	5	5	5	3	5	5	5	5	5	5	5	5	5	5	5	5	5	3	0.18
PIf-011 Yet Another 4	0.23	5	3	5	5	8	5	8	5	5	5	5	5	3	8	5	5	5	5	5	5	5	8	5	5	5	5	5	5	0.19
PIf-012 Complex Project 1	0.23	5	8	8	5	8	5	8	8	5	5	8	5	8	8	5	5	8	5	8	5	8	8	5	8	8	8	5	8	0.21
PIf-013 Complex Project 2	0.19	2	5	5	5	5	5	3	5	5	5	5	5	3	5	5	5	5	5	5	5	5	5	3	5	5	5	5	5	0.18
PIf-014 Complex Project 3	0.15	3	5	3	5	5	3	3	5	5	3	5	3	5	5	5	5	5	5	5	5	3	3	3	5	5	5	5	3	0.14
PIf-015 Complex Project 4	0.28	8	8	8	5	8	8	8	8	8	8	8	8	8	8	8	8	8	8	8	8	8	8	8	8	8	8	8	8	0.28
PIf-016 Complex Project 5	0.23	5	5	5	5	5	5	5	5	5	5	8	5	5	9	5	5	5	5	5	8	8	8	5	8	8	5	5	8	0.22
PIf-017 Complex Project 6	0.24	8	8	8	5	8	8	8	5	8	5	8	5	7	8	5	5	8	5	8	5	8	5	8	5	8	5	5	8	0.26
PIf-018 Complex Project 7	0.24	1	5	5	5	8	5	8	5	8	5	8	5	7	8	5	5	8	5	8	5	8	8	5	8	8	8	5	5	0.27
PIf-019 Complex Project 8	0.24	1	8	8	5	8	5	8	5	8	5	8	5	5	8	8	3	8	8	8	8	8	5	8	8	8	5	3	5	0.27
PIf-020 Complex Project 9	0.26	5	3	5	5	8	8	8	8	5	5	8	5	5	8	8	8	8	8	8	5	8	5	5	8	8	5	5	5	0.27
PIf-021 Minimal Project	0.12	0	3	3	3	3	3	3	3	5	3	5	3	3	3	3	3	3	5	5	5	3	5	5	5	5	3	3	3	0.11
PIf-022 Test Project	0.17	0	8	5	5	3	3	3	5	5	5	5	3	5	8	5	5	3	3	5	5	3	5	3	5	8	5	3	3	0.15
Solution Profile for Productivity Impact Factors		0.11	0.19	0.21	0.19	0.20	0.17	0.22	0.18	0.20	0.17	0.20	0.21	0.20	0.18	0.17	0.15	0.19	0.23	0.22	0.16	0.21	0.19	0.16	0.20	0.23	0.20	0.15	0.15	Convergence Gap 0.09

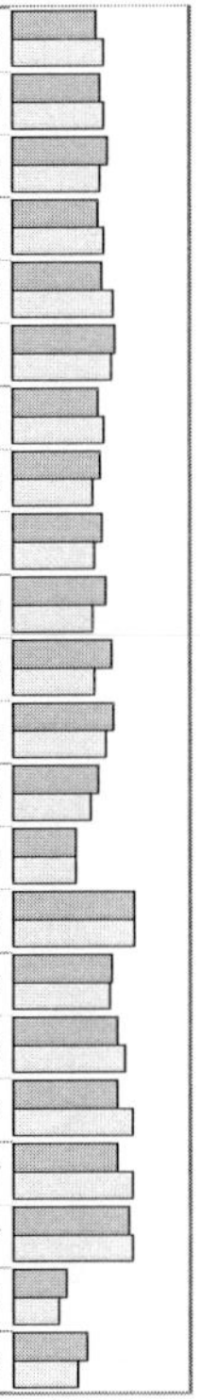

0.10 Convergence Range
0.20 Convergence Limit

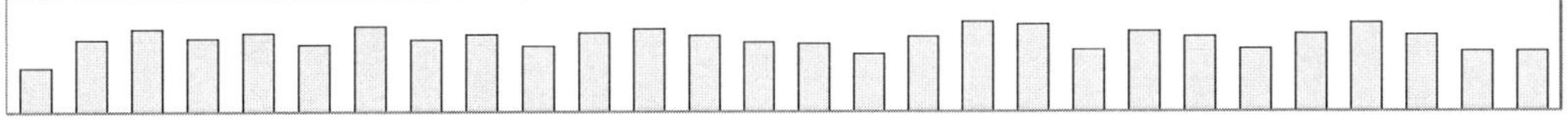

Table 12-11: Measured Efforts for 22 PIF Testing Projects

Measured Efforts Topics		Weight	Profile	
Pif-001 First Project	5.44	4%	0.19	
Pif-002 Next Project	5.67	4%	0.20	
Pif-003 Third Project	6.23	5%	0.22	
Pif-004 Fourth Project	5.61	4%	0.20	
Pif-005 Fifth Project	5.76	4%	0.20	
Pif-006 Large Project	6.60	5%	0.23	
Pif-007 Important Project	5.62	4%	0.20	
Pif-008 Yet Another 1	5.64	4%	0.20	
Pif-009 Yet Another 2	5.77	4%	0.20	
Pif-010 Yet Another 3	6.00	5%	0.21	
Pif-011 Yet Another 4	6.41	5%	0.23	
Pif-012 Complex Project 1	6.53	5%	0.23	
Pif-013 Complex Project 2	5.51	4%	0.19	
Pif-014 Complex Project 3	4.13	3%	0.15	
Pif-015 Complex Project 4	7.93	6%	0.28	
Pif-016 Complex Project 5	6.42	5%	0.23	
Pif-017 Complex Project 6	6.85	5%	0.24	
Pif-018 Complex Project 7	6.82	5%	0.24	
Pif-019 Complex Project 8	6.83	5%	0.24	
Pif-020 Complex Project 9	7.49	6%	0.26	
Pif-021 Minimal Project	3.47	3%	0.12	
Pif-022 Test Project	4.84	4%	0.17	

The 22 randomly chosen projects constitute another estimation stack. Since all projects contain the five effort types, the number of estimation items is $22 * 5 = 120$ in total, giving enough flexibility to align a cost model to this estimation stack. As before, the input value for the getting the cost profile are the logarithms of the efforts in days; consequently, here it means $5.44 = ln(231\ Days)$.

After calibration, this estimation stack listed in Table 12-11 was able to predict its constit-

uent projects with a convergence gap of only 0.09 (Figure 12-10), based on the mapping from PIF to measured efforts as the transfer function. The cell matrix reflects the a-parameters indicating how strong the impact of each PIF is on each project. The numbers are scaled down to range 0 to 9 using a suitable scaling factor. Table 12-12 is the bottom of Figure 12-10 turned clockwise. It shows the PIF as cost drivers with their priority profile; contrary to the R10 example, functional size is not dominant any longer. Other PIF are more prominently driving project cost.

Table 12-12: PIF Priority for the PIF Estimation Stack

	Productivity Impact Factors		Weight	Profile	
DFP COSMIC	Cfs	Functional Size	2%	0.11	
Pif-1 Personal	H1	Domain Knowhow	4%	0.19	
	H2	Personnel Capability	4%	0.21	
	H3	Technology Knowledge	4%	0.19	
	H4	Team Turnover	4%	0.20	
	H5	Management Capability	3%	0.17	
	H6	Team Size	4%	0.22	
Pif-2 Process	P1	Organization Maturity	3%	0.18	
	P2	Schedule Constraints	4%	0.20	
	P3	Requirement Completeness	3%	0.17	
	P4	Reuse	4%	0.20	
	P5	Project Type	4%	0.21	
	P6	Methodology	4%	0.20	
	P7	Stakeholder Cohesion	3%	0.18	
	P8	Project/Program Integration	3%	0.17	
	P9	Project Logistics	3%	0.15	
Pif-3 Product	S1	Product Size	4%	0.19	
	S2	Product Architecture	4%	0.23	
	S3	Product Complexity	4%	0.22	
	S4	Other Product Properties	3%	0.16	
	S5	Required Documentation	4%	0.21	
	S6	System Integration	4%	0.19	
	S7	Required Reusability	3%	0.16	
Pif-4 Technology	T1	Programming Language	4%	0.20	
	T2	Development Tools	4%	0.23	
	T3	Technical Environment	4%	0.20	
	T4	Technology Change	3%	0.15	
	T5	Technical Constraints	3%	0.15	

A transfer function such as Figure 12-10 with a small convergence gap predicts costing for new projects. For this, simply do the AHP for the PIF, and then use the PIF profile for determining their a-parameters. Using equation (12-6) applied to the respective estimation stack calculates a cost estimate. If your new project fits the characteristics of the projects from your history database that went into the estimation stack, the prediction accuracy of the PIF estimation stack is similarly good as the convergence gap indicates in Figure 12-7.

In practice, estimation stacks remain quite predictive and stable for a relatively long time within an organization. The more cost driver parameters are in use, the better the model fits the measured effort profiles. It is less clear whether these models work as well across organizations and technologies, although the known parametrical models always provide better estimations as experts can, even when used within an organization for the first time.

12-5.3 ESTIMATING ANOTHER PROJECT BASED ON THE PIF

Once an estimation stack is established with well-known projects that yields a good convergence gap, estimating new projects is straightforward. The recommended practice is using AHP as shown in *Chapter 3: What is AHP?* in section 3-5 to determine the cost driver profile using AHP for the new project, here called *Project X*, that shall be estimated based on the existing estimation stack (Figure 12-10).

To do this, first an AHP must calculate the profile. The AHP is like the process shown in section *3-5 Productivity Impact Factor Determination by AHP*. The decision process splits into the hierarchy of the four PIF groups; the top level AHP yields a preference for *D: Technology* (Figure 12-13).

Figure 12-13: Top Level AHP for Project X

AHP Priorities							
PIFs for Project X	A Personal	B Process	C Product	D Technology	Weight	Ranking	Profile
A Personal	1	3	3	1/5	24%	2	0.46
B Process	1/3	1	1/7	4	22%	4	0.43
C Product	1/3	7	1	1/2	23%	3	0.46
D Technology	5	1/4	2	1	32%	1	0.63

After breaking down the four hierarchy groups, it turns out that the top level cost drivers are again from that group *D: Technology*. In fact, the AHP process yields not only the top five but a profile for all 27 cost drivers. The full profile is used in Figure 12-15. The *D04: Technology Change* cost driver clearly outperforms all others in Project X, cost factors that depends from those cost drivers therefore will be heavily weighted, possibly even more than functional size.

Figure 12-14: The five Top Level Cost Drivers in Project X

Top Cost Drivers
PIFs for Project X

Top Cost Drivers		Attributes			Weight	Profile	
D04	Technology Change	Fast	Paradigm	Change	35%	0.74	
C03	Product Complexity	Changing	Inflationary	Challenging	18%	0.38	
A03	Technology Knowledge	Scarce Resource	Evolving		16%	0.34	
D03	Technical Environment	Agile	Knowledge	Acquisition	15%	0.32	
B03	Requirement Completeness	Even higher	Adapting	Flexible	15%	0.31	

Now the full profile $x_X = \langle x_1, x_2, \ldots, x_{27} \rangle$ can be filled into an estimation sheet for Project X using a-parameters from the PIF estimation stack. This calculates estimates for each of the five effort types. The profile x_X uses the power of eigenvectors from AHP to eliminate estimation bias. When adding this new estimation to the estimation stack, the convergence gap will remain as is.

Figure 12-15: Detail Estimations for Project X

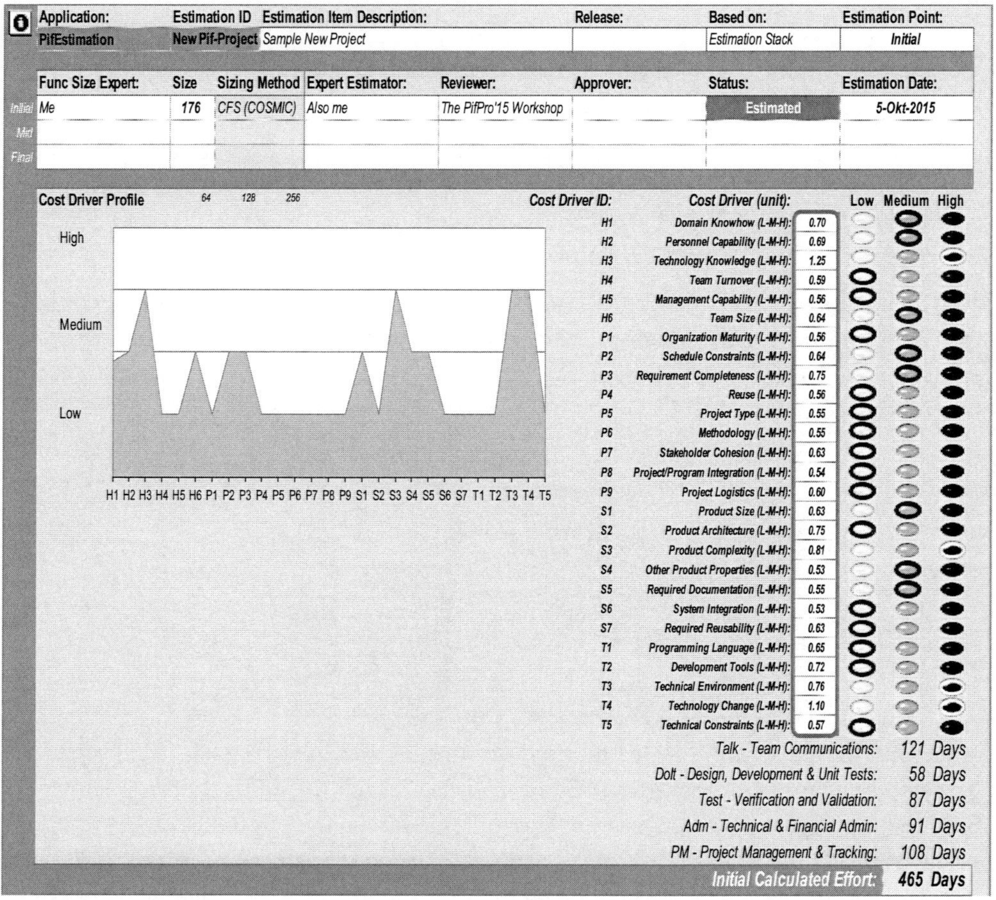

Application:	Estimation ID	Estimation Item Description:	Release:	Based on:	Estimation Point:
PifEstimation	New Pif-Project	Sample New Project		Estimation Stack	Initial

	Func Size Expert:	Size	Sizing Method	Expert Estimator:	Reviewer:	Approver:	Status:	Estimation Date:
Initial	Me	176	CFS (COSMIC)	Also me	The PifPro'15 Workshop		Estimated	5-Okt-2015
Mid								
Final								

Cost Driver Profile

Cost Driver ID:	Cost Driver (unit):		Low	Medium	High
H1	Domain Knowhow (L-M-H):	0.70			
H2	Personnel Capability (L-M-H):	0.69			
H3	Technology Knowledge (L-M-H):	1.25			
H4	Team Turnover (L-M-H):	0.59			
H5	Management Capability (L-M-H):	0.56			
H6	Team Size (L-M-H):	0.64			
P1	Organization Maturity (L-M-H):	0.56			
P2	Schedule Constraints (L-M-H):	0.64			
P3	Requirement Completeness (L-M-H):	0.75			
P4	Reuse (L-M-H):	0.56			
P5	Project Type (L-M-H):	0.55			
P6	Methodology (L-M-H):	0.55			
P7	Stakeholder Cohesion (L-M-H):	0.63			
P8	Project/Program Integration (L-M-H):	0.54			
P9	Project Logistics (L-M-H):	0.60			
S1	Product Size (L-M-H):	0.63			
S2	Product Architecture (L-M-H):	0.75			
S3	Product Complexity (L-M-H):	0.81			
S4	Other Product Properties (L-M-H):	0.53			
S5	Required Documentation (L-M-H):	0.55			
S6	System Integration (L-M-H):	0.53			
S7	Required Reusability (L-M-H):	0.63			
T1	Programming Language (L-M-H):	0.65			
T2	Development Tools (L-M-H):	0.72			
T3	Technical Environment (L-M-H):	0.76			
T4	Technology Change (L-M-H):	1.10			
T5	Technical Constraints (L-M-H):	0.57			

Talk - Team Communications:	121	Days
Dolt - Design, Development & Unit Tests:	58	Days
Test - Verification and Validation:	87	Days
Adm - Technical & Financial Admin:	91	Days
PM - Project Management & Tracking:	108	Days
Initial Calculated Effort:	**465**	**Days**

Thus, the estimation stack remains stable over several projects being estimated, and only if new paradigms, or observed deviations, indicate the need for recalibration, the estimation stack must be amended.

For practical purposes, the full profile x_X shifts from the range 0 to 1 onto 0.5 to 1.5; thus, the entries in the red encircled cost driver column in Figure 12-15 are actually the x_i that go into equation (12-1) and (12-6).

12-6 CONCLUSION

The concept of transfer functions is very powerful, and easily adaptable to software development cost estimations. We have laid the theoretical background for quality of estimations; the practical implementation is yet another challenge. The reward is huge: reliable project cost estimations, or at least estimations with a known accuracy. This is a crucial step to develop the information and communication technology, finally bringing the economic benefits that it promised long ago.

Although the method cannot verify that the selected cost drivers are correct, it can ascertain that their measurements are consistent and that the estimation model predicts project cost. The convergence gap between model prediction and actual project cost indicates how good the estimation stack used for cost prediction is.

CHAPTER 13: DYNAMIC SAMPLING OF TOPIC AREAS

One of the most difficult tasks in Six Sigma is separating the topics that influence the value chain. Requirements Engineering often does not address this difficulty properly, resulting on huge hierarchy-less lists of all-high important requirements.

Combinatorial Algebra deals with one operation: combining entities. The entities can be profiles of topic areas. Multiplication of profiles does not allow for combining an almost zero impact with a high impact – it would simply eliminate high impact. However, this is not factual. It brings combinatorial algebras into play; maybe linear transfer functions are just finite approximations to combinatorial algebras representing the complexity of today's world.

This chapter touches higher mathematics and is intended to readers interested in Theoretical Computer Science. It shows that QFD is part of a theoretical framework for knowledge representation. The proofs are based on Engeler 1995 (Engeler, 1995) *and mostly omitted here as they can be found in Fehlmann 2002* (Fehlmann, 2002).

13-1 INTRODUCTION

How to separate topics belonging to the control domain from the response domain? Sometimes, topics go through a process without being affected. In this case, they are not controls but still look like responses. They might originate from a previous value creation step.

Topics in the context of this chapter refer to responses and controls of Six Sigma transfer functions. Topics are the domain areas where transfer functions can be defined in between. The transfer functions implicitly define the topics of its domain areas.

The control topics of one domain Areas can become the response topics of another domain area. This creates the need to explain how to combine such topics.

13-2 COMBINATORY LOGIC

There is a mathematical theory called *Combinatory Logic* (Engeler, 1995) that explains quite generally how to combine topic areas. Combination is not only on the basic level possible; you can also explain how to combine topics on the second level; sometimes called meta-level. Intuitively, we would expect such a meta-level describing knowledge about how to deal with different topic areas.

To do this, the notion of topic areas must be enhanced somewhat. While up to here the description of topics used vector profiles, with the implicit assignment of a positive direction to every topic area, respectively its vector component, it is now necessary to make knowledge explicit. Thus, leaving the area of statistics, metrics, and the corresponding linear vector spaces, the focus shifts now to the knowledge about linking causes and effects – or control and responses - and go back to the basics, looking at propositional statements about the topic areas. Propositional calculus applies. For foundations of mathematical logic, see for instance Barwise (Barwise, et al., 1977).

Combinatory logic originates from mathematical logic. The issues addressed with combinatory logic is with the constructive variant of the axiom of choice. Informally, the axiom of choice says that given any collection of sets, each containing at least one object, it is possible to select exactly one object from each set, without requiring an algorithm saying how the selection is done. In the theory of *Complex Analysis*, such an algorithm seems an unnecessary condition; in fact, complex analysis proved to be very successful without requiring constructive selection algorithms.

In *Computer Science*, the existence of selection algorithms seems a natural condition for the applicability of the axiom of choice. Nothing exists resembling a program or process without an algorithm that constructs it. Whether such an algorithm ever stops with a result, or loops forever, is another question. Interestingly, this conditioning of the axiom of choice to mathematical logic has wide consequences.

13-2.1 THE GRAPH MODEL OF COMBINATORY LOGIC

Let $\mathcal{L}$ be the set of all propositions over a given domain. Examples include statements about customer's needs, solution characteristics, methods used, etc. These statements contain no free variables; i.e. they are propositions about the business domain we are going to model.

Denote by $\mathcal{G}(\mathcal{L})$ the power set containing all *Arrow Terms* of the form

$$\{a_1, ..., a_n\} \rightarrow b \tag{13-1}$$

The left-hand side is a finite set of arrow terms and the right-hand side is a single arrow term. This definition is recursive; thus, it is necessary to establish a base definition saying that every proposition itself is considered an arrow term. The arrows of the arrow terms are distinct from the logical imply that some authors also denote by an arrow. The arrows denote cause-effect, not logical imply.

The formal definition, written in set-theoretical language, is

$$\mathcal{G}_0(\mathcal{L}) = \mathcal{L}$$

$$\mathcal{G}_{n+1}(\mathcal{L}) = \mathcal{G}_n(\mathcal{L}) \cup \{\{a_1, \dots, a_m\} \to b \,|\, a_1, \dots a_m, b \in \mathcal{G}_n(\mathcal{L}), m = 0,1,2,3 \dots\} \tag{13-2}$$

$\mathcal{G}(\mathcal{L})$ is the set of all (finite and infinite) subsets of the union of all $\mathcal{G}_n(\mathcal{L})$:

$$\mathcal{G}(\mathcal{L}) = \bigcup_{n \in \mathbb{N}} \mathcal{G}_n(\mathcal{L}) \tag{13-3}$$

The elements of $\mathcal{G}_n(\mathcal{L})$ are arrow terms of level n. Terms of level 0 are *Topics*, terms of level 1 *Rules*. A *Rule Set* is an element of $\mathcal{G}_n(\mathcal{L})$ that consists of level 1 terms only and is finite; if it is infinite, we call it *Knowledge Base*. Hence, knowledge is a potentially unlimited set of rules about topics and rules. This definition is recursive, as before. As shown in Figure 13-1, the rules correspond to the cause/effect correlations in the QFD–matrix.

Arrow sets represent Six Sigma transfer functions in a way originally described by Ishikawa. The *Ishikawa Diagram* (Ishikawa, 1990) describes the cause-effect relations between topics and are considered the initial form of QFD matrices. Converting a series of Ishikawa diagrams into a QFD matrix is straightforward, see Figure 13-1 below.

Figure 13-1: Representing QFD Matrices as Rule Sets

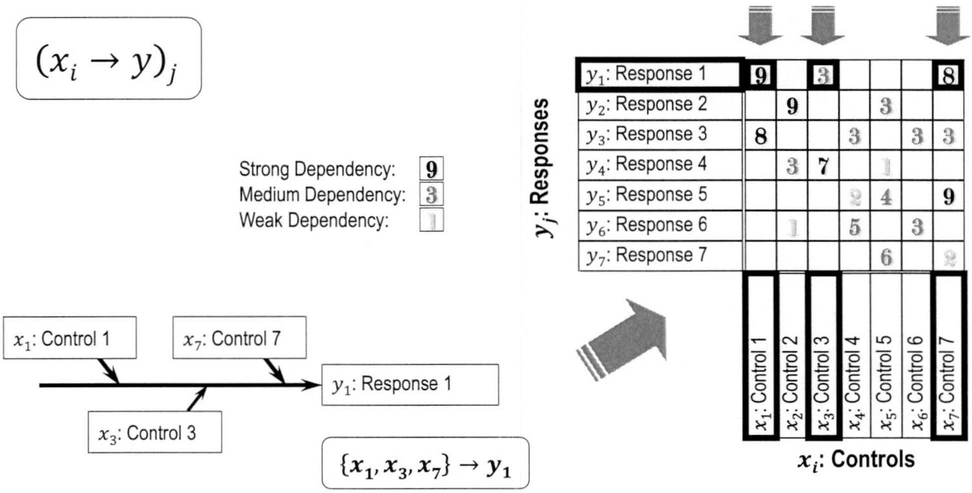

It is well known how to concatenate Ishikawa diagrams, when the response of one diagram becomes the control of another. This concatenation has been introduced in *Chapter 4: Quality Function Deployment*, in section *4-9 Comprehensive QFD – The Deming Chain.*

To avoid the many set-theoretical parenthesis, the following notations are applied:

- a_i for a finite set of arrow terms, i denoting some finite indexing function for arrow terms;
- a_1 for a singleton set of arrow terms; i.e. $a_1 = \{a\}$ where a is an arrow term;
- $\emptyset$ for the empty set, such as in the arrow term $\emptyset \rightarrow a$.

The indexing function cascades, thus $a_{i,j}$ denotes the union of a finite number of m arrow term sets

$$a_{i,j} = a_{i,1} \cup a_{i,2} \cup ... \cup a_{i,j} \cup ... \cup a_{i,m} \tag{13-4}$$

With these writing conventions, $(x_i \rightarrow y)_j$ (denotes a rule set; i.e., a finite set of arrow terms having at least one arrow. Thus, they are level 1 or higher.

Each element $x_i \rightarrow y$ of $(x_i \rightarrow y)_j$ denotes one Ishikawa diagram (Akao, 1990), which is a cause/effect constituent of a QFD deployment and stands at the origins of QFD in Japan. The matrix $(x_i \rightarrow y)_j$ represents the QFD deployment. This matrix obviously is a rule set within $\mathcal{G}(\mathcal{L})$. The union of all possible QFD matrices is infinite and therefore a knowledge base in $\mathcal{G}(\mathcal{L})$.

However, many elements of $\mathcal{G}(\mathcal{L})$ that do not resemble known QFD deployments, such as $((x_i \rightarrow y)_j \rightarrow z)_k$. This term consists of arrow terms whose left hands consist of finite rule sets, thus it is a nested rule. Another such example is $x_i \rightarrow (y_j \rightarrow z)$. This is a cascade of rules.

Subsequently, the association to the right for arrow terms is used:

$$x_i \rightarrow y_j \rightarrow z = x_i \rightarrow (y_j \rightarrow z) \tag{13-5}$$

13-2.2 THE APPLICATION OPERATION

Let $M, N \in \mathcal{G}(\mathcal{L})$. Then application of M to N is defined by

$$M \bullet N = \{a | \exists b_i \rightarrow a \in M, b_i \subset N\} \tag{13-6}$$

In case of M as a rule set, and N as a topic set, this represents the selection operation that chooses those rules $(b_i \rightarrow a)_j$ from rule set M that are applicable to the topic set N. However, the definition applies to all higher–level $\mathcal{G}(\mathcal{L})$ terms as well. As an example, consider $M = (b_i \rightarrow a)_j$ and another rule set $L = (d_k \rightarrow c)_p$, which represent two QFD deployments. Then,

$$M \bullet L = \{a | \exists (d_k \rightarrow c)_j \rightarrow a \in M, (d_k \rightarrow c)_j \subset L\} \tag{13-7}$$

Moreover,

$$(M \bullet L) \bullet N = \{f | \exists g_j \rightarrow f \in M \bullet L, f_j \subset N\} \tag{13-8}$$

and thus

$$
\begin{aligned}
(M &\bullet L) \bullet N \\
&= \{f | \exists (d_k \rightarrow c\ )_j \rightarrow g_i \rightarrow f \in M, (d_k \rightarrow c\ )_i \subset L, g_i \subset N\}
\end{aligned} \tag{13-9}
$$

Therefore, M selects those rules from L that are applicable to the topics in N.

Furthermore,

$$
\begin{aligned}
L \bullet (M \bullet N) &= \{a | \exists b_i \rightarrow a \in L, b_i \subset M \bullet N\} \\
&= \left\{a \middle| \exists b_i \rightarrow a \in L, \exists (c_j \rightarrow b\ )_i \subset M, c_{j,i} \subset N\right\}
\end{aligned} \tag{13-10}
$$

Thus, the result of $L \bullet (M \bullet N)$ are those elements of N that have both a rule $b_i \rightarrow a \in L$ and for each b_i there are rules $(c_j \rightarrow b\ )_i \subset M$. Applying a rule set selects those rules that apply to a given topic. Therefore, the application of graph terms represents the combination of QFD matrices as we use it in *Comprehensive QFD*, as observed by Prof. Akao (Akao, 1990). Thus, the application operation we defined is natural. In terms of QFD, this operation selects such cause/effect relations that apply to a given topic. Combinatory Logic provides a simple-to-use model for how to apply QFD. This observation was the reason for investigating models of Combinatory Logic.

For the application operator, association is to the right:

$$(M \bullet L) \bullet N = M \bullet L \bullet N \neq M \bullet (L \bullet N) \tag{13-11}$$

13-2.3 THE GRAPH MODEL CALCULUS

We can further considerably ease the set-theoretic representation introduced before when we omit to name sets and rather represent them by its "typical element". Thus, $b_i \rightarrow a$ represents a rule set, consisting of arrow terms of level 1 only. Omitting the subscript indicates that the set needs not to be finite. For instance, (t) is a – possibly infinite – set of arrow terms that contains topics and higher-level terms.

In short, the notation is

$$(b_i \rightarrow a) \bullet (t) = (a^{t \rightarrow a}) \tag{13-12}$$

Here, $(a^{t \rightarrow a})$ represents the set $\{a' | \exists b'_i \rightarrow a' \in (\ b_i \rightarrow a) \wedge b_i \subset (t)\}$, which is the subset of (a) that contain an element of (t) on the left-hand side of a graph term in $(\ b_i \rightarrow a)$.

This representation is particularly useful because it constitutes a rewrite rule algorithm to perform calculations. This is useful for automation and model building, because the rather complicated set-theoretic notation becomes unnecessary.

13-2.4 Sample Arrow Terms

Some examples might help to get used to the arrow term notation.

13-2.4.1 Example 1: Simple Application

Consider again $L \bullet (M \bullet N) = (b_i \to a) \bullet \big((c_j \to b) \bullet (t) \big)$. To apply $(c_j \to b)$ to (t), one needs to select the c_j such that they are members of (t). By applying the definition of application between graph terms, the rewrite formula looks as follows:

$$(c_j \to b) \bullet (t) = (t_j \to b) \bullet (t) = (b^{t \to b}) \tag{13-13}$$

where $b^{t \to b}$ selects those right-hand sides elements in $(t_j \to b)$ whose left elements fulfil $t_j \subset (t)$. Using the same variable names "t" in arrow terms means always $t_j \subset (t)$, as a writing convention.

Now this yields $(b_i \to a) \bullet (b^{t \to b})$, and in turn, applying equation (13-12), $a^{t \to b \wedge b \to a}$. If just interested in the type of the resulting terms, the selection superscripts are unnecessary.

Obviously, the aim is to get a combined matrix back, something like $(c_j \to b)$, to combine the matrix functions. This requires a deeper understanding of the graph model calculus that follows in section 13-3.2 : *The Lambda Theorem*. Nevertheless, before getting there, a few exercises help getting more at ease with this rewrite notation for arrow terms.

13-2.4.2 Example 2: Nested Application

In the next example, we combine two rule sets, denoted by $(b_i \to a)$ and $(d_k \to c)$. Then the result $(b_i \to a) \bullet (d_k \to c)$ is the solution of the equation $b_i = (d_k \to c)_p$, where b_i is some left-hand side within the set $(b_i \to a)$ and $(d_k \to c)_p$ is a finite subset of $(d_k \to c)$.

This implies $b = d_k \to c$ and $i = p$, thus

$$(b_i \to a) \bullet (d_k \to c)$$
$$= ((d_k \to c)_i \to a) \bullet (d_k \to c) \tag{13-14}$$
$$= a^{(d \to c) \to a}$$

This result represents the set $\{a' \in (a)|(d_k \to c)_i \to a' \in (b_i \to a)\}$. Note the observation that the combination of much more arrow terms is admissible than those corresponding to QFD deployments. Deployments can act on other rule sets instead of topics, and their impact investigated. Indeed, the power set structure $\mathcal{G}(\mathcal{L})$ is rich enough to describe any kind of application between arrow term sets.

13-2.4.3 EXAMPLE 3: TRADITIONAL FOUR–PHASE QFD DEPLOYMENT (ASI)

We can get comprehensive QFD deployments in the traditional manner; see Figure 4-26, by cascading the application operation between arrow terms as follows:

$$(b_i \to a) \bullet \Big((c_j \to b) \bullet \big((d_k \to c) \bullet ((e_s \to d) \bullet (e))\big)\Big) \qquad (13\text{-}15)$$

where $(b_i \to a)$ is the *House Of Quality*, $(c_j \to b)$ the *Solution Characteristics* deployment, $(d_k \to c)$ the *Methods & Tools* selection, $(e_s \to d)$ the *Vendor Selection*; and where (e) is the set of vendors, (d) the methods & tools, (c) the solution design, (b) the solution characteristics, and (a) the customer's needs.

A copy of Figure 4-26 is inserted here, for convenience:

Figure 13-2: The Four-House QFD Deployment for Simple Hardware – Part of 1990 ASI Training

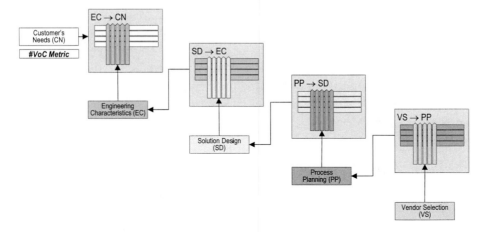

Using the abbreviations in Figure 13-2, equation (13-15) reads as

$$(EC_i \to CN) \bullet \Big((SD_j \to EC) \bullet \big((PP_k \to SD) \bullet ((VS_s \to PP) \bullet (VS))\big)\Big) \qquad (13\text{-}16)$$

Making clear that the vendor selection is what ultimately controls fulfilling the customer's needs (CN). For a training example, this is acceptable.

For more examples; e.g., for Figure 11-1: Deming Value Chain for Software Project Deployment or Figure 11-2: Deming Value Chain for Software Product Deployment, the graph model must be enhanced by additional operations; e.g., representing union of rule sets and arrow terms.

13-3 UNIVERSAL MODELS

Combinatory algebra is the simplest language for describing formal languages. The reason for this is that the two *Combinators* **S** and **K** fulfilling

$$\mathbf{K} \bullet M \bullet N = M \tag{13-17}$$

and

$$\mathbf{S} \bullet M \bullet N \bullet L = M \bullet L \bullet (N \bullet L) \tag{13-18}$$

are sufficient to describe any kind of construction needed in a process or procedure descriptions.

K is the *Projection Combinator* and **S** the *Substitution Combinator*. However, like an assembly language, the **S**- **K** terms become quite lengthy and are barely readable by humans, but they work fine as a foundation for computer science.

A *Model of Combinatory Algebra* (Engeler, 1995, p. 91) is a structure that contains an application operation between its members and has non-trivial implementations of the two combinators **S** and **K**. The elements of the graph model are the power set $\mathcal{G}(\mathcal{L})$ of the business cases $\mathcal{L}$ whose Deming chains shall become analyzed and measurable with QFD. If there is a definition for the combinators **S** and **K** in the graph model, this explains how to apply QFD for all kind of business cases. In that sense, the graph model is *universal* (Engeler, 1995, p. 8).

13-3.1 QFD DEPLOYMENTS ARE A MODEL OF COMBINATORY ALGEBRA

The following definitions demonstrate how the graph model implements the combinators **S** and **K** fulfilling equations (13-17) and (13-18). Therefore, the graph model is a model of combinatory algebra.

- $\mathbf{I} = (a_1 \to a)$ is the *Identification*; i.e. $(a_1 \to a) \bullet (b) = (b)$
- $\mathbf{K} = (a_1 \to \emptyset \to a)$ selects the 1st argument:
  $$\mathbf{K} \bullet (b) \bullet (c) = \left((b_1 \to \emptyset \to b) \bullet (b) \right) \bullet (c) = (\emptyset \to b) \bullet (c) = (b) \tag{13-19}$$
- $\mathbf{KI} = (\emptyset \to a_1 \to a)$ selects the 2nd argument:
  $$\mathbf{KI} \bullet (b) \bullet (c) = \left((\emptyset \to c_1 \to c) \bullet (b) \right) \bullet (c) = (c_1 \to c) \bullet (c) \\ = (c) \tag{13-20}$$
- $\mathbf{S} = \left(a_i \to (b_j \to c) \right)_1 \to \left((d_k \to b)_i \to (b_{j,i} \to c) \right)$

- 326 -

The proof that the latter definition fulfils equation (13-18) is somewhat more complex and therefore left to the interested reader who can find it in Engeler (Engeler, 1981, p. 389). With **S** and **K,** an abstraction operator can be constructed that builds new knowledge bases. This is the *Lambda Theorem.*

Obviously, it is impractical to use the **S-K** "assembly language" when doing QFD. However, combining arrow terms $(b_i \rightarrow a)_j$ is quite handy when planning the QFD approach, and can easily be automated.

13-3.1.1 APPLICABILITY OF THE GRAPH MODEL TO RULE–LEVEL DEPLOYMENTS

Rule-level deployments are straight QFD, as used by many authors including (Mazur & Bylund, 2009) and (Herzwurm, et al., 2000). There is always a metric assigned to combinators, called *Combinatory Metrics.* Its values are the vector profiles of the rule sets in the linear vector spaces.

This metric is recursively defined as a function

$$\|b\|_i = f\big(\|b\|_j\big) \text{ for a rule set } (b_i \rightarrow a)_j \tag{13-21}$$

In QFD, the function is linear. The term *Combinatory Metrics* refers to transfer functions represented by some rule set $(b_i \rightarrow a)_j$. It evaluates QFD deployments in view of their contribution to customer's needs, or project goals.

13-3.1.2 APPLICABILITY OF THE GRAPH MODEL TO HIGHER–LEVEL DEPLOYMENTS

Moreover, the multi-level QFD discipline opens; i.e., QFD not restricted to cause-effect on the level 1, but including higher levels. For instance, let $(b_i \rightarrow a)_j$ denote the cause-effect combinator between solution approaches $(b)_i$ and customer's needs $(a)_j$. Let $(d_k \rightarrow c)_m$ denote the cause-effect combinator that links ICT functionality $(d)_k$ to business processes $(c)_m$. Then, $((d_k \rightarrow c)_m \rightarrow a)_j$ is an example of second–level combinators describing the dependency of both customers' needs from how the ICT supports business processes. Another interesting example is $((d_k \rightarrow c)_m \rightarrow b)_i$. This combinator describes how the chosen solution approach depends from how ICT supports business processes. We call such higher–level deployments *Knowledge Deployment.*

13-3.2 THE LAMBDA THEOREM

The *Lambda Theorem* is the clue for working with knowledge deployments. Knowledge deployment – in conjunction with combinatory metrics – provides the means to measuring knowledge for organizations or individuals in terms of its applicability to topics such as customer's needs, business results, or project goals.

The Lambda Theorem. *Let* M *be an arrow term, and let* $x \in \mathcal{G}_0(\mathcal{L})$ *denote a topic; see equation (13-2). Then there is a term* $\lambda x.\,M$ *such that for all arrow terms* N *the combination* $(\lambda x.\,M) \bullet N = M[x/N]$ *holds, where* $M[x/N]$ *denotes the result of replacing all occurrences of* x *in* M *by* N.

Proof (Sketch). For the new term $\lambda x.\,M$, the form $(x_i \to M)$ will do exactly as postulated. Here, x_i is a variable running over arrow terms. If x does not occur in M, then $(\lambda x.\,M) \bullet N = M$. If M consists of the variable x only, then $(\lambda x.\,x) = \mathbf{I}$. If x occurs in M, then all references to the topic x in M are replaced by references to N.

For more details, see Fehlmann (Fehlmann, 1981).

It is now possible to represent the combination of two matrices M and N by a dedicated arrow term $(\lambda x.\,M) \bullet (N \bullet x)$. In case $M = (b_i \to a)_j$ and $M = (c_k \to b)_i$ this is

$$\left(\lambda x.\,(b_i \to a)_j\right) \bullet \left((c_k \to b)_i \bullet x\right) = \left(\lambda x.\,(b_i \to a)_j\right) \bullet b^{c \to b}$$
$$= \lambda x.\,a^{b \to a \wedge c \to b} = (c_k \to a^{b \to a \wedge c \to b})_j \tag{13-22}$$

Because of the application $(c_k \to b)_i \bullet x$ in $(\lambda x.\,M) \bullet (N \bullet x)$, the variable x runs by force only over the set c_k, because only elements of c_k can satisfy the condition $a^{b \to a \wedge c \to b}$.

Such structural replacement of topics in graph terms is helpful for investigating the effects of introducing new topics in an existing comprehensive QFD deployment. For instance, it was instrumental in developing new QFD deployments such as for an e-commerce site on the Web (Fehlmann, 2001) or Net Promoter® Score (*Chapter 5: Voice of the Customer by Net Promoter®*). In other words, existing QFD combinators always can combine with new topics. The structure of existing QFD deployments is reusable with new topics.

13-3.3 MEASURING POTENTIAL KNOWLEDGE DEPLOYMENT

The Lambda theorem does not preserve combinatory metrics. When replacing a topic with another, it becomes necessary to study the new cause/effect relationships. Thus, if we take M as the characterization of an existing organization – i.e., the comprehensive QFD deployment in an organization, or for a technology, that describes its full value chain (Fehlmann, 2003) – then the graph term $(\lambda x.\,M)$ represents the *potential* of that organization. When we apply $(\lambda x.\,M)$ to all possible new topics t, then the resulting combination $(\lambda x.\,M) \bullet (t)$ describes all possible applications to new topics of the organization's potential.

Using combinatory algebra and combinatory metrics, we have a means to optimize the possible new deployments of existing value chains. In theory, we can define the potential of an organization as the maximum combinatory metric when we let x vary along all possible topics. However, the number of such potential new topics is unlimited. In practice, it is crucial to select the most promising topics to get the best approximation, using known quality techniques. For the evaluation of the selection, we reuse the combinatory metrics based on the already existing comprehensive QFD deployment. Thus, combinatory algebra is a means to transform existing comprehensive QFD deployments into new ones, which may fit better into new market environments.

13-4 TESTING THE INTERNET OF THINGS

An important application of combinatory logic is in testing. Test cases have the same structure of arrow terms. The arrow terms represent tests; in $a_i \to b$, the a_i describe the test data and b the test response. Responses can be as simple as the amount of impact on the actuators in an *IoT Orchestra*. An IoT orchestra is a collection of "Things" in the *Internet of Things* that work together as a system. The goal of testing is measuring the *Schurr Radius*. The necessity for automatically produced test cases in IoT is apparent. There are no testers present when users connect a new sensor to their smart home network, or two autonomous cars meet each other. Behavior of the newly connected system still must remain safe. Test case automation is a long-standing need for testers; for IoT concerts, combinatory logic delivers automated test cases almost for free.

13-4.1 A SIMPLE IOT TESTING CASE

The mechanism in place are shown with a simplified IoT network. Consider a simple data retrieval application. The application meets two functional (FUR) and two non-functional (NFR) with the following goal profile, see Figure 13-3:

Figure 13-3: Customer's Needs' Priority Profile

	IoT Topics	Attributes			Weight	Profile	
FUR	y1 Extensible	Easy to extend IoT	Device independent	Flexible	31%	0.57	
	y2 Open	Open Source	Open Interfaces		20%	0.36	
NFR	y3 Reliable	Always correct	Always secure	Safe	39%	0.71	
	y4 Fast	No waiting			11%	0.20	

Only three user stories are needed to cover these requirements, see Figure 13-4:

Figure 13-4: User Stories' Priority Profile for Simple Data Retrieval Application

	User Stories Topics	Weight	Profile	
1)	Q001 Search Data	38%	0.65	
2)	Q002 Answer Questions	39%	0.66	
3)	Q003 Keep Data Safe	23%	0.39	

The priority profile reflects the number of data movements needed in the software to cope with the user requirements expressed in user stories.

The total functional size according ISO/IEC 19761 COSMIC is 6 CFP, i.e., six data movements only; thus this is a very small and simple application. The user stories' profiles reflect customer's needs for the IoT Topics shown in Figure 13-3 by transfer functions. User stories' priority profile is simply calculated by counting the number of data movements needed per user story to meet the customer's needs' priority profile.

The test coverage transfer function in Figure 13-5 is defined by the number of data movements in a test story delivering user stories.

Figure 13-5: Test Coverage for Simple Data Retrieval Application

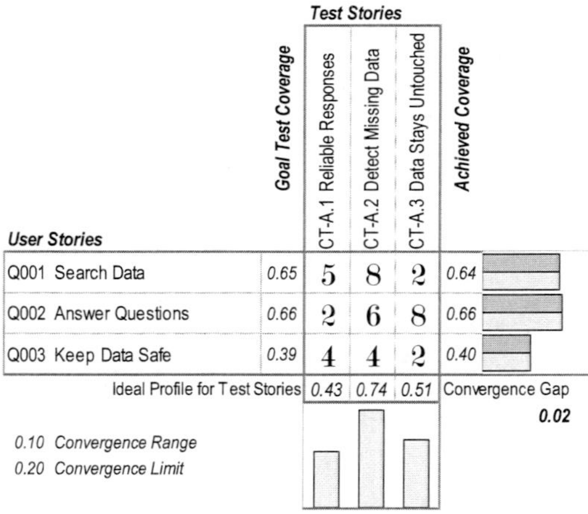

Coverage is fine with a convergence gap of 0.02. The total number of tested data movements per cell never exceeds eight.

13-4.2 CONNECTING IoT DEVICES TO THE DATABASE APPLICATION

Connecting IoT devices to a simple data retrieval application adds not only a continuous flow of searchable data but also considerable complexity. By adding one type of sensor and one type of actuator, the functional size almost triples and becomes 22 CFP. Security and safety risks increase with every data movement added to the IoT concert, as they can be misused or hacked, and cause unwanted and unsafe behavior.

Customer's needs remain the same – for simplicity, we do not consider additional needs that arise with IoT operations. Also, user stories remain the same, although data now refers not to static but to dynamic data and the priority profile now changes towards higher importance for *Q003: Keep Data Safe*. Test stories too remain the same but must cover many more data movements. Consequently, Figure 13-3 remains valid while Figure 13-6 changes its priority profile after connecting the database to the IoT concert because of the additional data movements between devices, database, sensors and actuators. Figure 13-4 transforms into Figure 13-6 with more focus on *Q003: Keep Data Safe*:

Figure 13-6: User Stories' Priority Profile for Full IoT Data Retrieval Application

User Stories Topics	Weight	Profile	
1) Q001 Search Data	30%	0.51	
2) Q002 Answer Questions	36%	0.63	
3) Q003 Keep Data Safe	34%	0.59	

Consequently, test cases increase in number. For instance, to keep data safe (*Q003: Keep Data Safe*), data transmissions to sensors and actuators must be tested against loss of data, or data transmission interference, e.g., by hackers. This increases test size but not the number of test stories.

The resulting test coverage (Figure 13-7) is expected to remain the same although test size increases considerably. This means that many more data movements are now under test; however, the test structure remains the same. The knowledge for testing the IoT is inherited from the original tests for the simple data retrieval test scenario.

Figure 13-7:. Test Coverage for Full IoT Data Retrieval Application

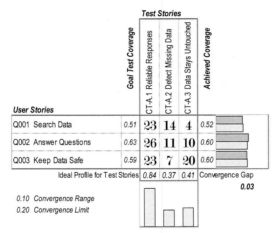

Clearly, both test coverage transfer functions remain within a safe Schurr Radius. Adding more types of IoT devices causes the cell counts grown in the test coverage matrix while the convergence gap remains within the Schurr Radius limits. This is what Combinatory Logic predicts. Thus, the original data retrieval application test serves as a model for the full IoT test. Adding more devices to the IoT concert makes the counts grown even further, but the matrix remains and the testing coverage convergence gap remains within the Schurr Radius. Only one rule set has been applied so far: $(x_3 \rightarrow y)_3$. If the IoT concert covers more user stories, say j, then this becomes $(x_3 \rightarrow y)_j$; what in turn most likely requires i more test stories: $(x_i \rightarrow y)_j$.

The importance of the original three test stories changed between the data retrieval application and the full IoT concert, see Table 13-8.

Table 13-8: Test Priority Change when Adding IoT Concert

	Data Retrieval			Full IoT Concert		
Test Story	**Weight**	**Profile**		**Weight**	**Profile**	
CT-A.1 Reliable Responses	25%	0.43	11	49%	0.84	72
CT-A.2 Detect Missing Data	44%	0.74	18	21%	0.37	32
CT-A.3 Data Stays Untouched	30%	0.51	12	24%	0.41	34

While test priority initially focused on *CT-A.2: Detect Missing Data*, focus moved to *CT-A.1: Retrieve Previous Responses* after adding sensors and actuator to the system. This is reflecting the additional effort needed to protect data movements between sensors and database from interferences, e.g., data loss or compromise.

The following table (Table 13-9) shows a comparison of test sizes between the original data retrieval application test, and the full IoT concert test.

Table 13-9: Simple Data Retrieval Application Test Size vs. IoT Test Size

Data Retrieval		Full IoT Concert	
Test Size in CFP:	41	Test Size in CFP:	138
Test Intensity in CFP:	6.8	Test Intensity in CFP:	6.3
Defect Density:	17%	Defect Density:	18%
Test Coverage:	100%	Test Coverage:	100%

The key indicator for tests is the *Test Intensity*, the ratio between *Test Size* and *Functional Size*. *Defect Size* in turn is the percentage of defective data movements in the software. All size measurements are according the international standard ISO/IEC 19761 COSMIC (ISO/IEC 19761:2011, 2011). There are no limits for neither functional size nor test size.

13-5 AUTOMATED TEST CASE GENERATION

Thanks to the test priority goal profile, derived from the original customer's needs and carried through user stories to test stories finally, it is possible to generate test cases automatically because the convergences gap serves as guidance which test cases to add. The principle behind artificial intelligence are heuristics, i.e., metrics telling which search branches to follow and which to avoid.

Because of the guidance, artificial intelligence adds only test cases that pertain to the functionality of the implemented user stories, notwithstanding whether the IoT concert now features additional but untested functionality. The data retrieval approach does not cover additional requirements that might come with the IoT concert, such as whether window stores close when sun shine is strong, or avoid car collisions when needed. It is therefore necessary to extend the model, for instance from the simple data retrieval application to something more sophisticated such as a smart home, or autonomous cars. Then, combinatory logic allows to extend the test suite from the model that still is remains relatively simple, to the full-blown the reality covering a much wider range of functionality and that is much more difficult to manage.

Automated testing is a must for IoT systems especially for autonomous cars. For them, allotted testing time can be very small. For instance, in case of an encounter with another car from a different manufacturer that wants to connect and whose behavior is hardly predictable, testing time allowance might be reduced to a few milliseconds.

13-5.1 The Future of Software Testing

We demonstrated with an example how testing scenarios carry over from simple applications to complex IoT concerts, using the original test cases as testing patterns for automatically extending the test to the full IoT application. Using combinatory logic, testing scenarios designed for the original model carry over to its extended IoT implementation, and this is already an important saving, enabling safe IoT concertation.

Combinatory logic paves a way to testing complex IoT concerts and networked systems, based on the solid ground of existing testing experiences. It has been shown how the quality of testing can be maintained even after moving to automated testing. For testing the IoT, this approach offers significant savings; however, the full potential of combinatory logic in organizing knowledge is significantly greater.

Using combinatory logic for testing obviously is a high hurdle for testers that neither understand eigenvectors nor combinatory logic. Moreover, the approach relies on using customer's needs profiles and software metrics for sizing and evaluating tests. Metrics are not common practices in today's software communities. To help with this, tools must be constructed that make the theoretical background available for practitioners. A prototype tool is available to the readers of tis book. Other applications of the theory will become apparent with the ongoing cosmic inflation, that Information & Communication Technology (ICT) currently undergoes. This opens new fields, and new practices expand at an astonishing pace.

13-6 Conclusions

The cosmic inflation that the ICT is currently undergoing creates new needs for managing complexity. While restoring health to a balance sheet of a company did require handling knowledge at a basic level, typically align spending with revenue – a rule set at level 1 – keeping today's networks of intelligent things healthy, sane and benevolent to humans requires much more knowledge at different abstraction levels. Security, safety, privacy and human autonomy are a few of many rule sets that interfere when creating these huge networks, from which our life depends more and more. For instance, the autonomous car with all its legal, security and safety implication is impossible to handle without a sound theory of knowledge that relates to such a complex machine that once was a car.

The arrow sets represent knowledge in a most general way. They constitute a theory suitable for explaining what knowledge is and what can be done with knowledge. Metrics about knowledge – such as software metrics – are obtainable as exercises in the in applying arrow set theory; see for instance the example shown in section *6-6 Metrology and Measurement Accuracy* of *Chapter 6: Functional Sizing*. In most application areas for this theory, the rule sets are yet unexplored. To apply the theory to practice will take generations to come.

The first step in most application areas within this cosmic inflation is making knowledge coherent; the next step is learning how to combine knowledge. A domain area is *Coherent* if their arrow term representation does not mix left and right sides; i.e., all formulas describing left-hand causes do not appear in right sides. Most attempts to requirements engineering of today simply ignore coherence. QFD still plays an ephemeral role within the requirement engineering domain. Obviously, coherence imposed by QFD should be central to requirements engineering, already by now.

In practice, coherence is difficult to assess, as most domain areas are not described in a formal way but rather in natural language, and viewpoints affect the meaning of words. A simple verbatim analysis of phrases describing the domain area is not good enough; even with predictive analysis relying on contextual relationship (Siegel, 2013) requirements mix up among left and right sides in arrow terms.

Another important conclusion is about combining different profiles. Akao invented this practice as *Comprehensive QFD* (Akao, 1990); however, without understanding the nature of QFD matrices as Six Sigma transfer functions, this was not very well adopted. Today, with the advent of the ISO 16355 standard, it will become common practice.

Combinatorial Algebra provides the operation of combining profiles in a very natural and intuitive way. Combining profiles is needed when different transfer functions yield contributions to some focus topic that uses both; e.g., business drivers that are supported by functions, qualities, and architecture. Combinatorial algebras are not finite. Finite dimensions do not limit combinatorial algebras. There is no limit set when modeling knowledge about the reality. In some way, Six Sigma transfer functions are finite approximations to combinatorial algebras representing the complexity of today's world.

APPENDICES

APPENDIX A:
LINEAR ALGEBRA IN A NUTSHELL

Concepts of linear algebra play a prominent role in this book. This appendix presents basic definitions and results of these concepts such as vectors, matrices, and eigenvalues and eigenvectors. Since only those concepts are described which are relevant to the content of this book, the reader is referred to the excellent book of Meyer (Meyer, 2000) for further details.

A-1 VECTOR SPACES

A set $\mathcal{V}$ over the real numbers $\mathbb{R}$ in conjunction with the algebraic addition

$$+: \mathcal{V} \times \mathcal{V} \to \mathcal{V}, \qquad (x, y) \mapsto x + y \tag{A-1}$$

and scalar multiplication

$$*: \mathbb{R} \times \mathcal{V} \to \mathcal{V}, \qquad (\alpha, y) \mapsto \alpha y \tag{A-2}$$

is called a *Vector Space*, if the following properties of the operation *Addition* hold:

(A1) *Closure Property*: $x + y \in \mathcal{V}$ for all $x, y \in \mathcal{V}$.

(A2) The addition is *associative*; i.e.,

$$(x + y) + z = x + (y + z), \quad \text{for all } x, y, z \in \mathcal{V} \tag{A-3}$$

(A3) The addition is *commutative*; i.e.,

$$x + y = y + x, \qquad \text{for all } x, y \in \mathcal{V} \tag{A-4}$$

(A4) There exists a *Neutral Element* $0 \in \mathcal{V}$ with $x + 0 = x$ for all $x \in \mathcal{V}$.

(A5) For each $x \in \mathcal{V}$, there exists an *Inverse Element* $-x \in \mathcal{V}$ with $x + (-x) = 0$.

Furthermore, the properties of the operation *Scalar Multiplication* hold:

(M1) *Closure Property*: $\alpha x \in \mathcal{V}$ for all $\alpha \in \mathbb{R}$ and $x \in \mathcal{V}$.

(M2) The scalar multiplication is *associative*; i.e.,

$$(\alpha \beta) x = \alpha(\beta x), \qquad \text{for all } \alpha \in \mathbb{R} \text{ and each } x \in \mathcal{V} \tag{A-5}$$

(M3) The unit $1 \in \mathbb{R}$ also behaves neutral in $\mathcal{V}$:

$$1x = x \text{ for each } x \in \mathcal{V} \tag{A-6}$$

(M4) The scalar multiplication is *distributive*; i.e.,

$$\alpha(x + y) = \alpha x + \alpha y, \qquad \text{for all } \alpha \in \mathbb{R} \text{ and each } x, y \in \mathcal{V}. \tag{A-7}$$

(M5) This holds also for the addition in the underlying body of real numbers $\mathbb{R}$:

$$(\alpha + \beta)x = \alpha x + \beta x, \qquad \text{for all } \alpha \in \mathbb{R} \text{ and each } x \in \mathcal{V}. \tag{A-8}$$

The elements $x \in \mathcal{V}$ are termed *Vectors*, and the addition is called *Vector Addition*.

A nonempty subset $\mathcal{U} \subset \mathcal{V}$ is called a *subspace of the vector space* $\mathcal{V}$, if and only if the properties A1 and M1 are satisfied for each $x, y \in \mathcal{U}$ and $\alpha \in \mathbb{R}$; i.e.,

$$x, y \in \mathcal{U} \implies x + y \in \mathcal{U}, \quad x \in \mathcal{U} \implies \alpha x \in \mathcal{U} \tag{A-9}$$

Obviously, if $\mathcal{U} \subset \mathcal{V}$, then $\mathcal{U}$ satisfies all vector space properties of $\mathcal{V}$ except the properties (A1), (A4), (A5), and (M1). It can be shown that (A1) in conjunction with (M1) implies the properties (A4) and (A5). By the closure property (M1), $(-x) = (-1)x \in \mathcal{U}$ for all $x \in \mathcal{U}$ so that (A5) holds. Since x and its inverse element $-x$ are both elements of $\mathcal{U}$, (A1) asserts that $x + (-x) \in \mathcal{U}$, hence $\mathbf{0} \in \mathcal{U}$.

Let $\mathcal{V}$ be a vector space and $x_1, x_2, \ldots, x_n \in \mathcal{V}$ be vectors. For scalars α_j with $1 \leq j \leq n$, a vector $x \in V$ is a *linear combination* of the vectors $x_1, x_2, \ldots, x_n$, if

$$x = \alpha_1 x_1 + \alpha_2 x_2 + \cdots + \alpha_n x_n = \sum_{j=1}^{n} \alpha_j x_j \tag{A-10}$$

For a vector space $\mathcal{V}$ with elements $x_1, x_2, \ldots, x_n \in \mathcal{V}$, the vector space

$$span(x_1, x_2, \ldots, x_n) = \left\{ x \in \mathcal{V} \; \middle| \; x = \sum_{j=1}^{n} \alpha_j x_j \right\} \tag{A-11}$$

is termed the *vector space spanned by* $x_1, x_2, \ldots, x_n$. If $\mathcal{V} = span(x_1, x_2, \ldots, x_n)$ then $\mathcal{V}$ is said to be *generated by* $x_1, x_2, \ldots, x_n$.

Vectors $x_1, x_2, \ldots, x_n \in \mathcal{V}$ are said to be *linearly independent*, if

$$x = \sum_{j=1}^{n} \alpha_j x_j = 0 \implies \alpha_j = 0 \tag{A-12}$$

for all $j = 1, \ldots, n$.

A *Basis* $\mathcal{B}$ of a vector space $\mathcal{V}$ is a subset of linearly independent vectors of $\mathcal{V}$ that span $\mathcal{V}$. Every vector space has a basis. All bases of a vector space $\mathcal{V}$ have the same number of elements. This unique number is termed the *dimension* of V, denoted by $dim(\mathcal{V})$. The proofs are left to the reader.

A vector space $\mathcal{V}$ over the real numbers $\mathbb{R}$ of dimension n is abbreviated by $\mathbb{R}^n$. Let $\boldsymbol{b}_1, \boldsymbol{b}_2, \dots, \boldsymbol{b}_n \in \mathbb{R}^n$ be a basis for a vector space $\mathcal{V} = \mathbb{R}^n$. Then a vector $\boldsymbol{x} \in \mathbb{R}^n$ has a unique representation as a linear combination of its basis:

$$\boldsymbol{x} = x_1 \boldsymbol{b}_1 + x_2 \boldsymbol{b}_2 + \cdots + x_n \boldsymbol{b}_n = \sum_{j=1}^{n} x_j \, \boldsymbol{b}_j \tag{A-13}$$

Thus, the *Components* of a vector $\boldsymbol{x}$ uniquely represent by its components

$$\boldsymbol{x} = \langle x_1, x_2, \dots, x_n \rangle \text{ with scalar components } x_i \in \mathbb{R}. \tag{A-14}$$

A *Norm* for a vector space $\mathcal{V} = \mathbb{R}^n$ is a function $\| \quad \| : \mathcal{V} \to \mathbb{R}$ that fulfills the following conditions:

$$\|\boldsymbol{x}\| > 0 \text{ and } \|\boldsymbol{x}\| = 0 \Leftrightarrow \boldsymbol{x} = \boldsymbol{0} \tag{A-15}$$

$$\|\alpha \boldsymbol{x}\| = |\alpha| \|\boldsymbol{x}\| \text{ for all scalars } \alpha \in \mathbb{R} \tag{A-16}$$

$$\|\boldsymbol{x} + \boldsymbol{y}\| \le \|\boldsymbol{x}\| + \|\boldsymbol{y}\| \tag{A-17}$$

The $\boldsymbol{\mathcal{L}p}$-*norm* $(p \ge 1)$ of a vector $\boldsymbol{x} \in \mathbb{R}^n$ with $\boldsymbol{x} = \langle x_1, x_2, \dots, x_n \rangle$ is defined as

$$\|\boldsymbol{x}\|_p = \left(\sum_{j=1}^{n} |x_j|^p \right)^{1/p} \tag{A-18}$$

In general, three forms of the $\boldsymbol{p}$-norm are applied:

$$\|\boldsymbol{x}\|_1 = \sum_{j=1}^{n} |x_j|, \|\boldsymbol{x}\|_2 = \left(\sum_{j=1}^{n} x_j^2 \right)^{1/2} \text{ and } \|\boldsymbol{x}\|_\infty = max(|x_j|) \tag{A-19}$$

which are called the *Grid Norm* or $\boldsymbol{\mathcal{L}_1}$-*norm*, the *Euclidean Norm* or $\boldsymbol{\mathcal{L}_2}$-*norm*, and the *Maximum Norm* or $\boldsymbol{\mathcal{L}_\infty}$-*norm*, respectively.

To construct a *Normalized Vector* x with $x \neq 0$ one defines a vector y such that $y = x/\|x\|$. Apparently, in view of (A-15) and (A-16):

$$\|y\| = \left\|\frac{x}{\|x\|}\right\| = \frac{1}{\|x\|}\|x\| = 1 \tag{A-20}$$

Hence, the normalized vector y points in the same direction as x, but its length is equal to one.

If $x = \langle x_1, x_2, \dots, x_n \rangle$ and $y = \langle y_1, y_2, \dots, y_n \rangle$ are vectors in $\mathbb{R}^n$, then the real-valued term

$$x \star y = \sum_{j=1}^{n} x_j y_j \tag{A-21}$$

is called the *Standard Inner Product* for $\mathbb{R}^n$. Setting $y = x$ in (A-21) results in

$$x \star x = \sum_{j=1}^{n} x_j x_j = \sum_{j=1}^{n} x_j{}^2 \tag{A-22}$$

implying $\|x\|_2 = \sqrt{x \star x}$ in case of the Euclidean norm.

If the inner product of two vectors $x, y \in \mathbb{R}^n$ is equal to zero; i.e., $x \star y = 0$, the vectors are said to be orthogonal to each other. Note that $x \star y = 0$, if and only if

$$\|x + y\|^2 = \|x\|^2 + \|y\|^2 \text{ and } \|x - y\|^2 = \|x + y\|^2 \tag{A-23}$$

The Euclidean Distance between vectors x and y in $\mathbb{R}^n$ is defined by

$$\|x - y\|_2 = \left(\sum_{j=1}^{n} |x_j - y_j| \right)^{1/2} \tag{A-24}$$

A-2 LINEAR TRANSFORMATIONS

Let $\mathcal{D}$ and $\mathcal{R}$ be vector spaces over $\mathbb{R}$. A function f with $f: \mathcal{D} \to \mathcal{R}$ is called a *Linear Transformation* with *Domain* $\mathcal{D}$ and *Range* $\mathcal{R}$, if the following conditions hold:

$$f(x + y) = f(x) + f(y) \qquad\qquad\qquad \text{(A-25)}$$

and

$$f(\alpha x) = \alpha f(x) \qquad\qquad\qquad \text{(A-26)}$$

for all vectors $x, y \in \mathcal{D}$ and all scalars $\alpha \in \mathbb{R}$.

Merging the conditions (A-25) and (A-26) results in an alternative definition of a linear function: f is a *Linear Transformation*, if

$$f(\alpha x + y) = \alpha f(x) + f(y) \qquad\qquad\qquad \text{(A-27)}$$

for all vectors $x, y \in \mathcal{D}$ and all scalars $\alpha \in \mathbb{R}$.

The set of all linear transformations $f: \mathcal{V} \to \mathcal{W}$ is denoted by $\mathfrak{L}(\mathcal{V}, \mathcal{W})$, where $\mathcal{V}$ and $\mathcal{W}$ are vector spaces. $\mathfrak{L}(\mathcal{V}, \mathcal{W})$ is a vector space: For $f, g \in \mathfrak{L}(\mathcal{V}, \mathcal{W})$, the addition is defined as

$$(f + g)(x) = f(x) + f(x) \qquad\qquad\qquad \text{(A-28)}$$

for all vectors $x \in \mathcal{V}$, and for all scalars $\alpha \in \mathbb{R}$ and $f \in \mathfrak{L}(\mathcal{V}, \mathcal{W})$, the scalar multiplication is defined as

$$(\alpha f)(x) = \alpha f(x) \qquad\qquad\qquad \text{(A-29)}$$

for all vectors $x \in \mathcal{V}$.

The following result shows that linear transformations are completely determined if their values on basis vectors are specified.

Theorem. *Let* $\mathcal{B}_{\mathcal{V}} = \{v_1, v_2, \dots, v_n\}$ *be a basis of* $\mathcal{V}$ *and* $w_1, w_2, \dots, w_n$ *be arbitrary vectors in* $\mathcal{W}$. *There exists a unique linear transformation* $f \in \mathfrak{L}(\mathcal{V}, \mathcal{W})$ *such that*

$$f(v_j) = w_j \qquad\qquad\qquad \text{(A-30)}$$

for all $j = 1, 2, \dots, n$.

Proof. Since $\mathcal{B}_V$ is a basis of V there exist unique scalars $\alpha_1, \alpha_2, ..., a_n$ such that any vector $v \in V$ has a representation as

$$v = \sum_{j=1}^{n} \alpha_j v_j \tag{A-31}$$

In view of linearity,

$$f(v) = f\left(\sum_{j=1}^{n} \alpha_j v_j\right) = \sum_{j=1}^{n} \alpha_j f(v_j) = \sum_{j=1}^{n} \alpha_j w_j \tag{A-32}$$

Hence, $f(v)$ is completely determined.

The existence of the linear transformation f is verified as follows. Consider the map

$$f\left(\sum_{j=1}^{n} \alpha_j v_j\right) = \sum_{j=1}^{n} \alpha_j w_j \tag{A-33}$$

compare (A-32). Setting

$$\sum_{j=1}^{n} \alpha_j v_j = 1 \cdot v_j + \sum_{\substack{i=1 \\ i \neq j}}^{n} 0 \cdot v_i \tag{A-34}$$

and

$$\sum_{j=1}^{n} \alpha_j w_j = 1 \cdot w_j + \sum_{\substack{i=1 \\ i \neq j}}^{n} 0 \cdot w_i \tag{A-35}$$

and replacing the corresponding terms shows that $f(v_j) = w_j$.

The linearity of the map f defined in (A-33) is proved as follows. At first it is checked, whether the map f satisfies $f(\alpha v) = \alpha f(v)$ for all $v \in V$ and scalars $\alpha \in \mathbb{R}$, compare (A-29). By (A-31) and (A-33)

$$f(\alpha v) = f\left(\sum_{j=1}^{n} \alpha(\alpha_j v_j)\right) = \sum_{j=1}^{n} \alpha(\alpha_j w_j) = \alpha\left(\sum_{j=1}^{n} \alpha_j w_j\right) = \alpha f(v) \tag{A-36}$$

Thus, $f(\alpha v) = \alpha f(v)$ for all $v \in V$ and all scalars $\alpha \in \mathbb{R}$.

Secondly, it is checked whether the map f in (A-33) satisfies $f(\boldsymbol{v} + \boldsymbol{v}') = f(\boldsymbol{v}) + f(\boldsymbol{v}')$ for arbitrary vectors $\boldsymbol{v}, \boldsymbol{v}' \in V$. The vector $\boldsymbol{v}'$ that can be written as $\boldsymbol{v}' = \sum_{j=1}^{n} \beta_j \boldsymbol{v}_j$ with coefficients $\beta_1, \beta_2, \dots, \beta_n \in \mathbb{R}$. By (A-33)

$$
f(\boldsymbol{v} + \boldsymbol{v}') = f\left(\sum_{j=1}^{n}(\alpha_j + \beta_j)\boldsymbol{v}_j\right) = \sum_{j=1}^{n}(\alpha_j + \beta_j)\boldsymbol{w}_j
$$

$$
= \sum_{j=1}^{n}\alpha_j \boldsymbol{w}_j + \sum_{j=1}^{n}\beta_j \boldsymbol{w}_j = f(\boldsymbol{v} + \boldsymbol{v}')
$$

(A-37)

Hence, the map f defines a linear transformation.

It remains to prove the uniqueness of the map f defined in (A-33). Consider a map $g \in \mathfrak{L}(V, W)$ with $g(\boldsymbol{v}_j) = \boldsymbol{w}_j$ for all $j = 1, 2, \dots, n$. Then $(f - g)(\boldsymbol{v}_j) = \boldsymbol{0}$ holds for all $j = 1, 2, \dots, n$.

By (A-31) and the linearity of $(f - g)$

$$
(f - g)(\boldsymbol{v}) = (f - g)\left(\sum_{j=1}^{n}\alpha_j \boldsymbol{v}_j\right) = \sum_{j=1}^{n}\alpha_j(f - g)(\boldsymbol{v}_j) = 0
$$

(A-38)

Thus, $(f - g) = 0$ implies $f = g$. $\qquad\qquad\qquad\qquad\qquad\qquad\qquad\qquad$ □

A linear transformation $f \in \mathfrak{L}(V, W)$ is said to be *injective*, if

$$
f(x) = f(y) \Longrightarrow x = y
$$

(A-39)

for all $x, y \in V$.

A linear transformation $f \in \mathfrak{L}(V, W)$ is called *surjective*, if for every $y \in W$ there exists an $x \in V$ such that $f(x) = y$, and it is called *invertible* (synonyms: *bijective*, *isomorph*), if and only if the transformation f is injective and surjective.

A linear transformation $id_V \in \mathfrak{L}(V, V)$ is termed the *Identity Transformation* or an *Endomorphism*. The linear map $0 \colon V \to W$ mapping every $\boldsymbol{v} \in V$ to $\boldsymbol{0} \in W$ is called the *Zero Transformation*.

If $f \in \mathfrak{L}(U, V)$ and $g \in \mathfrak{L}(V, W)$ are linear transformations between their respective vector spaces, then the *Composition* $f \circ g \colon U \to W$ defined by $(f \circ g)(x) = f(g(x))$ for all $x \in U$ is a linear transformation; i.e., $(f \circ g) \in \mathfrak{L}(U, W)$. The composition of linear transformations is both associative; i.e., $(f \circ g) \circ h = f \circ (g \circ h)$ and distributive; i.e., $(f + g) \circ h = (f \circ h) + (g \circ h)$ and $f \circ (g + h) = (f \circ g) + (f \circ h)$.

For the identity transformations hold

$$id_V \circ f = id_W \circ f = f \text{ for all } f \in \mathfrak{L}(V, W) \tag{A-40}$$

The *Kernel* (also known as *Null Space*) of a linear map $f: V \to W$ between two vector spaces V and W, is the set of all elements $v \in V$ for which $f(v) = \mathbf{0}$, where $\mathbf{0}$ denotes the zero vector in W. That is, in set-builder notation,

$$ker(f) = \{v \in V \mid f(v) = \mathbf{0}\} \tag{A-41}$$

The following properties of the kernel are of interest for the further sections. The kernel of f is a linear subspace of the domain V. In the linear map $f: V \to W$, two elements of V are mapped to the same vector element of W if and only if their difference lies in the kernel of f:

$$f(v_1) = f(v_2) \iff f(v_1 - v_2) = \mathbf{0} \tag{A-42}$$

It follows that the image $im(f)$ of f is isomorphic to the *Quotient* of V by the kernel:

$$im(f) \cong V/ker(f) \tag{A-43}$$

This implies the *rank-nullity theorem*:

$$dim\big(ker(f)\big) + dim\big(im(f)\big) = dim(f) \tag{A-44}$$

where "rank" denotes the dimension of the image of f, and "nullity" that of the kernel of f.

Let V be a vector space over $\mathbb{R}$. The vector space $V^* := \mathfrak{L}(V, \mathbb{R})$ is called the *Dual Vector Space* of V V and consists of all linear transformations $f \in \mathfrak{L}(V, \mathbb{R})$. To define a *Dual Linear Transformation*, consider a linear transformation $f \in \mathfrak{L}(V, \mathbb{R})$ and a linear transformation $g \in W^* = \mathfrak{L}(V, \mathbb{R})$. Then the composition $(g \circ f)$ is a transformation with $(g \circ f) \in \mathfrak{L}(V, \mathbb{R})$; i.e., $(g \circ f) \in V^*$. Hence, the *Dual of f* is the transformation

$$f^*: W^* \to V^* \text{ with } f^*(g) = g \circ f \tag{A-45}$$

Lemma. *If $f \in \mathfrak{L}(V, W)$, then f^* is linear; i.e. $f^* \in \mathfrak{L}(W^*, V^*)$.*

Proof. Let $g, h \in W^*$ and $\alpha \in \mathbb{R}$ be a scalar. Analogous to (A-27), it will be proved that for $f^*(\alpha \cdot g + h), \alpha f^*(g), f^*(h) \in V^*$ the linearity property holds:

$$f^*(\alpha \cdot g + h) = \alpha f^*(g) = f^*(h) \tag{A-46}$$

Let $v \in V$ be an arbitrary vector. Then,

$$
\begin{aligned}
f^*(\alpha \cdot g + h)(v) &= ((\alpha \cdot g + h) \; ^\circ \; f)(v) \\
&= \alpha \cdot (g^\circ f)(v) + (h^\circ f)(v) \\
&= (\alpha f^*(g))(v) \; + \; (f^*(h))(v) \\
&= (\alpha f^*(g) \; + \; f^*(h))(v)
\end{aligned}
\tag{A-47}
$$

Hence, (A-46) is verified.

A-3 MATRICES

Consider a linear transformation $f \in \mathfrak{L}(V, W)$, where V and W are vector spaces. Assume that $\mathcal{B}_V = \{v_1, v_2, \ldots, v_n\}$ is a basis of V, and $\mathcal{B}_W = \{w_1, w_2, \ldots, w_n\}$ is a basis of W. As proved before, the linear transformation f is uniquely determined when vectors $f(v_1), f(v_2), \ldots, f(v_n) \in W$ can be specified. Since $\mathcal{B}_W$ is a basis of W, there exists (real-valued) scalars a_{ij} such that

$$
f(v_j) = \sum_{i=1}^{m} a_{ij} \, w_j
\tag{A-48}
$$

for $1 \le j \le n$, compare (A-32). For simplicity, the scalars a_{ij} where $1 \le i \le m$ and $1 \le j \le n$ are arranged in a rectangular array with m rows and n columns:

$$
A = \begin{pmatrix}
a_{11} & a_{12} & \cdots & a_{1n} \\
a_{21} & a_{22} & \cdots & a_{2n} \\
\vdots & \vdots & \ddots & \vdots \\
a_{m1} & a_{m2} & \cdots & a_{mn}
\end{pmatrix}
\tag{A-49}
$$

which is called an $m \times n$ *Matrix*, where each a_{ij} denotes an *Entry* of the matrix in the ith row and the jth column. To omit naming matrix coefficients, sometimes $(A)_{ij}$ replaces the coefficient a_{ij}. No commas are set between matrix indices.

If the vector spaces V and W are equal to $\mathbb{R}^n$ and $\mathbb{R}^m$, respectively, then there exists a bijective correspondence between the linear transformations in $\mathfrak{L}(\mathbb{R}^n, \mathbb{R}^m)$ and matrices in $\mathbb{R}^{m \times n}$. If $f \in \mathfrak{L}(\mathbb{R}^n, \mathbb{R}^m)$, then the matrix A is defined by (A-48) and (A-49), respectively. Conversely, for any matrix $A \in \mathbb{R}^{m \times n}$ as in (A-49), a linear transformation $f \in \mathfrak{L}(\mathbb{R}^n, \mathbb{R}^m)$ exists by setting $f(v_j)$ as in (A-48).

A matrix $A \in \mathbb{R}^{n \times n}$ is a *Square Matrix*. A square matrix D with $(D)_{ii} \ne 0$ and all other entries are zero is termed a *Diagonal Matrix*. If $(D)_{ii} = 1$ for a diagonal matrix, D is the *Identity Matrix*, denoted by I. If all entries are zero, then 0 denotes the *Zero Matrix*.

In view of the bijective correspondence between matrices and linear transformations, the matrix addition of any two matrices $A \in \mathbb{R}^{m \times n}$ and $B \in \mathbb{R}^{m \times n}$ is another matric in $\mathbb{R}^{m \times n}$ denoted by $(A + B) \in \mathbb{R}^{m \times n}$. It is calculated component-wise by

$$(A + B)_{ij} = (A)_{ij} + (B)_{ij} \tag{A-50}$$

where $1 \leq i \leq m, 1 \leq j \leq n$, and the scalar multiplication $\alpha A \in \mathbb{R}^{m \times n}$ by

$$(\alpha A)_{ij} = \alpha (A)_{ij} \tag{A-51}$$

running over the same range of indices as above.

To derive the multiplication of any two matrices, consider linear transformations $f \in \mathfrak{L}(\mathcal{U}, \mathcal{V})$ and $g \in \mathfrak{L}(\mathcal{V}, \mathcal{W})$ and the composition $(f \circ g) \in \mathfrak{L}(\mathcal{U}, \mathcal{W})$. Assume that $\mathcal{B}_\mathcal{U} = \{u_1, u_2, \dots, u_n\}$, $\mathcal{B}_\mathcal{V} = \{v_1, v_2, \dots, v_p\}$, and $\mathcal{B}_\mathcal{W} = \{w_1, w_2, \dots, w_m\}$ are the bases of the vector spaces U, V, and W, respectively. Let A, B, and C be the matrix associated with f, g and $f \circ g$, respectively, and with coefficients $(A)_{ik} = a_{ik}$, $(B)_{kj} = b_{kj}$ and $(C)_{ij} = c_{ij}$. Then for each $j = 1, 2, \dots, n$

$$
(f \circ g)(u_j) = f\left(\sum_{k=1}^{p} b_{kj} v_k\right) = \sum_{k=1}^{p} b_{kj} f(v_k)
$$

$$
= \sum_{k=1}^{p} b_{kj} \left(\sum_{i=1}^{m} a_{ik} w_i\right) = \sum_{i=1}^{m} \left(\sum_{k=1}^{p} a_{ik} b_{kj}\right) \tag{A-52}
$$

Hence,

$$c_{ij} = \sum_{k=1}^{p} a_{ik} b_{kj} \tag{A-53}$$

for all $1 \leq i \leq m$, $1 \leq j \leq n$.

Therefore, the *Product of Matrices* $A \in \mathbb{R}^{m \times p}$ and $B \in \mathbb{R}^{p \times n}$, denoted by $AB \in \mathbb{R}^{m \times n}$, is determined by (A-53):

$$(C)_{ij} = (AB)_{ij} = \sum_{k=1}^{p} a_{ik} b_{kj} \tag{A-54}$$

In general, $AB \neq BA$; i.e., the matrix multiplication is not commutative.

Since a vector can be regarded as matrix in $\mathbb{R}^{n \times 1} = \mathbb{R}^n$, thus with exactly one column, the product of a matrix $A \in \mathbb{R}^{m \times n}$ and a vector $x \in \mathbb{R}^n$ is calculated by

$$(Ax)_i = \sum_{j=1}^{n} a_{ij} y_j \tag{A-55}$$

for all $1 \leq i \leq m$, compare (A-54).

Obviously, $A(\alpha x + y) = \alpha A x + A y$ with $A \in \mathbb{R}^{m \times n}$, $x, y \in \mathbb{R}^n$ and $\alpha \in \mathbb{R}$. This implies that the map $f_A \colon \mathbb{R}^n \to \mathbb{R}^m$ with $x \longmapsto Ax$ is a linear transformation.

Consider now the linear transformation $f \in \mathfrak{L}(\mathbb{R}^n, \mathbb{R}^m)$ with its associated matrix A. Let $\mathfrak{B}_{\mathbb{R}^m} = \{v_1, v_2, \dots, v_m\}$, and $\mathfrak{B}_{\mathbb{R}^n} = \{w_1, w_2, \dots, w_n\}$ denote the standard basis of $\mathbb{R}^m$ and $\mathbb{R}^n$, respectively. The dual linear transformation is defined by the linear mapping $f^* \in \mathfrak{L}((\mathbb{R}^n)^*, (\mathbb{R}^m)^*)$ with the two corresponding standard vector bases $\mathfrak{B}_{(\mathbb{R}^m)^*} = \{v_1^*, v_2^*, \dots, v_m^*\}$, and $\mathfrak{B}_{(\mathbb{R}^n)^*} = \{w_1^*, w_2^*, \dots, w_n^*\}$. To determine the matrix of f^*, the map $f^*(v_j^*) \colon \mathbb{R}^n \to \mathbb{R}$ is analyzed.

By definition of a dual linear transformation,

$$\left(f^*(v_j^*)\right)(w_i^*) = v_j^* \circ f(w_i^*) = v_j^* \left(\sum_{k=1}^{m} a_{ki} v_k\right) = a_{ji} \tag{A-56}$$

for all $1 \leq j \leq m$ and $1 \leq i \leq n$; see (A-45). Hence,

$$f^*(v_j^*) = \sum_{i=1}^{n} a_{ji} w_i \tag{A-57}$$

for all $1 \leq j \leq m$. That means that an (i, j)-entry of the matrix of the dual linear transformation f^* is identical to the (j, i)-entry of the matrix A of the linear transformation f.

The *Transpose* of a matrix $A \in \mathbb{R}^{m \times n}$ is the matrix $A^\mathsf{T} \in \mathbb{R}^{n \times m}$ defined componentwise by $(A^\mathsf{T})_{ij} = (A)_{ji}$, $1 \leq i \leq m$, $1 \leq j \leq n$. By definition, $(A^\mathsf{T})^\mathsf{T} = A$. Furthermore, if both $A, B \in \mathbb{R}^{m \times n}$, $(A + B)^\mathsf{T} = A^\mathsf{T} + B^\mathsf{T}$, and $(\gamma A)^\mathsf{T} = \gamma A^\mathsf{T}$, $\gamma \in \mathbb{R}$.

Transposing the product of matrices A and B has an interesting result:

Lemma. *For matrices* $A \in \mathbb{R}^{m \times p}$ *and* $B \in \mathbb{R}^{p \times n}$,

$$(AB)^\mathsf{T} = B^\mathsf{T} A^\mathsf{T} \tag{A-58}$$

Proof. By definition,

$$((AB)^\mathsf{T})_{ij} = (AB)_{ji} = \sum_{k=1}^{p} a_{jk} b_{ki} \tag{A-59}$$

and

$$(B^\mathsf{T} A^\mathsf{T})_{ij} = \sum_{k=1}^{p} b_{ki} a_{jk} = \sum_{k=1}^{p} a_{jk} b_{ki} \tag{A-60}$$

Hence, $((AB)^\mathsf{T})_{ij} = (B^\mathsf{T} A^\mathsf{T})_{ij}$ for all i and j, and thus $(AB)^\mathsf{T} = B^\mathsf{T} A^\mathsf{T}$. □

A square matrix $A \in \mathbb{R}^{n \times n}$ for which $A = A^\mathsf{T}$ is called a *Symmetric Matrix*.

Lemma. *For a matrix $A \in \mathbb{R}^{m \times n}$, the products $AA^\mathsf{T} \in \mathbb{R}^{m \times m}$ and $A^\mathsf{T} A \in \mathbb{R}^{n \times n}$ are symmetric matrices.*

Proof. In view of the previous lemma,

$$(AA^\mathsf{T})^\mathsf{T} = (A^\mathsf{T})^\mathsf{T} A^\mathsf{T} = AA^\mathsf{T} \tag{A-61}$$

and

$$(A^\mathsf{T} A)^\mathsf{T} = A^\mathsf{T} (A^\mathsf{T})^\mathsf{T} = A^\mathsf{T} A \tag{A-62}$$

By definition of the matrix transposition and matrix multiplication, $AA^\mathsf{T} \in \mathbb{R}^{m \times m}$ and $A^\mathsf{T} A \in \mathbb{R}^{n \times n}$. □

A symmetric matrix $A \in \mathbb{R}^n$ with $a_{ij} \in \mathbb{R}$ is said to be *positive definite*, if $x^\mathsf{T} Ax > 0$ for every vector $x \in \mathbb{R}^n \setminus \{0\}$. If $x^\mathsf{T} Ax \geq 0$, A is termed *positive semidefinite*.

An equivalent characterization of a positive definite matrix is based on the uniquely determined *Cholesky Factorization* LDL^T of a symmetric matrix A, where L is a lower triangular and D a diagonal matrix with

$$(D)_{ij} = \begin{cases} d_{jj} & \text{for } i = j \\ 0 & \text{for } i \neq j \end{cases} \qquad (L)_{ij} = \begin{cases} 0 & \text{for } i < j \\ 1 & \text{for } i = j \\ l_{ij} & \text{for } i > j \end{cases} \tag{A-63}$$

where the entries of the corresponding matrix are calculated recursively:

$$d_{jj} = a_{jj} - \sum_{k=1}^{j-1} l_{jk} d_{kk} \qquad l_{ij} = \frac{1}{d_{jj}} \left(a_{ij} - \sum_{k-1}^{j-1} l_{ik} l_{jk} d_{kk} \right) \tag{A-64}$$

Thus, a symmetric matrix A, which can be decomposed by the Cholesky factorization into $A = LDL^\mathsf{T}$ with $(D)_{jj} = d_{jj} > 0$, is said to be *positive definite*.

A square matrix $A \in \mathbb{R}^{n \times n}$ is termed a *Non-Negative Matrix* if each entry is non-negative; i.e., $a_{ij} \geq 0$, denoted by $A \geq 0$. The matrix A is said to be *positive*, if all entries $a_{ij} > 0$, denoted by $A > 0$.

Lemma. *If A is a square matrix and $x, y \in \mathbb{R}^n$, the following statements are true:*

$$
\begin{aligned}
A > 0, x \geq 0, x \neq 0 &\implies Ax > 0 \\
A \geq 0, x \geq y \geq 0 &\implies Ax \geq Ay \\
A \geq 0, x > 0, Ax = 0 &\implies A = 0 \\
A > 0, x > y > 0 &\implies Ax > Ay
\end{aligned}
\tag{A-65}
$$

Proof. To show the first implication, let $A > 0$ and $x > 0$. Since $x > 0$, there exists an index $k \in \{1, 2, \dots, n\}$ such that $x_k > 0$. Hence, for all $i = 1, 2, \dots, n$

$$
(Ax)_i = \sum_{j=1}^{n} a_{ij} x_j \geq a_{ik} x_k > 0
\tag{A-66}
$$

implying $Ax > 0$.

To show the second implication, let $A \geq 0$ and $x \geq y \geq 0$. Since $x \geq y$, there must be some $j \in \{1, 2, \dots, n\}$ such that $x_j \geq y_j$, thus

$$
(Ax)_i = \sum_{j=1}^{n} a_{ij} x_j \geq \sum_{j=1}^{n} a_{ij} y_j = (Ay)_i
\tag{A-67}
$$

for all $i = 1, 2, \dots, n$. Hence, $Ax \geq Ay$.

To show the third implication, let $A \geq 0$, $x > 0$, and $Ax = 0$. For all $i = 1, 2, \dots, n$,

$$
0 = (Ax)_i = \sum_{j=1}^{n} a_{ij} x_j
\tag{A-68}
$$

Since each $a_{ij} x_j$ is non-negative, it must be true that $a_{ij} x_j = 0$ for all $j = 1, 2, \dots, n$. This implies that all entries $a_{ij} = 0$, since $x_j > 0$. Hence, $A = 0$.

To show the fourth implication, let $A > 0$ and $x > y > 0$. Then, for each $i \in \{1, 2, \dots, n\}$

$$
(Ax)_i = \sum_{j=1}^{n} a_{ij} x_j > \sum_{j=1}^{n} a_{ij} y_j = (Ay)_i
\tag{A-69}
$$

since all entries $a_{ij} > 0$ and $x_j > y_j$. Hence, $Ax > Ay$. $\square$

The *Trace* of a square matrix is equal to the sum of the diagonal entries $(Ax)_{ii}$; i.e.,

$$trace(A) = \sum_{i=1}^{n} a_{ii} \qquad \text{(A-70)}$$

The trace function is a linear mapping: $trace(A + B) = trace(A) + trace(B)$, and $trace(\alpha A) = \alpha \cdot trace(A), \ \alpha \in \mathbb{R}$.

As stated above, matrix multiplication is generally not commutative. However, the trace function is one of the few cases in which the order of the matrices can be changed.

Lemma. *For* $A \in \mathbb{R}^{m \times n}$ *and* $B \in \mathbb{R}^{m \times n}$,

$$trace(AB) = trace(BA) \qquad \text{(A-71)}$$

Proof. Since $AB \in \mathbb{R}^{m \times m}$ and $BA \in \mathbb{R}^{n \times n}$,

$$trace(AB) = \sum_{i=1}^{m} (AB)_{ii} = \sum_{i=1}^{m} \left(\sum_{j=1}^{n} a_{ij}b_{ji} \right) \qquad \text{(A-72)}$$

$$= \sum_{j=1}^{n} \left(\sum_{i=1}^{m} b_{ji}a_{ij} \right) = \sum_{j=1}^{n} (BA)_{jj} = trace(BA)$$

$\square$

If I is the $n \times n$ identity matrix, then $trace(I) = n$. *(A-73)*

A square matrix A is said to be *nonsingular* if there exists a square matrix B such that $AB = I$ and $BA = I$. The matrix B is termed the *inverse* of the A, denoted by $B = A^{-1}$. A square matrix A that does not have an inverse is called a *singular matrix*.

Note that a system of linear equations $Ax = b$, where the square matrix A is nonsingular, has a unique solution.

The *determinant* of a square matrix A, denoted by $det(A)$, is defined by the following sum of products:

$$det(A) = \sum_{\sigma \in S_n} sgn(\sigma) \prod_{i=1}^{n} a_{i\sigma_i} \qquad \text{(A-74)}$$

The sum is computed over all permutations σ of the set $\{1, 2, \dots n\}$. A permutation is a function that reorders this set of integers. The value in the ith position after the reordering σ is denoted σ_i. The set of all such permutations – the *Symmetric Group* on n elements – is denoted S_n. For each permutation σ, $sgn(\sigma)$ denotes the signature of σ, a value that is $+1$, whenever the reordering given by σ can be achieved by successively interchanging two entries an even number of times; and -1, whenever it can be achieved by an odd number of such interchanges. (A-74) is known as *Leibniz Formula*.

For square matrices A and B, $det(AB) = det(A) \cdot det(B)$, $det(A) = det(A^\mathsf{T})$, and $det(A^{-1}) = det(A)^{-1}$.

The determinant of a lower triangular matrix L is equal to the product of its diagonal entries:

$$det(L) = det(L^\mathsf{T}) = \prod_{j=1}^{n} l_{jj} \tag{A-75}$$

The determinant of a diagonal matrix D is equal to the product of all d_{jj}, $1 \le j \le n$.

Clearly, if a symmetric matrix A is decomposed by the Cholesky factorization, then

$$det(A) = det(LDL^\mathsf{T}) = det(L)det(D)det(L^\mathsf{T}) = det(D) = \prod_{j=1}^{n} d_{jj} \tag{A-76}$$

Lemma. *Let A be a square matrix, and $x \in \mathbb{R}^n$. If the system of linear equations $Ax = 0$ has a solution in which not all the unknowns are zero, then $det(A) = 0$.*

Proof. Obviously, the system $Ax = 0$ has the trivial solution $x = 0$. Suppose now that A is nonsingular, then $x = 0$ is the unique solution (as stated above). However, this contradicts the hypothesis that not all entries of x are zero. ☐

Lemma. *A square matrix A is nonsingular if and only if $det(A) \ne 0$.*

Proof.

($\Rightarrow$) Recall that $det(I) = 1$.

Then, $det(I) = det(AA^{-1}) = det(A)det(A^{-1}) = det(A)\big(det(A)\big)^{-1}$. Hence, $det(A)det(A)^{-1} = 1$ implies that $det(A) \ne 0$.

($\Leftarrow$) $det(A) \ne 0 \Rightarrow \big(det(A)\big)^{-1} \ne 0$. ☐

Lemma. *A square matrix A is singular if and only if there exists an $x \in \mathbb{R}^n \backslash \{0\}$ such that $Ax = 0$.*

Proof. See Fiedler (Fiedler, 2008, p. 24). ☐

Square matrices A and B are said to be *similar*, if a nonsingular square matrix P exists such that $B = P^{-1}AP$. The product $P^{-1}AP$ is termed a *similarity transformation* on A. A square matrix A is said to be *diagonalizable*, if A is like a diagonal matrix.

APPENDIX B:
EIGENVALUES AND EIGENVECTORS

Eigenvalues and Eigenvectors are probably the most prominent mathematical concepts in the beginning of the information age. Engineers have used them since their inception in the early 20ᵗʰ century. Initially, the computational power needed to find the eigenvalues was a major hurdle against widespread use. This hurdle is gone.

B-1 BASIC DEFINITIONS AND RESULTS

For a square matrix A, the scalar $\lambda \in \mathbb{R}$ and the vector $x \in \mathbb{R}^n \backslash \{0\}$ that satisfy $Ax = \lambda x$ is termed the *Eigenvalue* (or characteristic value or root) and the *Eigenvector* (or characteristic vector) for the matrix A, respectively.

The pair (λ, x) is called an *Eigenpair* of A. The set of distinct eigenvalues denoted by σ_A is termed the *Spectrum* of A. The maximum of the absolute values in the spectrum of the matrix A is called the *Spectral Radius* and denoted as $\rho(A) = max\{|\lambda| | \lambda \in \sigma_A\}$.

Geometrically, eigenvectors determine directions that are invariant under multiplication with a square matrix A. The eigenvalue λ reflects the amount the eigenvector x is stretched or shrunk when it is transformed by the matrix A. The eigenvalue λ for a square matrix A may be zero, but the associated eigenvector x may be not equal to the zero vector 0, by definition.

It is easy to check whether a given square matrix A is singular or not.

Lemma. *A square matrix A is singular if and only if $\lambda = 0$ is an eigenvalue for A.*

Proof. The assertion of the lemma is shown by the following equivalences:

$$
\begin{aligned}
A \ \ is \ \ singular &\Leftrightarrow there \ \ exists \ \ x \neq 0 \ \ with \ \ Ax \neq 0 \\
&\Leftrightarrow there \ \ exists \ \ x \neq 0 \ \ with \ \ Ax = 0x \\
&\Leftrightarrow \lambda = 0 \ \ is \ \ an \ \ eigenvalue \ \ for \ \ A
\end{aligned} \tag{B-1}
$$

$\square$

Lemma. *An eigenvector for a square matrix A cannot be associated with two distinct eigenvalues for A.*

Proof. If there exists a non-zero vector x satisfying $Ax = \lambda_1 x$ and $Ax = \lambda_2 x$ with $\lambda_1 \neq \lambda_2$, then

$$0 = Ax - Ax = \lambda_1 x - \lambda_2 x = (\lambda_1 - \lambda_2)x \qquad \text{(B-2)}$$

This implies $x = 0$, which contradicts the assumption that $x \neq 0$. Thus, such a zero vector x cannot exist. $\qquad \square$

Lemma. *Let $\lambda_1, \lambda_2, \ldots \lambda_n$ be the pairwise different eigenvalues for the matrix A and $x_1, x_2, \ldots, x_n$ the associated eigenvectors. Then $x_1, x_2, \ldots, x_n$ are linearly independent.*

Proof. (By contradiction). Assume that $x_1, x_2, \ldots, x_n$ are linearly dependent. Hence, there exist a k with $1 \leq k \leq n$ and non-zero scalars $c_1, c_2, \ldots, c_k$ such that

$$0 = \sum_{i=1}^{k} c_i x_i \qquad \text{(B-3)}$$

Then,

$$
\begin{aligned}
0 &= (A - \lambda_k I)0 \\
&= (A - \lambda_k I) \sum_{i=1}^{k} c_i x_i = \sum_{i=1}^{k} c_i (A - \lambda_k I) x_i \\
&= \sum_{i=1}^{k} c_i (\lambda_i x_i - \lambda_k x_i) = \sum_{i=1}^{k} c_i (\lambda_i - \lambda_k) x_i = \sum_{i=1}^{k-1} c_i (\lambda_i - \lambda_k) x_i
\end{aligned}
\qquad \text{(B-4)}
$$

For $1 \leq i \leq k - 1$, $(\lambda_i - \lambda_k) \neq 0$, since it is assumed that the eigenvectors x_i have pairwise different eigenvalues. Hence, $c_i(\lambda_i - \lambda_k) \neq 0$ for $1 \leq i \leq k - 1$. At this point, the linear dependence relation

$$0 = \sum_{i=1}^{k-1} c_i (\lambda_i - \lambda_k) x_i \qquad \text{(B-5)}$$

is obtained, which involves the eigenvectors x_i, $1 \leq i \leq k - 1$, where $c_i(\lambda_i - \lambda_k) \neq 0$ for the same indices.

Repeating the described procedure in view of the matrices $A - \lambda_{k-1}I$, $A - \lambda_{k-2}I, \ldots$ results finally in

$$0 = c_1 x_1, \qquad c_1 \neq 0, \qquad \text{(B-6)}$$

which implies $x_1 = 0$. However, this is a contradiction, since eigenvectors are never equal to the zero vector 0. Hence, the eigenvectors $x_1, x_2, \ldots, x_n$ are linearly independent. $\qquad \square$

Since eigenvectors are linearly independent, they build a basis. A basis of eigenvectors for a square matrix A is termed the *Eigenbasis* for A.

Since $Ax = \lambda x \Leftrightarrow (A - \lambda I)x = 0$, vectors of interest are the non-zero vectors x in the kernel $ker(A - \lambda I)$. Obviously, the kernel contains non-zero vectors if and only if the matrix $A - \lambda I$ is singular. Hence, those eigenvalues λ are of interest for which $det(A - \lambda I) = 0$. Or, alternatively

$$
\begin{aligned}
\lambda \in \sigma_A \;&\Leftrightarrow\; Ax = \lambda x \text{ for some } x \neq 0 \\
&\Leftrightarrow\; (A - \lambda I)x = 0 \text{ for some } x \neq 0 \\
&\Leftrightarrow\; A - \lambda I \text{ is singular} \\
&\Leftrightarrow\; det(A - \lambda I) = 0
\end{aligned}
\tag{B-7}
$$

Hence,

Lemma. $\lambda \in \mathbb{R}$ *is an eigenvalue of a square matrix* A *if and only if the matrix* $(A - \lambda I)$ *is singular; i.e., if and only if* $det(A - \lambda I) = 0$.

The set $\mathfrak{E}_A(\lambda) = \{x \in \mathbb{R}^n \backslash \{0\} | x \in ker(A - \lambda I)\}$ contains all eigenvectors associated with the eigenvalue λ and is called the *Eigenspace* of the matrix A.

The polynomial $p_A(\lambda) = det(A - \lambda I)$ is of degree n and is called the *characteristic polynomial*. Hence, the eigenvalues for A are the *characteristic roots* of $p_A(\lambda)$, calculated by solving $p_A(\lambda) = 0$. Note that eigenvalues may be repeated.

Lemma. *Let* A *be a square matrix with eigenvalues* λ_i, $1 \leq i \leq n$. *Then,* $trace(A) = \sum_{i-1}^{n} \lambda_i$ *and* $det(A) = \prod_{i=1}^{n} \lambda_i$.

Proof. Consider the characteristic polynomial

$$
p_A(\lambda) = (-1)^n \lambda^n - trace(A)\lambda^{n-1} + \cdots + (-1)^n det(A)
\tag{B-8}
$$

and the characteristic polynomial

$$
p_A(\lambda) = \prod_{i-1}^{n}(\lambda - \lambda_i) = \lambda^n - \lambda^{n-1}\sum_{i=1}^{n} \lambda_i + \cdots + (-1)^n \prod_{i=1}^{n} \lambda_i
\tag{B-9}
$$

where λ_i are the eigenvalues of the matrix A. A comparison of the coefficients shows that $trace(A) = \sum_{i-1}^{n} \lambda_i$ and $det(A) = \prod_{i=1}^{n} \lambda_i$. $\qquad\square$

The *algebraic multiplicity* of an eigenvalue λ for a square matrix A the multiplicity of the eigenvalue λ as a root of the characteristic polynomial; i.e., $p_A(\lambda) = 0$. The eigenvalue λ is termed a *simple eigenvalue*, if $alg \; mult_A(\lambda) = 1$.

The *geometric multiplicity* of an eigenvalue λ for the square matrix A is equal to the number of linearly independent eigenvectors associated with the set of eigenvalues λ_i, $i = 1, 2, \ldots, n$. Formally, this means $geom \; mult_A(\lambda) = dim(ker(A - \lambda I))$.

Note that $geom \; mult_A(\lambda) \leq alg \; mult_A(\lambda)$ is always valid; see e.g., Meyer's textbook (Meyer, 2000, p. 511). In addition, since every has at least one eigenvector, $geom \; mult_A(\lambda) \geq 1$.

Let λ_i, $i = 1, 2, \ldots, n$ be the (not necessarily distinct) eigenvalues of a square matrix $\boldsymbol{A}$. λ_1 is called a *dominant eigenvalue* for $\boldsymbol{A}$, if $|\lambda_1| > |\lambda_i|$ for $i = 2, 3, \ldots, n$. A dominant eigenvalue is denoted by λ_{max}. The eigenvectors corresponding to λ_{max} are termed the *dominant eigenvectors* for A.

Determining the value of λ_{max} is important in practice. The most applied bounds for the dominant eigenvalue for a square matrix $\boldsymbol{A}$ are as follows. Denote by

$$r_i = \sum_{j=1}^{n} |a_{ij}| \qquad\qquad c_{ji} = \sum_{i=1}^{n} |a_{ij}| \qquad\qquad \text{(B-10)}$$

the ith row sum and j^{th} column sum of $\boldsymbol{A} = (a_{ij})$, respectively, for $i, j \in \{1, 2, \ldots, n\}$. Then,

$$|\lambda_{max}| \leq min\left(\max_{1 \leq i \leq n} r_i, \max_{1 \leq j \leq n} c_j \right) \qquad\qquad \text{(B-11)}$$

B-2 EIGENVALUES & EIGENVECTORS FOR SYMMETRIC MATRICES

Recall that for a matrix $\boldsymbol{A} \in \mathbb{R}^{m \times n}$ the matrices $\boldsymbol{A}^\mathsf{T}\boldsymbol{A}$ and $\boldsymbol{A}\boldsymbol{A}^\mathsf{T}$ are symmetric.

Lemma. *The eigenvalues of the symmetric matrices $\boldsymbol{A}\boldsymbol{A}^\mathsf{T}$ and $\boldsymbol{A}^\mathsf{T}\boldsymbol{A}$ are real and non-negative, and the eigenvalues are strictly positive, if $\boldsymbol{A}\boldsymbol{A}^\mathsf{T}$ and $\boldsymbol{A}^\mathsf{T}\boldsymbol{A}$ are nonsingular.*

Proof. If $(\lambda, \boldsymbol{x})$ is an eigenpair for $\boldsymbol{A}\boldsymbol{A}^\mathsf{T}$, then

$$\boldsymbol{A}\boldsymbol{A}^\mathsf{T}\boldsymbol{x} = \lambda\boldsymbol{x} \Leftrightarrow \boldsymbol{x}^\mathsf{T}\boldsymbol{A}\boldsymbol{A}^\mathsf{T}\boldsymbol{x} = \lambda\boldsymbol{x}^\mathsf{T}\boldsymbol{x} \Leftrightarrow \frac{(\boldsymbol{A}^\mathsf{T}\boldsymbol{x})^\mathsf{T}(\boldsymbol{A}^\mathsf{T}\boldsymbol{x})}{\boldsymbol{x}^\mathsf{T}\boldsymbol{x}} \Leftrightarrow \lambda = \frac{\|\boldsymbol{A}\boldsymbol{x}\|_2^2}{\|\boldsymbol{x}\|_2^2} \qquad \text{(B-12)}$$

It follows that $\lambda \in \mathbb{R}$ and $\lambda \geq 0$. As stated previously, $\lambda = 0$ if and only if $\boldsymbol{A}\boldsymbol{A}^\mathsf{T}$ is singular. Hence, $\lambda > 0$ if and only if $\boldsymbol{A}\boldsymbol{A}^\mathsf{T}$ is nonsingular.

The lemma is completely shown when similar arguments are applied to $\boldsymbol{A}^\mathsf{T}\boldsymbol{A}$. $\quad\square$

Lemma. *Let $\boldsymbol{A}$ be a symmetric matrix. Then, the eigenvectors associated with distinct eigenvalues for $\boldsymbol{A}$ are orthogonal.*

Proof. Assume that $\boldsymbol{x}$ and $\boldsymbol{y}$ are eigenvectors for $\boldsymbol{A}$ associated with distinct eigenvalues λ and μ, respectively. Then,

$$\lambda(\boldsymbol{x}^\mathsf{T}\boldsymbol{y}) = (\lambda\boldsymbol{x}^\mathsf{T})\boldsymbol{y} = (\boldsymbol{x}^\mathsf{T}\boldsymbol{A}^\mathsf{T})\boldsymbol{y} = \boldsymbol{x}^\mathsf{T}(\boldsymbol{A}^\mathsf{T}\boldsymbol{y}) = \boldsymbol{x}^\mathsf{T}(\mu\boldsymbol{y}) = \mu(\boldsymbol{x}^\mathsf{T}\boldsymbol{y})$$

Hence, $(\lambda - \mu)(\boldsymbol{x}^\mathsf{T}\boldsymbol{y}) = 0$. Since $\lambda \neq \mu$, it follows that $(\boldsymbol{x}^\mathsf{T}\boldsymbol{y}) = 0$; i.e., $\boldsymbol{x}$ and $\boldsymbol{y}$ are orthogonal eigenvectors. $\quad\square$

As derived in the previous section, a square matrix A is positive definite if it can be decomposed by the Cholesky factorization; i.e., $A = LDL^{\mathsf{T}}$ where the lower triangular matrix L is uniquely defined and non-singular, and the entries of the diagonal matrix D are strictly positive.

Lemma. *All eigenvalues of a positive definite matrix A are real and strictly positive.*

Proof. Since the matrix A is assumed to be positive definite, it can be decomposed by the Cholesky factorization; i.e., $A = LDL^{\mathsf{T}}$. If (λ, x) is an eigenpair for A, then

$$LDL^{\mathsf{T}}x = \lambda x \Leftrightarrow \left(D^{\frac{1}{2}}L^{\mathsf{T}}x\right)^{\mathsf{T}}\left(D^{\frac{1}{2}}L^{\mathsf{T}}x\right) = \lambda x^{\mathsf{T}}x \Leftrightarrow \lambda = \frac{\left\|D^{\frac{1}{2}}L^{\mathsf{T}}x\right\|_2^2}{\|x\|_2^2} \tag{B-13}$$

It follows that $\lambda \in \mathbb{R}$ and $\lambda \geq 0$. $\lambda = 0$ if and only if $D^{\frac{1}{2}}L^{\mathsf{T}}x = 0$ or, equivalently, $x = 0$, but this is a contradiction to the definition of an eigenvector. Hence, $\lambda > 0$. $\square$

Lemma. *All eigenvalues of a positive semidefinite matrix A are real and non-negative.*

Proof. A symmetric matrix A is positive semidefinite if $x^{\mathsf{T}}Ax \geq 0$ for eigenvectors of A. If (λ, x) is an eigenpair for A, then

$$\lambda = \frac{x^{\mathsf{T}}Ax}{x^{\mathsf{T}}x} = \frac{x^{\mathsf{T}}B^{\mathsf{T}}Bx}{x^{\mathsf{T}}x} = \frac{\|Bx\|_2^2}{\|x\|_2^2} \geq 0 \tag{B-14}$$

and the assertion is proved. $\square$

Lemma. *Let A be a symmetric matrix. Then all eigenvalues for A real numbers. If these eigenvalues are distinct, then the associated eigenvectors are orthogonal.*

Proof. Recall that a square matrix A diagonalizable, if A is like a diagonal matrix D; i.e., if a nonsingular, thus invertible, matrix P exists such that $D = P^{-1}AP$. $\square$

A *complete set of eigenvectors* for a matrix $A \in \mathbb{R}^{n \times n}$ is a set of n linearly independent eigenvectors of A. A matrix A does not have a complete set of eigenvectors is termed a *defective matrix*.

Lemma. *The columns of P build a complete set of eigenvectors and λ_{ij}, $j = 1,2,\dots,n$ are the associated eigenvalues if and only if $P^{-1}AP = D = diag(\lambda_1, \lambda_2, \dots \lambda_n)$.*

Proof. See textbooks on linear algebra; e.g., (Meyer, 2000, p. 506) and (Nipp & Stoffer, 2002, p. 156). $\square$

A square matrix A is said to be *orthogonally diagonalizable*, if there exists an orthogonal square matrix P such that $P^{-1}AP = P^{\mathsf{T}}AP$ is a diagonal matrix with real entries.

Lemma. *Assume that a square matrix* A *is orthogonally diagonalizable. Then the matrix* A *is symmetric.*

Proof. By assumption, the matrix A is orthogonally diagonalizable. Then there exists a diagonal square matrix P and a diagonal matrix D such that $P^\mathsf{T}AP = D$. By definition of an orthogonal matrix, $P^\mathsf{T}P = PP^\mathsf{T} = I$, and for a diagonal matrix holds $D = D^\mathsf{T}$. Hence,

$$A = P^\mathsf{T}AP = (P^\mathsf{T})^\mathsf{T}D^\mathsf{T}P^\mathsf{T} = (PD^\mathsf{T}P^\mathsf{T})^\mathsf{T} = (PDP^\mathsf{T})^\mathsf{T} = A^\mathsf{T} \qquad \text{(B-15)}$$

and the implication is proved. $\qquad\qquad\qquad\qquad\qquad\qquad\qquad\qquad\qquad\qquad$ $\square$

Without the obvious proof:

Lemma. *A matrix* A *is orthogonally diagonalizable if and only if* A *is symmetric.*

B-3 THE POWER METHOD

Theorem. *If* x *is an eigenvector of a square matrix* A, *then its corresponding eigenvalue is given by*

$$\lambda = \frac{x^\mathsf{T}Ax}{x^\mathsf{T}x} \qquad \text{(B-16)}$$

This quotient is called the *Rayleigh Quotient.*

Proof. Since $x > 0$ is an eigenvector of the matrix A, one has $Ax = \lambda x$. Then

$$Ax = \lambda x \Longleftrightarrow x^\mathsf{T}Ax = \lambda x^\mathsf{T}x \Longleftrightarrow \lambda = \frac{x^\mathsf{T}Ax}{x^\mathsf{T}x} \qquad \text{(B-17)}$$

$$\square$$

Hence, in cases for which the power method generates a good approximation of a dominant eigenvector, the Rayleigh Quotient delivers a good approximation of a dominant eigenvalue.

APPENDIX C:
THE PERRON-FROBENIUS THEORY

The Perron–Frobenius Theorem – stated and proved by Perron in 1907 and Frobenius in 1912 – asserts that a real square matrix with positive entries has a unique largest real eigenvalue and that the corresponding eigenvector has strictly positive components, and asserts a similar statement, valid for certain classes of non-negative matrices. The Perron–Frobenius Theorem describes the properties of the leading eigenvalue and of the corresponding eigenvectors when A is a non-negative real square matrix.

Although results and implications of this theorem are now well understood, proofing it is still a matter of advanced mathematics. Not all details will be given in this appendix, only the major concepts for readers who need an in-depth understanding of the basics for surviving in today's complex world.

C-1 PRELIMINARIES

A square matrix A is *reducible*, if there exists a permutation matrix P such that

$$B = PAP^{\mathsf{T}} = \begin{bmatrix} A_{11} & A_{12} \\ 0 & A_{22} \end{bmatrix} \tag{C-1}$$

where $A_{11} \in \mathbb{R}^{k \times k}$, $A_{12} \in \mathbb{R}^{k \times (n-k)}$, and $A_{22} \in \mathbb{R}^{(n-k) \times (n-k)}$ with $0 < k < n$.

On the other hand, a matrix A is said to be *irreducible*, when such a permutation matrix P does not exist.

Keener (Keener, 1993) states two alternative definitions of an irreducible matrix:

1) A non-negative square matrix A is *irreducible*, if for any eigenvector $x \geq 0$, $Ax > 0$.

2) A non-negative square matrix A is *irreducible*, if for any two integers i and j there exists an integer $p \geq 0$ and a sequence of integers $k_1, k_2, \ldots, k_p$ such that the product $a_{ik_1} \cdot a_{k_1 k_2} \cdot \ldots \cdot a_{k_p j} \neq 0$, where a_{ij} denotes an entry of A.

A question arises: How can the non-negativity and the irreducibility converted to positivity? An answer gives

Lemma. *If a non-negative square matrix A is irreducible, then $(I + A)^{n-1} > 0$.*

Proof. Note that it is sufficient to show that $(I + A)^{n-1}x > 0$ for any $x \geq 0,\ x \neq 0$. To accomplish this, define the sequence

$$x_{k+1} = (I + A)x_k \geq 0 \qquad \text{(C-2)}$$

for $k = 0, 1, \dots, n - 2$, with $x_0 = x$. Since $x_{k+1} = x_k + Ax_k$, the vector x_{k+1} has no more zero components than x_k.

Suppose that x_{k+1} and x_k have the same zero components. Then there exists a permutation matrix P such that

$$P x_k = \begin{bmatrix} y \\ 0 \end{bmatrix},\ P x_{k+1} = \begin{bmatrix} z \\ 0 \end{bmatrix} \qquad \text{(C-3)}$$

with $y, z \in \mathbb{R}^m,\ y, z > 0,\ 1 \leq m < n$.

Then,

$$
\begin{aligned}
P x_{k+1} = \begin{bmatrix} z \\ 0 \end{bmatrix} &= P(x_k + Ax_k) \\
&= P x_k + PA(P^\mathsf{T}P)x_k \\
&= P x_k + (PAP^\mathsf{T})P x_k \\
&= \begin{bmatrix} y \\ 0 \end{bmatrix} + \begin{bmatrix} A_{11} & A_{12} \\ A_{21} & A_{22} \end{bmatrix} \begin{bmatrix} y \\ 0 \end{bmatrix}
\end{aligned}
\qquad \text{(C-4)}
$$

Hence, $A_{12} = 0$, which contradicts the assumption that A is irreducible.

It follows that the initial iteration vector x_0 has at most $n - 1$ zero components, and that x_k has at most $n - k - 1$ zero components. Thus,

$$x_{n-1} = (I + A)^{n-1}x_0 > 0 \qquad \text{(C-5)}$$

$\square$

Let A be non-negative. For a fixed index i, the *Period of Index* i is the greatest common divisor of all natural numbers m such that $(A^m)_{ii} > 0$. When A is irreducible, the period of every index is the same and is called the *Period of* A. The period is also called the *Index of Imprimitivity* (Meyer, 2000, p. 674).

If the period is 1, A is *aperiodic*.

A matrix A is *primitive* if it is non-negative and its mth power is positive for some natural number m. It can be proved that primitive matrices are the same as irreducible aperiodic non-negative matrices.

A positive square matrix is primitive and a primitive matrix is irreducible. All statements of the Perron–Frobenius theorem for positive matrices remain true for primitive matrices. However, a general non-negative irreducible matrix A may possess several eigenvalues whose absolute value is equal to the spectral radius $\rho(A)$ of A, so the statements need to be correspondingly modified. The number of such eigenvalues is exactly equal to the period. Results for non-negative matrices were first obtained by Frobenius in 1912 (Wikipedia - Perron-Frobenius Theorem, 2015).

C-2 THE PERRON THEOREM

Cairns published the following elegant proof (Cairns, 2014). Recall the definitions of spectral radius $\rho(A)$ of a square matrix A.

Theorem. (*Perron's Theorem.*) *Let A be a positive square matrix. Then*

   a) $\rho(A)$ *is an eigenvalue, and it has a positive eigenvector.*
   b) $\rho(A)$ *is the only eigenvalue with $|\lambda| = \rho(A)$.*
   c) $\rho(A)$ *has geometric multiplicity* 1.
   d) $\rho(A)$ *has algebraic multiplicity* 1.

Preliminaries. The proof of the Perron-Frobenius theorem will rely on the following simple positivity trick. For the following, denote matrix and vector components always by $(A)_{ij} = a_{ij}$, $v = \langle v_1, v_2, \dots, v_m \rangle$ and $w = \langle w_1, w_2, \dots, w_m \rangle$.

Trick. Let A be a positive square matrix. If v and w are two non-equal column vectors with $v \geq w$, then $Av > Aw$. There is some positive $\varepsilon > 0$ with $Av > (1 + \varepsilon)Aw$.

Proof. The vector $A(v - w) = \sum a_{ij}(v_j - w_j)$, so

$$\left(A(v - w) \right) \geq \sum_j \min_{ij}(a_{ij})\,(v_j - w_j) = \min_{ij}(a_{ij}) \sum_j (v_j - w_j) > 0 \qquad \text{(C-6)}$$

So, $(Av)_i - (Aw)_i > 0$, and by definition $Av > Aw$. This proves the first statement. If we change the vector $A(v - w)$ by a small amount, then it will still be positive, so there is some $\varepsilon > 0$ with $A(v - w) - \varepsilon Aw > 0$, or $Av > (1 + \epsilon)Aw$. That proves the second statement.

Theorem. (*Gelfand's formula*) *The spectral radius of a matrix A can be written in terms of the norms of its powers*:

$$\rho(A) = \lim_{n \to \infty} (\|A\|^n)^{1/n} \qquad \text{(C-7)}$$

Proof. Consult textbooks or Wikipedia (Wikipedia - Spectral Radius, 2015).

With these tools, the Perron-Frobenius theorem can easily be proved.

Proof. (a) A *has* λ *as an eigenvalue, and* λ *has a positive eigenvector.*

An eigenvalue λ exists with $|\lambda| = \rho(A)$. Let $w = \langle w_1, w_2, \dots, w_m \rangle$ be an eigenvector. Let $\hat{w}$ be the vector with $\hat{w}_j = |w_j|$. Then

$$(A\hat{w})_i = \sum_j a_{ij}|w_j| \geq \left| \sum_j a_{ij}w_j \right| = |\lambda w_i| = \rho(A)w_i \tag{C-8}$$

so $A\hat{w} \geq \rho(A)w$.

If they are not equal, then by the positivity trick, $A^2\hat{w} > \rho(A)Aw$, and there is some positive $\varepsilon > 0$ with $A^2\hat{w} \geq (1 + \varepsilon)\rho(A)Aw$.

The matrix A is non-negative, and so are all its powers A^n, so multiplying both sides of the equation by A^n preserves the inequality:

$$A^{n+2}\hat{w} \geq (1 + \varepsilon)\rho(A)A^{n+1}w \tag{C-9}$$

This is true for every n, so

$$\begin{aligned}
A^{n+1}\hat{w} &\geq (1 + \varepsilon)\rho(A)A^n w \\
&\geq \big((1 + \varepsilon)\rho(A)\big)^2 A^{n-1}w \\
&\qquad \vdots \\
&\geq \big((1 + \varepsilon)\rho(A)\big)^2 Aw
\end{aligned} \tag{C-10}$$

Gelfand's formula proofs that

$$\rho(A) = \lim_{n\to\infty} (\|A\|^n)^{1/n} \geq (1 + \varepsilon)\rho(A) \tag{C-11}$$

which is a contradiction, thus the assumption must have been mistaken. This means that the two vectors were equal: $A\hat{w} = \rho(A)w$.

So, $|w|$ is also an eigenvector with eigenvalue $\rho(A)$. It is not only non-negative but positive, because $A|w| = \rho(A)|w|$ is positive.

(b) *The only eigenvalue with* $|\lambda| = \rho(A)$ *is* $\rho(A)$.

Suppose $\lambda \neq \rho(A)$ is some other eigenvalue with $|\lambda| = \rho(A)$. This will yield a contradiction. Repeat the reasoning from part (a) again: if $\hat{w} = |w|$, then $A\hat{w} = \rho(A)w$, or

$$\sum a_{ij}|w_j| = \left| \sum a_{ij}w_j \right| \tag{C-12}$$

for every $1 \le j \le n$.

Lemma. *Let A A be a positive $n \times n$ matrix. Let w be a vector in $\mathbb{R}^n$. Choose an index j. If*

$$\sum a_{ij}|w_j| = \left|\sum a_{ij}w_j\right| \tag{C-13}$$

then there is some $\gamma \in \mathbb{R}$ with $\gamma \ne 0$, so that the product γw is a non-negative vector.

Proof. If $w = 0$, then $\gamma = 1$ works. Suppose w is not zero. Square both sides and write $\left|\sum a_{ij}w_j\right|^2 = (\sum a_{ij}w_j)(\sum a_{kl}w_l)$. This results in the equality

$$\sum a_{ij}a_{kl}|w_j||w_l| = \sum a_{ij}a_{kl}w_j w_l \tag{C-14}$$

Subtract the right-hand side from the left:

$$\sum a_{ij}a_{kl}(|w_j||w_l| - w_j w_l) = 0 \tag{C-15}$$

All coefficients $a_{ij}a_{kl}$ are positive real numbers. Thus, an index l must exist such that $w_l \ne 0$. Setting $\gamma = |w_l|$ proves the lemma. $\square$

The conditions of the lemma hold, so $\gamma w > 0$ for some real number $\gamma \ne 0$. However, this means that

$$\lambda(\gamma w) = \gamma(\lambda w) = \gamma(A w) = A(\gamma w) \ge 0 \tag{C-16}$$

is positive, and $\gamma w > 0$, so λ must be non-negative.

However, $\rho(A)$ is the only non-negative number that meets the condition $|\lambda| = \rho(A)$.

(c) $\rho(A)$ *has geometric multiplicity* 1.

Suppose w is the positive eigenvector above and $\acute{w}$ is a linearly independent eigenvector of the eigenvalue $\rho(A)$.

Let $\gamma > 0$ be chosen so that $w - \gamma \acute{w}$ is non-negative and at least one entry is zero. It is not the zero vector, because $w, \acute{w}$ are linearly independent. Thus

$$w - \gamma \acute{w} = \frac{A(w - \gamma \acute{w})}{\rho(A)} > 0 \tag{C-17}$$

We chose γ so that at least one entry was zero, so this is a contradiction. Therefore, there cannot be two linearly independent eigenvectors, so $\rho(A)$ has geometric multiplicity 1.

(d) $\rho(A)$ *has algebraic multiplicity* 1.

As in previous parts, let w be a right eigenvector of $\rho(A)$. We now know that there is only one eigenvector and that it is positive.

Let u be a positive left eigenvector of $\rho(A)$. We can get such an eigenvector by applying (a) to A^T.

This pair of eigenvectors allows us to decompose $\mathbb{R}^n$ as a direct sum.

Lemma. *The space* $u^\perp = \{x \in \mathbb{R}^n | ux = 0\}$ *is invariant under* A.

Proof. If $ux = 0$, then $uAx = (uA)x = ux = 0$. $\qquad\qquad\qquad\qquad\qquad\square$

This space has dimension $n - 1$, and w is a vector that is not in $u^\perp$, because

$$ux = \sum u_j w_j > 0 \qquad\qquad\qquad\text{(C-18)}$$

Therefore, $\mathbb{R}^n$ is the direct sum of $span(w)$ and $u^\perp$.

Let $w_2, w_3, \ldots, w_n$ be a basis for the space $u^\perp = \{x \in \mathbb{R}^n | ux = 0\}$. Let X be the matrix

$$X = \begin{bmatrix} \cdots & \cdots & \cdots & \cdots \\ w & w_2 & \cdots & w_n \\ \cdots & \cdots & \cdots & \cdots \end{bmatrix} \qquad\qquad\text{(C-19)}$$

Then, XAX^{-1} leaves invariant the spaces $X^{-1}span(u) = span(e_1)$ and $X^{-1}u^\perp = span(e_2, e_3, \ldots, e_n)$, so this change of basis turns A into a block diagonal matrix

$$X^{-1}AX = \begin{bmatrix} \rho(A) & 0 \\ 0 & Y \end{bmatrix} \qquad\qquad\text{(C-20)}$$

for some matrix Y.

Suppose λ has algebraic multiplicity more than one. Then it must also be an eigenvalue of Y. However, then Y must have an eigenvector for $\rho(A)$ too, which makes two eigenvectors, which contradicts (b).

Therefore, $\rho(A)$ has algebraic multiplicity one. This proves part (d).

C-3 THE PERRON-FROBENIUS THEOREM

Finally, we can easily extend the theorem to the case where A is non-negative and has a positive power A^m.

Theorem. *(Perron-Frobenius theorem.) The statements (a), (b), (c), (d) are also true for non-negative matrices* A *so that some power* A^m *is positive.*

Proof. Let $\lambda_1, \lambda_2, \dots, \lambda_n$ be the eigenvalues of A, counted with algebraic multiplicity. Suppose λ_1 has the largest absolute value.

Then the eigenvalues of A^m are $\lambda_1^m, \lambda_2^m, \dots, \lambda_n^m$, where λ_1^m is the largest in absolute value. Perron's Theorem implies that λ_1^m is positive, and it has a positive eigenvector w, and that all the other eigenvalues of A^m are smaller.

However, if $|\lambda_1|^m > |\lambda_j|^m$ for $j \geq 2$, then $|\lambda_1| > |\lambda_j|$ for $j \geq 2$. Therefore, λ_1 is the largest eigenvalue in absolute value. It has algebraic multiplicity one. Its eigenvector must be w, because no other eigenvector of A has the right eigenvalue. So, conditions (b), (c), (d) hold, and (a) except for the positivity of λ_1. This follows from the fact that w is positive, and $\lambda w = Aw \geq 0$.

It turns out that this is equivalent to a condition on the directed graph $\mathfrak{G}(A)$ on $\{1, 2, \dots, n\}$ with an edge from i to j if $a_{ij} > 0$.

There is some m with $A^m > 0$ if and only if

a) Every vertex in $\mathfrak{G}(A)$ can be reached from every other vertex
b) The *Greatest Common Divisor* of the lengths of the loops in $\mathfrak{G}(A)$ is one.

The length of a path in a graph is equal to the number of edges traversed. The above statement can be proved by the methods of graph theory. For graph theory, consult the textbooks or Wikipedia (Wikipedia - Path (Graph Theory), 2015). A sketch will do for a proof.

Sketch. There is a path of length m from i to j in $\mathfrak{G}(A)$ if and only if $(A^m)_{ij} > 0$. The first condition guarantees that for every i, j there are infinitely many m with $(A^m)_{ij} > 0$, and the second condition ensures there exist loops at any vertex of any sufficiently long length, obtainable by chaining the shorter loops together.

Suppose that it is possible to go from any vertex i to any other vertex j in time at most s, and that there are loops at every vertex of every length at least r. Let $m = s + r$. Then for every pair i, j, one can go from i to j in time $t \leq s$ and pick a loop at j which has length $(m - t) \geq r$. This is a path of length m from i to j, so $(A^m)_{ij} > 0$.

C-4 PERRON-FROBENIUS FOR SYMMETRIC & POSITIVE MATRICES

Recall that a square matrix A is symmetric if and only if $A = A^\mathsf{T}$. For the components of positive and symmetric square matrices, $a_{ij} = a_{ji}$ holds. Thus, it is easy to see that there is always an m ensuring $A^m > 0$. Even integer numbers will do the job: $m = 2, 4, \dots$; compare also with the simple proof of the Perron-Frobenius theorem for positive symmetric matrices, published by Ninio (Ninio, 1976). Consequently, the Perron-Frobenius theorem holds for QFD matrices.

BIBLIOGRAPHY

Abran, A., 2010. *Software Metrics and Software Metrology.* Hoboken, NJ(New Jersey): John Wiley & Sons, Inc..

Akao, Y., ed., 1990. *Quality Function Deployment - Integrating Customer Requirements into Product Design.* Portland, OR: Productivity Press.

Albrecht, A. J., 1979. *Measuring Application Development Productivity.* Monterey, California, IBM Corporation, p. 83–92.

Ambler, S. W., 2004. *The Object Primer, 3rd Edition – Agile Model–Driven Development With UML 2.0.* New York, NY: Cambridge University Press.

American Association of Cost Estimators, 2008-1. *Recommended Practice No. 41R-08,* Morgantown, WV: AACE, Inc..

American Association of Cost Estimators, 2008-2. *Risk Analysis and Contingency Determination Using Range Estimating,* Morgantown, WV: AACE International Recommended Practices.

American Association of Cost Estimators, 2014. *Basis of Estimate - as Applied for the Software Services Industries,* Morgantown, WV: AACE Recommended Practices.

Arduino Community, 2005. *Arduino.* [Online]
Available at: http://www.arduino.cc/
[Accessed 25 June 2014].

Bagriyanik, S., Karahoca, A. & Ersoy, E., 2014. *Selection of a Functional Sizing Methodology: A Telecommunications Company Case Study.* Rome, Italy, Elsevier.

Bakalova, Z., 2014. *Towards Understanding the Value-Creation in Agile Projects.* Enschede: CTIT Dissertation Series.

Bana e Costa, C. A. & Vansnick, J.-C., 2008. A Critical Analysis of the Eigenvalue Method used to derive priorities in AHP. *European Journal of Operational Research ,* Volume 187, p. 1422–1428.

Barwise, J. et al., 1977. *Handbook of Mathematical Logic.* Studies in Logic and the Foundations of Mathematics ed. Amsterdam, NL: North-Holland Publishing Company.

Beck, K., 2000. *eXtreme Programming Explained.* Boston : Addison-Wesley.

Bektas, S. & Sisman, Y., 2010. The comparison of L1 and L2-norm minimization methods. *International Journal of the Physical Sciences*, 18 September, 5(11), pp. 1721-1727.

Bell, D., 2004. *UML basics: The Sequence Diagram – Introductory Level,* Armonk, NY: IBM DeveloperWorks.

Beni, I. et al., 2011. *A Taxonomy of Software Projects Productivity Impact Factors V1.1,* Rome, Italy: GUFPI-ISMA.

Bhushan, N. & Rai, K., 2004. *Strategic Decision Making: Applying the Analytic Hierarchy Process.* London: Springer Verlag.

Boehm, B. & et.al., 2000. *Software Cost Estimation with COCOMO II.* Upper Saddle River, NJ: Prentice Hall.

Bourbaki, N., 1989. *Elements of Mathematics, Algebra I..* Heidelberg, Germany: Springer-Verlag.

Buglione, L. et al., 2011. *Guideline for the Use of COSMIC FSM to manage Agile projects,* Montréal, Canada: The COSMIC Consortium.

Buglione, L. & Cencel, C., 2008. *Impact of Base Functional Component Types on Software Size Based Effort Estimation.* Berlin Heidelberg, PROFES 2008, LNCS 5089, pp. 75-89.

Buglione, L. & Trudel, S., 2010. *Guideline for Sizing Agile Projects with COSMIC.* Stuttgart, Germany, DASMA e.V.

Cairns, H., 2014. *A short proof of Perron's theorem.* [Online]
Available at: http://www.math.cornell.edu/~web6720/Perron-Frobenius_Hannah%20Cairns.pdf
[Accessed 25 August 2015].

Clausing, D., 1994. *Total Quality Development.* ASME Press Series on International Advances in Design Productivity ed. New York, NY: The American Society of Mechanical Engineers.

Cohn, M., 2005. *Agile estimating and planning.* New Jersey, NJ: Prentice Hall.

Cooley, J. W. & Tukey, J. W., 1964. Cooley, James W.; Tukey, An algorithm for the machine calculation of complex Fourier series". Math. Comput. 19 (90): 297–301. *Mathematics of Computation,* 17 August, Volume 19, pp. 297-301.

COSMIC and IFPUG, 2015. *Glossary of terms for NF and Project Requirements.* [Online]
Available at: http://cosmic-sizing.org/publications/glossary-of-terms-for-nf-and-project-requirements/
[Accessed 18 November 2015].

COSMIC Consortium, 2015. *Guideline on Non-Functional & Project Requirements V1.03.* [Online]
Available at: http://cosmic-sizing.org/publications/guideline-on-non-functional-project-requirements/
[Accessed 18 November 2015].

COSMIC Measurement Practices Committee, 2015. *The COSMIC Functional Size Measurement Method – Version 4.0.1 – Measurement Manual,* Montréal: The COSMIC Consortium.

Creveling, C., Slutsky, J. & Antis, D., 2003. *Design for Six Sigma.* New Jersey, NJ: Prentice Hall.

Damm, L.-O., Lundberg, L. & Wohlin, C., 2006. Faults-slip-through – A Concept for Measuring the Efficiency of the Test Process. *Software Process Improvement And Practice,* Volume 11, p. 47–59.

Damm, L.-O., Lundberg, L. & Wohlin, C., 2008. A model for software rework reduction through a combination of anomaly metrics. *The Journal of Systems and Software,* Volume 81, pp. 1968-1982.

Dasgupta, A., Gencel, C. & Symons, C., 2015. *A Process to Improve the Accuracy of MkII FP to COSMIC Size Conversions: Insights into the COSMIC Method Design Assumptions.* Kraków, Poland, Springer Lecture Notes in Business Information Processing.

Deming, W., 1986. *Out of the Crisis.* Center for Advanced Engineering Study ed. Boston, MA: Massachusetts Institut of Technology.

Denney, R., 2005. *Succeeding with Use Cases – Working Smart to Deliver Quality.* Booch–Jacobson–Rumbaugh – Series ed. New York, NY: Addison-Wesley.

Denniston, B., 2006. Capabilities Indices and Conformance for Specification: The Motivation for Using Cpm. *Quality Engineering,* 18(1), pp. 79-88.

El-Haik, B. S., 2005. *Axiomatic Quality - Integrating Axiomatic Design with Six-Sigma, Reliability, and Quality Engineering.* Hoboken, NJ: John Wiley & Sons.

El-Haik, B. S. & Shaout, A., 2010. *Software Design for Six Sigma – A Roadmap for Excellence.* Hoboken, NJ: Wiley&Sons, Inc..

Engeler, E., 1981. Algebras and Combinators. *Algebra Universalis,* pp. 389-392.

Engeler, E., 1995. *The Combinatory Programme.* Basel, Switzerland: Birkhäuser.

Fagan, M. E., 1974. Design and Code inspections to reduce errors in program development. *IBM Systems Journal,* 15(3), p. 182–211.

Fagg, P. & Rule, G., 2010. *Sizing User Stories with the COSMIC FSM method.* [Online] Available at: http://sms.amemory.co.uk/wp-content/uploads/20100408-COSMICstories-article-v0c1.pdf [Accessed 27 February 2015].

Fehlmann, T. M., 1981. *Theorie und Anwendung der Kombinatorischen Logik,* Zürich, CH: ETH Dissertation 3140-01.

Fehlmann, T. M., 1999. *Quality Function Deployment for the full Business Life Cycle - Towards the "Mafia Proposal".* Vienna, Austria, STEV Austria.

Fehlmann, T. M., 2000. *Measuring Competitiveness in Service Design.* Novi, MI, The 12th Symposium on QFD.

Fehlmann, T. M., 2001. *Risk Exposure Measurements on Web Sites.* Heidelberg, Germany, 4th European Conference on Software Measurement and IT Control (FESMA).

Fehlmann, T. M., 2002. *QFD as Algebra of Combinators.* Tokyo, Japan, 8th International QFD Symposium, ISQFD 2002.

Fehlmann, T. M., 2003. *Linear Algebra for QFD Combinators.* Orlando, FL, International Council for QFD (ICQFD).

Fehlmann, T. M., 2003. Strategic Management by Business Metrics: An application of Combinatory Metrics. *International Journal of Quality & Reliability,* 20(1), pp. 134-145.

Fehlmann, T. M., 2005-1. *Six Sigma in der SW-Entwicklung.* Wiesbaden: Vieweg.

Fehlmann, T. M., 2005-2. The Impact of Linear Algebra on QFD. *International Journal of Quality & Reliability Management,* 22(9), pp. 83-96.

Fehlmann, T. M., 2006-1. *Statistical Process Control for Software Development – Six Sigma for Software revisited.* Joenssu, FI, University of Joensuu, Dep.CSS, International Proceedings Series 6.

Fehlmann, T. M., 2008. *New Lanchester Theory for Requirements Prioritization.* Barcelona, Spain, 2nd International Workshop on Software Product Management, ISWPM'08.

Fehlmann, T. M., 2009-1. *Defect Density Prediction with Six Sigma.* Roma, Italia, DPO s.r.l.

Fehlmann, T. M., 2009-2. *Using Six Sigma for Software Project Estimations.* Kaiserslautern, Germany, MetriKon 2009 – Praxis der Software-Messung.

Fehlmann, T. M., 2011-1. *Measuring and Estimating Ongoing Agile Projects in Real-Time.* London, UK, United Kingdom Software Metrics Association (UKSMA).

Fehlmann, T. M., 2011-2. *Agile Software Projects with Six Sigma.* Glasgow, UK, Strathclyde Institute for Operations Management.

Fehlmann, T. M., 2011-3. *Understanding Business Drivers for Software Products from Marketing.* Kaiserslautern, Germany, Tagungsband des DASMA Software Metrik Kongresses - MetriKon 2011.

Fehlmann, T. M., 2011. When use COSMIC FFP? When use IFPUG FPA? A Six Sigma View. In: R. Dumke & A. Abran, eds. *COSMIC Function Points - Theory and Advanced Practices.* Boca Raton, FL: CRC Press, pp. 260-274.

Fehlmann, T. M. & Kranich, E., 2011-1. *COSMIC Functional Sizing based on UML Sequence Diagrams.* Kaiserslautern, DASMA e.V.

Fehlmann, T. M. & Kranich, E., 2011-2. *Transfer Functions, Eigenvectors and QFD in Concert.* Stuttgart, Germany, QFD Institut Deutschland e.V.

Fehlmann, T. M. & Kranich, E., 2012-1. *Quality of Estimations.* Assisi, Italy, IEEE Computer Society 2012, pp. 8-14.

Fehlmann, T. M. & Kranich, E., 2012-2. *Social Media Metrics for Embedded Software.* Stuttgart, Germany, DASMA e.V.

Fehlmann, T. M. & Kranich, E., 2012-3. *Using Six Sigma Transfer Functions for Analysing Customer's Voice.* Glasgow, UK, Strathclyde Institute for Operations Management.

Fehlmann, T. M. & Kranich, E., 2013-1. *Customer-Driven Software Product Development - Software Products for the Social Media World. A Case Study..* Berlin Heidelberg, Springer-Verlag, pp. 300-312.

Fehlmann, T. M. & Kranich, E., 2013-2. *Experiments with Short-Run Control Charts for Monitoring the Software Development Process.* Kaiserslautern, DE, DASMA e.V.

Fehlmann, T. M. & Kranich, E., 2014-1. *Defect Density Measurements Using COSMIC - Experiences with Mobile Apps and Embedded Systems.* Rotterdam, IWSM Mensura.

Fehlmann, T. M. & Kranich, E., 2014-2. Early Software Project Estimation the Six Sigma Way. *Lecture Notes in Business Information Processing,* Volume 199, pp. 193-208.

Fenton, N. E., Neil, M. & Marquez, D., 2008. Using Bayesian networks to predict software defects and reliability. *Journal of Risk and Reliability,* Proceedings of the Institution of Mechanical Engineers(Part O), pp. 701-712.

Fiedler, M., 2008. *Special Matrices and Their Applications in Numerical Mathematics.* Mineola, NY: Dover Publications, Inc. Mineola.

Frohnhoff, S., 2009. *Use Case Points 3.0: Implementation of a Use Case Based Estimating Method for the Software Engineering of Business Information Systems,* Paderborn, Germany: Universität Paderborn, Institut für Informatik.

Gallardo, P. F., 2007. Google's Secret and Linear Algebra.. *EMS Newsletter,* Volume 63, pp. 10-15.

Gencel, Ç. & Bideau, C., 2012. *Exploring the Convertibility between IFPUG and COSMIC Function Points: Preliminary Findings.* Assisi, Italy, IEEE Computer Society 2012, pp. 170-177.

George, M. L., 2002. *Lean Six Sigma: Combining Six Sigma Quality with Lean Production Speed.* New York, NY: McGraw-Hill.

George, M. L., 2010. *The Lean Six Sigma Guide to Doing More with Less.* Hoboken, NJ: John Wiley & Sons.

Gigerenzer, G., 2007. *Gut Feelings. The Intelligence of the Unconscious..* New York, NY: Viking.

Girod, B., Rabenstein, R. & Stenger, A., 2001. *Signals and Systems, 2nd Edition.* Hoboken, NJ: John Wiley & Sons.

Herzwurm, G. & Pietsch, W., 2009. *Management von IT-Produkten.* Wirtschaftswissenschaften ed. Heidelberg, Germany: dpunkt.verlag.

Herzwurm, G. & Schockert, S., 2006. *What are the Best Practices of QFD?.* Tokyo, Japan, Transactions from the 12th Int. Symposium on Quality Function Deployment.

Herzwurm, G., Schockert, S. & Mellis, W., 2000. *Joint Requirements Engineering. QFD for Rapid Customer Focused Software and Internet-Development.* Braunschweig: Vieweg.

Hill, P., ed., 2010. *Practical Software Project Estimation 3rd Edition.* New York, NY: McGraw-Hill.

Hu, M. & Antony, J., 2007. Enhancing Design Decision-Making through Development of Proper Transfer Function in Design for Six Sigma Framework. *International Journal of Six Sigma and Competitive Advantage,* Issue 3, pp. 33-55.

IFPUG Counting Practice Committee, 2010. *Function Point Counting Practices Manual - Version 4.3.1,* Princeton Junction, NJ: International Function Point User Group (IFPUG).

IFPUG Non-Functional Sizing Standards Committee, April 2013. *Software Non-functional Assessment Process (SNAP) - Assessment Practices Manual,* Princeton Junction, NJ: International Function Point Users Group (IFPUG).

Imai, M., 1997. *Gemba Kaizen: A Commonsense, Low-Cost Approach to Management.* New York, NY: McGraw-Hill.

Ishikawa, K., 1990. *Introduction to Quality Control.* Translated by J. H. Loftus; distributed by Chapman & Hall, London ed. Tokyo, Japan: JUSE Press Ltd.

ISO 16355-1:2015, 2015. *ISO 16355-1:2015, 2015. Applications of Statistical and Related Methods to New Technology and Product Development Process - Part 1: General Principles and Perspectives of Quality Function Deployment (QFD), Geneva, Switzerland: ISO TC 69/SC 8/WG 2 N 14,* Geneva, Switzerland: ISO TC 69/SC 8/WG 2 N 14.

ISO/IEC 14143-1:2007, 2007. *Information technology - Software measurement - Functional size measurement - Part 1: Definition of concepts,* Geneva, Switzerland: ISO/IEC JTC 1/SC 7.

ISO/IEC 19761:2011, 2011. *Software engineering - COSMIC: a functional size measurement method,* Geneva, Switzerland: ISO/IEC JTC 1/SC 7.

ISO/IEC 20926:2009, 2009. *Software and systems engineering - Software measurement - IFPUG functional size measurement method,* Geneva, Switzerland: ISO/IEC JTC 1/SC 7.

ISO/IEC 20968:2002, 2002. *Software engineering - Mk II Function Point Analysis - Counting Practices Manual,* Geneva, Switzerland: ISO/IEC JTC 1/SC 7.

ISO/IEC 24570:2005, 2005. *Software engineering - NESMA functional size measurement method version 2.1 - Definitions and counting guidelines for the application of Function Point Analysis,* Geneva, Switzerland: ISO/IEC JTC 1/SC 7.

ISO/IEC 29881:2010, 2010. *Information technology - Systems and software engineering - FiSMA 1.1 functional size measurement method,* Geneva, Switzerland: ISO/IEC JTC 1/SC 7.

ISO/IEC 31010:2009, 2009. *Risk Management - Risk Assessment Techniques,* Geneva, Switzerland: ISO/TC 262.

ISO/IEC CD Guide 98-3, 2015. *Evaluation of measurement data - Part 3: Guide to uncertainty in measurement (GUM),* Geneva, Switzerland: TC/SC: ISO/TMBG.

ISO/IEC Guide 99:2007, 2007. *International vocabulary of metrology – Basic and general concepts and associated terms (VIM)*, Geneva, Switzerland: TC/SC: ISO/TMBG.

Jenner, M. S., 2011. Automation of Counting of Functional Size Using COSMIC FFP in UML. In: *COSMIC Function Points - Theory and Advanced Practices*. Boca Raton, FL: CRC Press - Auerbach, pp. 276-283.

Kan, H., 2004. *Metrics and Models in Software Quality Engineering*. 2nd edition ed. Boston, MA: Addison-Wesley.

Kano, N., Seraku, N., Takahashi, F. & Tsuji, S., 1984. Attractive quality and must-be quality. *Journal of the Japanese Society for Quality Control (in Japanese)*, April, 14(2), p. 39–48.

Keener, J. P., 1993. The Perron-Frobenius Theorem and the Ranking of Football Teams. *SIAM Review*, 35(1), pp. 80-93.

Kressner, D., 2005. Numerical Methods for General and Structured Eigenvalue Problems. *Lecture Notes in Computational Science and Engineering*, Volume 46.

Lang, E., 1973. *Linear Algebra*. 3rd ed. New York, NY, USA: Springer-Verlag New York Inc..

Langville, A. N. & Meyer., C. D., 2006. *Google's PageRank and Beyond: The Science of Search Engine Rankings*. Princeton, NJ: Princeton University Press.

Levesque, G., Bevo, V. & Cao, D., 2008. *Estimating Software Size with UML Models*. Montréal, Canadian Conference on Computer Science & Software Engineering, pp. 81-87.

Levie, R. d., 2012. *Advanced Excel for Scientific Data Analysis*. 3rd Edition ed. Orrs Island, ME: Atlantic Academic LLC.

López, T. S., Ranasinghe, D. C., Harrison, M. & McFarlane, D., 2013. Using Smart Objects to build the Internet of Things. *IEEE Internet Computing (to appear)*.

Maass, E. & McNair, P. D., 2010. *Applying Design for Six Sigma to Software and Hardware Systems*. Upper Saddle River, NJ 07458, USA: Prentice Hall.

Marín, B., Giachetti, G. & Pastor, O., 2008. *Measurement of Functional Size in Conceptual Models: A Survey of Measurement Procedures Based on COSMIC*. Heidelberg, Springer, pp. 170-183.

Mazur, G., 2014. *QFD and the New Voice of Customer (VOC)*. Istanbul, Turkey, International Council for QFD (ICQFD), pp. 13-26.

Mazur, G. & Bylund, N., 2009. *Globalizing Gemba Visits for Multinationals*. Savannah, GA, USA, Transactions from the 21st Symposium on Quality Function Deployment.

Meyer, C. D., 2000. *Matrix Analysis and Applied Linear Algebra*. Philadelphia, PA: SIAM.

Michell, J., 1986. Measurement scales and statistics: a clash of paradigms. *Psychological Bulletin*, Volume 3, p. 398–407.

Miller, G., 1956. The Magical Number Seven, Plus or Minus Two: Some Limits on Our Capacity for Processing Information. *The Psychological Review*, Volume 63, pp. 81-97.

Mizuno, S. & Akao, Y., 1994. QFD: The Customer-Driven Approach to Quality Planning and Deployment, translated by Glenn Mazur. In: S. Mizuno & Y. Akao, eds. *Quality Function Deployment*. Tokyo: Asian Productivity Institute.

Montgomery, D. C., 2009. *Introduction to Statistical Quality Control*. 6th ed. New York, NY, USA: John Wiley & Sons, Inc..

Morgenshtern, O., Raz, T. & Dvir, D., 2007. Factors Affecting Duration and Effort Estimation Errors in Software Development Projects. *Information and Software Technology*, 49(8), p. 827–837.

Myers, R. H., Montgomery, D. C. & Anderson-Cook, C. V., 2009. *Response Surface Methodology: Process and Product Optimization Using Designed Experiments.* New York, NY: John Wiley & Sons.

Ninio, F., 1976. A Simple Proof of the Perron-Frobenius Theorem for Positive Symmetric Matrices. *Journal of Physics A: Mathematical and General,* 9(8), pp. 1281-1282.

Nipp, K. & Stoffer, D., 2002. *Lineare Algebra: eine Einführung für Ingenieure unter besonderer Berücksichtigung numerischer Aspekte.* Zürich: vdf, Hochschul-Verlag an der ETH.

Oriou, A., 2014. *Manage the automotive embedded software development with automation of COSMIC.* Rotterdam, NESMA.

Oudrhiri, R., 2005. Six Sigma and DFSS for IT and Software Engineering. *TickIt International,* Issue 2Q05.

Owen, R. & Brooks, L., 2009. *Answering the Ultimate Question – How Net Promoter Can Transform Your Business.* San Francisco, CA, USA: Jossey-Bass.

Pande, P., Neuman, R. & Cavanagh, R., 2002. *The Six Sigma Way – Team Fieldbook.* New York, NY: McGraw-Hill.

Petersen, K. & Wohlin, C., 2010. Software process improvement through the Lean Measurement (SPI-LEAM) method. *The Journal of Systems and Software,* Volume 83, p. 1275–1287.

Phadke, M. S., 1989. *Quality Engineering Using Robust Design.* Englewood Cliffs, NJ: Prentice Hall.

Poppendieck, M. & Poppendieck, T., 2007. *Implementing Lean Software Development.* New York, NY: Addison-Wesley.

Quesenberry, C. P., 1997. *SPC Methods for Quality Improvement.* New York, NY, USA: John Wiley & Sons, Inc.

Rahman, M. & Mitchell, J., 1994. *A Guide to Quality Function Deployment.* ASI Press ed. Richmond Hill New York, NY: Mukhles Rahman.

Refflinghaus, R., Winzer, P., Schlüter, N. & Esser, C., 2014. *Customer-integrated QFD by Using System Engineering.* Istanbul, Turkey, 20th International Symposium on Quality Function Deployment.

Reichheld, F., 2001. *The Loyalty Effect: The Hidden Force Behind Growth, Profits, and Lasting Value..* Paperback ed. Boston, MA, USA: Harvard Business School Press.

Reichheld, F., 2007. *The Ultimate Question: Driving Good Profits and True Growth.* Boston, MA: Harvard Business School Press.

Roman, S., 2007. *Advanced Linear Algebra.* 3rd ed. New York, NY: Springer Verlag.

Roriz Filho, H., 2010. *Can Scrum Support Six Sigma?,* Indianapolis, IN: Scrum Alliance.

Russo, L., 2004. *The Forgotten Revolution - How Science Was Born in 300 BC and Why It Had to Be Reborn.* Berlin Heidelberg New York: Springer-Verlag.

Saaty, T. & Alexander, J., 1989. *Conflict Resolution: The Analytic Hierarchy Process.* New York, NY: Praeger, Santa Barbara, CA.

Saaty, T. L., 1990. *The Analytic Hierarchy Process – Planning, Priority Setting, Resource Allocation.* Pittsburgh, PA : RWS Publications.

Saaty, T. L., 2003. Decision-making with the AHP: Why is the principal eigenvector necessary?. *European Journal of Operational Research,* Volume 145, pp. 85-91.

Saaty, T. L. & Özdemir, M., 2003. Negative Priorities in the Analytic Hierarchy Process. *Mathematical and Computer Modelling*, 37(9–10), pp. 1063-1075.

Saaty, T. L. & Peniwati, K., 2008. *Group Decision Making: Drawing out and Reconciling Differences*. Pittsburg, PA: RWS Publications.

Santillo, L., 2011. Early and Quick COSMIC FFP Analysis Using Analytic Hierarchy Process. In: R. Dumke & A. Abran, eds. *COSMIC Function Points - Theory and Advanced Practices*. Boca Raton, FL: CRC Press, pp. 176-191.

Satmetrix Systems, Inc., 2007. *Net Promoter® Economics: The Impact of Word of Mouth*, San Mateo, CA, USA: Satmetrix Systems, Inc..

Schneider, D., Berent, M., Thomas, R. & Krosnick, J., 2008. *Measuring Customer Satisfaction and Loyalty: Improving the 'Net-Promoter' Score*. New Orleans, Louisiana, Annual Meeting of the American Association for Public Opinion Research.

Schurr, S., 2011. *Evaluating AHP Questionnaire Feedback with Statistical Methods*. Stuttgart, Germany, 17th International QFD Symposium, ISQFD 2011.

Schwaber, K. & Beedle, M., 2002. *Agile Software Development with Scrum*. Upper Saddle River, NJ: Prentice Hall PTR.

Siegel, E., 2013. *Predictive Analytics - The Power to Predict who will Click, Buy, Lie, or Die*. Hoboken, NJ: John Wiley & Sons.

Siviy, M. J., Penn, M. L. & Stoddard, R. W., 2008. *CMMI and Six Sigma - Partners in Process Improvement*. Amsterdam: Addison-Wesley Longman.

Taoka, N., 1997. *Lanchester Strategy – An Introduction*. Sunnyvale, CA: Lanchester Press Inc..

The R Foundation, 2015. *The R Project for Statistical Computing*. [Online]
Available at: http://www.r-project.org
[Accessed 16 April 2015].

Trudel, S., 2012. *Using the COSMIC Functional Size Measurement Method (ISO 19761) as a Software Requirements Improvement Mechanism*, Montréal, Canada: École de Technologie Supérieure - Université du Québec.

UKSMA, 2015. *Software Defect Measurement and Analysis Handbook*, London, UK: United Kingdom Software Metrics Association.

Vandermerwe, S. & Rada, J., 1988. Servitization of business: Adding value by adding services. *European Management Journal*, 6(4), pp. 314-324.

Vogelezang, F., 2015. *Best Practices in Software Cost Estimation*. [Online]
Available at: http://www.slideshare.net/frankvogelezang/best-practices-in-software-cost-estimation-metrikon-2015-frank-vogelezang
[Accessed 26 November 2015].

Volpi, L. & Team, 2007. *Matrix.xla*. [Online]
Available at: http://www.bowdoin.edu/~rdelevie/excellaneous/matrix.zip
[Accessed 29 March 2015].

Wikipedia - Path (Graph Theory), 2015. *Path (Graph Theory)*. [Online]
Available at: https://en.wikipedia.org/wiki/Path_(graph_theory)
[Accessed 26 August 2015].

Wikipedia - Perron-Frobenius Theorem, 2015. *Perron–Frobenius Theorem*. [Online]
Available at: https://en.wikipedia.org/wiki/Perron–Frobenius_theorem
[Accessed 25 August 2015].

Wikipedia - Spectral Radius, 2015. *Spectral Radius.* [Online]
Available at: https://en.wikipedia.org/wiki/Spectral_radius
[Accessed 26 August 2015].

Womack, J., 2013. *Gemba Walks - Expanded 2nd Edition.* Barters Island, ME: Lean Enterprise Institute, Inc..

Wu, H.-H., 2004. Using target costing concept in loss function and process capability indices to set up goal control limits. *The International Journal for Advanced Manufacturing Technology*, Volume 24, p. 206–213.

Yang, K. & El-Haik, B. S., 2009. *Design for Six Sigma.* 2nd ed. New York, NY, USA: Mc Graw-Hill.

Yan, X. & Su, X. G., 2009. *Linear Regression Analysis – Theory and Computing.* Singapore: World Scientific Publishing Co. Pte. Ltd.

Zultner, R. E., 1995. Blitz QFD: Better, Faster, and Cheaper Forms of QFD. *American Programmer*, October '95, p. 24–36.

REFERENCE INDEX